W0257543

FELSMECHANIK UND INGENIEURGEOLOGIE
ROCK MECHANICS AND ENGINEERING GEOLOGY
SUPPLEMENTUM I

Grundfragen auf dem Gebiete der Geomechanik

Principles in the Field of Geomechanics

XIV. Kolloquium der Österreichischen Regionalgruppe (i. Gr.)
der Internationalen Gesellschaft für Felsmechanik

14th Symposium of the Austrian Regional Group (i. f.)
of the International Society for Rock Mechanics
Salzburg, 27. und 28. September 1963

Herausgegeben von / Edited by
L. Müller, Salzburg

Unter Mitwirkung von / In Cooperation with
C. Fairhurst, Minneapolis

Mit 136 Textabbildungen

With 136 Figures

SPRINGER-VERLAG WIEN GMBH 1964

ISBN 978-3-662-23624-6   ISBN 978-3-662-25703-6 (eBook)
DOI 10.1007/978-3-662-25703-6

Titel Nr. 9129

Inhaltsverzeichnis

Some Problems on Failure of Rock Masses

By

Manuel Rocha*

With 3 Figures

Summary — Zusammenfassung — Résumé

Some Problems on Failure of Rock Masses. After emphasizing the difference between joints and faults, from the standpoint of mechanical behaviour in rock masses, the author presents ideas on how to account for the effect of joints on their shear strength. The rock mass is considered as a set of blocks defined by its geometrical parameters, and by the shearing properties of the material of the blocks and their boundary surfaces. The author presents values of these properties and discusses the conditions in which low-strength fillings of joints and faults actually contribute towards reducing shear strength in rock masses.

The concept of the factor of safety to be adopted in failure studies of rock masses is also discussed in the paper. Attention is drawn to the fact that the concept usually followed is unsuited to those problems in which the forces applied to the rock mass must, by their very nature, be considered constant and it is suggested that, in such cases, the factor of safety be defined with respect to the degree of weakening of the property of the rock mass that will bring about failure. After emphasizing that less correct statements about the factor of safety of rock masses are very often presented, the author discusses the criteria to be followed to ascribe values to these factors and presents recommended figures.

Einige Probleme betreffend Brucherscheinungen in Gebirgen. Der Unterschied zwischen Kluft und Verwerfung, vom mechanischen Verhalten des Gebirges aus gesehen, wird hervorgehoben und Vorschläge gemacht, wie der Einfluß der Klüfte auf die Scherfestigkeit berücksichtigt werden könnte. Demnach wird das Gebirge einer Blockgruppe angeglichen, welche durch geometrische Parameter und durch die Schubeigenschaften des Blockmaterials und der begrenzenden Flächen definiert wird. Es werden Werte bezüglich dieser Eigenschaften angegeben. Die Bedingungen, unter welchen die Kluft- und Verwerfungsfüllungen schwachen Widerstandes tatsächlich für die Schubwiderstandsminderung beitragen, werden erörtert.

Der Verfasser behandelt auch bei dieser Arbeit den Begriff des bei den Bruchuntersuchungen des Gebirges anzuwendenden Sicherheitsgrades. Es wird betont, daß der übliche Begriff sich nicht denjenigen Problemen anpaßt, bei denen die angreifenden Kräfte, ihrer physischen Natur wegen, als konstant angesehen werden müssen, und es wird vorgeschlagen, daß bei solchen Fällen die Definition des Sicherheitsgrades auf der Berücksichtigung der Schwächung derjenigen Gebirgseigenschaften, die den Bruch hervorriefen, fußt.

Nachdem hervorgehoben wird, daß öfters nicht ganz einwandfreie Behauptungen über die Sicherheitsgrade der Gebirge gemacht werden, wird das Problem der Kriterien bezüglich der Bewertung dieser Sicherheitsgrade diskutiert. Es werden einige für ratsam gehaltene Werte vorgeschlagen.

Quelques problèmes concernant la rupture des massifs rocheux. Après avoir souligné la différence entre les joints et les failles en ce qui concerne le comportement mécanique des massifs rocheux, l'auteur indique comment tenir compte des joints par leur résistance

* Ing. Manuel R o c h a, Director, Laboratório Nacional de Engenharia Civil, Avenida do Brasil, Lisboa, Portugal.

au cisaillement. Dans cette conception le massif rocheux est assimilé à un assemblage de blocs défini par ses paramètres géométriques et par les propriétés de cisaillement du matériau des blocs et des surfaces qui les limitent. L'auteur donne des valeurs de ces propriétés et discute dans quelles conditions les remplissages à faible résistance des joints et des failles contribuent effectivement à diminuer la résistance au cisaillement du massif rocheux.

On discute aussi dans cet article la notion de coefficient de sécurité à adopter dans les études de rupture des massifs rocheux. On attire l'attention sur le fait que la notion habituelle ne convient pas aux problèmes dans lesquels les forces appliquées au massif rocheux doivent être, par leur nature physique, considérées comme constantes. On suggère de définir dans de tels cas le coefficient de sécurité d'après l'affaiblissement de la propriété du massif rocheux qui amènera la rupture. Après avoir souligné que l'on présente très souvent des conceptions moins correctes sur le coefficient de sécurité des massifs rocheux, l'auteur discute les critères à adopter pour fixer les valeurs de ces coefficients et en recommande quelques valeurs.

I. Introduction

In accordance with the theme of this Colloquium — Fundamental Questions in the Field of Geomechanics — two basic problems concerning the failure of rock masses will be dealt with in the present paper: how to account for joints and faults, and the concept of factor of safety to be followed.

II. Ideas on how to Take Account of Joints and Faults in the Shear Strength of Rock Masses

The usual influence of joints and faults on the failure of rock masses is widely recognized. Nevertheless, the way in which these geological features should be considered in the prediction of failure of foundations or of rock masses solely under the action of their own weight has not been systematically discussed.

From the standpoint of the mechanical behaviour of rock masses, the wide difference between joints and faults should be emphasized. The latter, having by their very nature already undergone displacements along their surfaces, are usually able by their shape to move without considerable fracturing of the rock material and, in addition, the mechanical characteristics of the joint filling materials are frequently poor. Thus, under the loadings applied by the structure, the rock mass can slide easily along the fault surfaces. In seismic areas faults present an additional problem, as displacements disturbing the structure can occur along them.

As for joints, it should be noted in the first place that they are generally arranged more or less regularly as regards orientation, forming parallel families, and spacing. Rock masses are often cut by families of joints with different, frequently three, orientations. Furthermore, the superposition of families with the same orientation but with different characteristics, notably as regards spacing, is sometimes observed. Thus, rock masses are more or less perfectly divided into blocks, which sometimes are imbricated i. e., contrary to faults, joints can consist of fractures without continuity. For the purpose in view, regularly arranged low-cohesion surfaces, e. g. certain sedimentation or schistosity surfaces, can be considered as joints.

The geometry of families of joints has a decisive influence on the strength of the rock mass. In particular the more imbricated the blocks, the higher their shear strength, as rupture surfaces have not only to follow the joints but also to cut the material of the blocks.

There is a marked difference between the fillings of faults and joints. The latter is usually thinner, joints without any filling being very frequent. That is why shear strength is, as a rule, higher in joints than in faults.

The foregoing considerations on joints and faults are schematic, their purpose being merely to establish a simplified model of reality, which is indispensable in any attempt to develop a scientific approach to the behaviour of rock masses. These should be assimilated to a set of more or less imbricated blocks, cut very often by some faults.

The volumes of rock masses involved in the failure of foundations of large structures, particularly of dams, or in failure under their weight are usually considerably larger than the blocks defined by joints. This makes it possible to consider the rock mass as a medium with certain average global characteristics, i. e. consideration of the individual joints can be avoided, which is a simplification of the utmost interest. On the contrary, faults, because of their irregularity, their smaller number and consequently wider spacings, and their possible great importance for the structure must, as a rule, be considered one by one as regards their influence on the strength of the rock mass. The difference just mentioned between joints and faults is of basic importance from both practical and conceptual standpoints. It is obvious that joints may have to be considered one by one if the dimensions of the blocks into which they divide the rock do not meet the condition stated above.

Assuming the rock mass can be considered as a set of blocks, let us see the patterns of joints to be considered. The most current one, shown in fig. 1 a, is characterized by the spacings d_1 and d_2, by the angle α between the two families of joints and by the imbrication i. As a rule spacings d_1 and d_2 range from a few decimeters to a few meters. Very frequent special cases of this model are those corresponding to $i = 0$ (fig. 1 b), blocks without imbrication, to $i = 0$ and $\alpha = 90^0$ (fig. 1 c), rectangular blocks, and to a sole family of joints (fig. 1 d). Models with imbrication in two directions, instead of one, can also be considered, but it seems they will not be needed except in very special cases. When the problem to be solved requires three-dimensional models, those just mentioned will have to be extended.

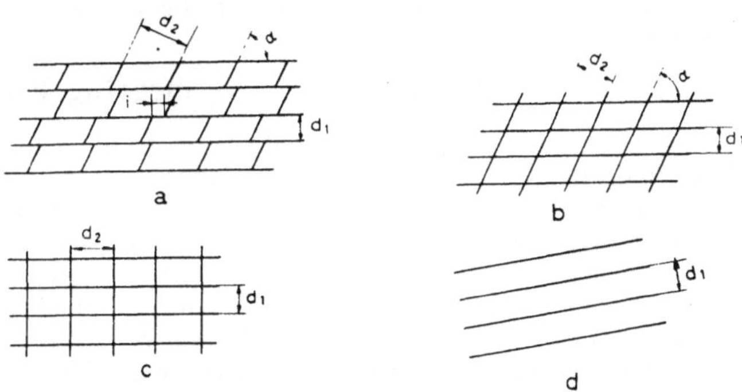

Fig. 1. Joint patterns of rock masses
Modelle der im Gebirge vorkommenden Klüfte

The mechanical behaviour of a medium cut by joints depends on the parameters defining their geometry, such as those of fig. 1, and also on the mechanical properties of the joints and of the material of the blocks. If the problem is the failure of the medium, and if this takes place by shear, the properties to be considered are the cohesion and the angle of friction of joints, c_j and φ_j, and of the material of the blocks, c_b and φ_b. It may also be necessary to ascribe different properties to the different families of joints and to take the anisotropy of the material of the blocks into account, but the number of parameters must be reduced to a minimum, which requires a global assessment of each problem. When joints have no filling or when, as usual, this has poor mechanical properties, it can be assumed that $c_j = 0$.

In the present state of our knowledge no theory is available permitting the determination of the shear strength of the medium just defined, for the sizes considered. That is why "in situ" tests would be the best way to determine shear strength, but difficulties arise for testing samples large enough to reproduce the influence of the network of joints. In fact such a shear test should be carried out on samples (fig. 2) with a thickness d sufficiently large in comparison with the spacings of the families of joints, which as a rule is impossible. It is important to note that in the directions along which imbrication can be neglected the test becomes radically simpler, as it is then sufficient to determine the shear strength along the joints, which can be done with much smaller samples. Such is the case for the direction of joints considered in fig. 1, except those spaced d_2 in fig. 1 a.

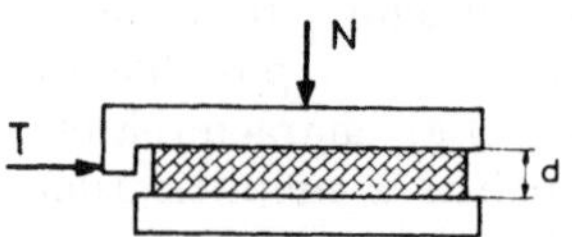

Fig. 2. In-situ shear test

Scherversuche in situ

On account of the difficulties mentioned it is often necessary to consider the rock mass as isotropic with the same angle of friction as the joints, assuming that cohesion is zero. It should be noted that if it can be assumed that $\varphi_b = \varphi_j$, that simplification amounts to neglecting c_b and c_j alone. Notice that the method of characterizing the rock mass by the pair of parameters c and φ results from the fact that the analytical methods available can assess the load carrying capacity of a rock mass only if this is assimilated to a homogeneous isotropic solid defined by the parameters referred to.

In the general case where it is desired to determine the shear strength of the medium in any direction, the best way, we consider, is to make laboratory tests on models where the geometry of the joints and their mechanical properties are reproduced.

III. Values of the Shearing Properties

As has just been shown, prediction of the behaviour of a rock mass under shear requires the knowledge of the shear characteristics of the joints and the rock, i. e. the material of the blocks defined by joints.

Our experience of more than 200 "in situ" shear tests, carried out on samples 70×70 cm in cross section, of rocks of very different kinds has shown that cohesion values, c_b, are as a rule very low and that those of the angle of friction, φ_b, are usually above 55^0, although many of the rocks tested showed marked alteration. The results of these tests are described in detail in a paper to be presented at the next Congress on Large Dams[1].

As for the properties of joints, as mentioned, cohesion c_j is as a rule negligible. In the fairly frequent case of rocks displaying no alteration near joints which present no filling material, the angle of friction is not likely to be much different in the joints, φ_j, and the rock, φ_b.

As for the influence of filling materials, often of a clay nature, on the shear properties of joints, and also of faults, care should be taken to determine whether these materials are continuous or whether the blocks of rock are in direct contact. Our experience shows that, frequently, clay filling materials appear to impair the safety of the rock mass, whereas in fact they do not since it behaves roughly as a rockfill with its voids filled with clayey materials. This scheme, nevertheless, should be considered a two-phase system in which, as in soil mechanics, the applied loads can be transferred from one to the other phase according to their relative deformability. A frequent situation, which corresponds to the scheme in reference, is the case of joints or faults filled with clayey materials brought from the surface by seepage water. Thus, in some cases, the presence of filling material can be favourable, as it increases the watertightness of the rock mass.

The preceding considerations show the importance of following the joints and faults by means of galleries in order to check the evolution of their characteristics, particularly to investigate whether there is direct contact between the blocks of the rock.

In order to investigate the influence of filling material on the behaviour of rock masses, it is of interest to determine the state of stress existing in it. In fact, if these stresses are very low compared to the overburden stresses, it can safely be concluded that the influence of the filling material may be disregarded.

If it is concluded that filling material does not influence the shearing characteristics, joints and faults must be characterized by tests in zones where the blocks are in direct contact.

From the preceding considerations a conclusion of great practical significance can be drawn, concerning the very frequent case where joints are closed for certain lengths and filled with weak material in others. This material need not be taken in consideration if it does not occur continuously to a considerable extent compared to the dimensions of the foundation surface.

The low values of rock cohesion observed make it advisable to neglect cohesion in rock masses, unless the rock mass is of very high quality and the shearing strength to be considered concerns directions for which the imbrication of the blocks is efficient. In this case, if one wishes to determine cohesion with a reasonable accuracy it will be necessary to resort to model tests reproducing joints, which indicate, for each direction, the cohesion in the rock mass as a function of the cohesion in the rock, which must be proportional to each other. The problem becomes obviously simpler when the results of "in situ" tests can be assumed to characterize the rock mass. Only in particularly favourable cases do we consider the sometimes recommended values of 20 to 50 kg/cm^2, multiplied by a factor of safety, an acceptable value for cohesion in the rock mass[2].

As for the angles of friction in rock masses, the results obtained and the considerations previously mentioned show that they can usually be ascribed values above 55^0, which amounts to a coefficient of friction of 1.4. This excludes cases in which joints with continuous fillings of low-strength materials have to be taken into account.

IV. The Concept of Factor of Safety

In the study of the failure of a rock mass a very delicate problem arises, viz. the definition of a factor of safety. As a rule, following the definition usually adopted in other fields, the factor of safety is the number n by which the magnitudes of the forces F applied at the rock mass should be multiplied for failure to occur. That is, $n = \dfrac{F_r}{F}$, F_r being the intensity of the forces causing failure. If the problem under consideration is the safety of a foundation, e. g. of a dam, F represents the forces transmitted by the foundations to the rock mass. If it is the safety of a rock mass under its own weight, for instance a slope, F represents the weight.

Sometimes a coefficient of safety is defined in which the magnitude of only some of the forces increases. Such is usually the case of gravity dams when a factor of safety is defined as the ratio of the intensity of hydrostatic pressure for which rupture would occur — as if the specific gravity of water increased — to the pressure actually applied at the dam, the weight of this remaining constant. In the definition of the factor of safety with respect to sliding along a plane surface, it can in general be assumed, similarly, that the resultant of forces normal to the plane remains constant and that only the resultant of the forces parallel to the plane varies, i. e. $n = \dfrac{T_r}{T}$, where T_r is the value of this resultant which causes failure and T its actual value. In the study of safety against sliding along the foundation surface of a gravity dam or buttress in which the foundation is usually characterized by its angle of friction alone, a factor of safety thus defined has to be used if we require it to remain a ratio of forces. In fact in this case, as also in the analysis of safety against sliding along a joint or fault defined by an angle of friction, if the angle between the resultant of the hydrostatic pressure and the weight, and the normal to the foundation, or the joint or the fault, is less than the angle of friction, no failure will occur whatever the magnitude of the resultant, whereas there must be failure if the angle of friction is exceeded.

Nevertheless, the safety criteria just defined are unsuited to the physical nature of the majority of Rock Mechanics problems, notably those connected to dams and stability of slopes. In fact, in these cases, the forces involved — hydrostatic pressure and weight — are ascribed design values which are never exceeded in service conditions, and consequently to define factors of safety in terms of higher values has no meaning. The uncertainty involved results from precarious knowledge of, and the wide variation in the properties of the rock mass. Therefore, within the modern probability concepts of safety, one should adopt a factor of safety which characterizes, by means of these properties, how far the conditions anticipated in the rock mass are from failure.

Thus, if the rock mass is defined by its angle of friction, it is possible to take as factor of safety, η_1, the ratio between the angle of friction φ which can reasonably be ascribed to the rock mass, and the value φ_R for which rupture will occur, or preferably the ratio of the corresponding tangents, i. e. $n_1 = \dfrac{\tan \varphi}{\tan \varphi_R}$.
As an illustration, let us consider the problem of a foundation that can be considered as a half-space, characterized by $\varphi = 50^0$, subjected to a normal stress $\sigma = 50 \, \text{kg/cm}^2$, acting on a strip, and a stress $\sigma_0 = 1.5 \, \text{kg/cm}^2$ outside the strip. The normal stress causing rupture, σ_R, can be computed by the well-known Prandtl-Caquot formula

$$\sigma_R = \sigma_0 \, e^{\pi \tan \varphi} \tan^2 \left(\frac{\pi}{4} + \frac{\varphi}{2} \right),$$

from which results $\sigma_R = 500 \, \text{kg/cm}^2$. On the other hand this same expression enables the value of the angle of friction φ_R for which rupture would occur under a stress $\sigma = 50 \, \text{kg/cm}^2$ to be computed. This value is $\varphi_R = 35^0$. It is thus possible to define two factors of safety, $n = \dfrac{\sigma_R}{\sigma} = 10$, and $n_1 = \dfrac{\tan \varphi}{\tan \varphi_R} = 1.7$. If σ cannot exceed $50 \, \text{kg/cm}^2$, the factor 10 says little about the safety of the foundation. In fact, although the factor 10 may seem satisfactory — in comparison with the values currently adopted in other fields — the other factor of safety n_1 may lead to the conclusion that safety is unsatisfactory, if the probability of occurrence of an angle of friction of 35^0 is not considered sufficiently low, which depends on the knowledge available on the properties of the rock mass.

Notice that, in the analysis of safety of the foundation against sliding along a plane, the above-considered factors of safety $n = \dfrac{T_R}{T}$ and $n_1 = \dfrac{\tan \varphi}{\tan \varphi_R}$ coincide.
In fact, N being the component normal to the surface, $T_R = N \tan \varphi$ and, as $T = N \tan \varphi_R$, it follows that $n = \eta_1$.

Let us now assume that the rock mass is defined by a cohesion alone. It is possible, likewise, to consider as the factor of safety the ratio between the value, c, of the cohesion ascribed to the rock mass and the value, c_R, for which rupture occurs. In this particular case, the value thus obtained for the safety factor, $n_1 = \dfrac{c}{c_R}$, coincides with the value $n = \dfrac{F_R}{F}$ obtained assuming that the intensities of applied forces increase until rupture takes place.

In the most general case of a rock mass characterized by a cohesion c and an angle of friction φ, it is still possible to characterize safety by a sole coefficient, since as a rule in a given rock mass the values of c and φ are correlated. Thus, let us consider the analysis of the foundation of a dam resting on granites with different degrees of alteration, defined by a quality index i, for which the Coulomb straight lines were obtained by means of *"in situ"* tests (fig. 3). The quality index adopted is the ratio of the weight of water absorbed by the rock dried at 105^0 C to its dry weight[3]. Due to the correlation mentioned, the factor of safety can be defined from the cohesion, from the angle of friction or even from the quality index. If, as is usually the case, no range of values of c and φ — observed in different points of the rock mass — is available up to values corresponding to rupture, a correlation between c and φ has to be ascribed in order to enable the determination of the factor of safety n_1, by computation or by model tests. Given that the angle of friction is more significant, it is preferable to consider n_1 as the ratio of coefficients of friction instead of the ratio of cohesions.

Summing up, we consider that in most cases the factor of safety to be used in the analysis of rock masses should represent the weakening which the rock mass must undergo for failure to take place under the action of the constant loads applied.

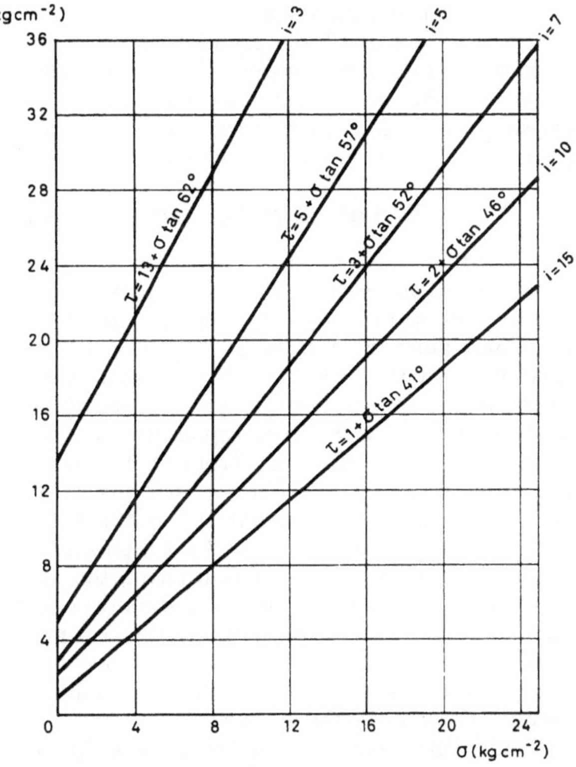

Fig. 3. Coulomb straight lines obtained by in-situ tests in granites with different degrees of alteration

Coulombsche Gerade, erhalten durch in-situ-Versuche mit Graniten verschiedenen Verwitterungsgrades

The preceding considerations clearly show how delicate it is to speak of the safety factor of rock masses without previously defining it. Sometimes it is even attempted to choose the values to be adopted for the factor of safety, without taking this basic precaution. These values and their significance depend to a decisive extent on the concept adopted for the factor of safety. Furthermore it should be noted that, whatever this concept, a given value of this factor does not correspond to a given margin of safety, which depends on the probability of occurrence of forces or properties of the material, according to the concept adopted, for which rupture takes place. Thus, let us consider again a foundation on the granite formation to which fig. 3 relates, and let us admit that in the area under consideration a value $i = 3$, which corresponds to $c = 13$ kg/cm^2 and $\varphi = 62^0$, was ascribed. Let us assume that computations or model tests showed that failure would occur for values of c and φ corresponding to $i = 10$, i. e. for 2 kg/cm^2 and 46^0 respectively, and that these were obtained in the formation under study where the granite is entirely decomposed. The factor of safety defined in terms of the coefficients of friction is only $\frac{\tan 62^0}{\tan 46^0} = 1.8$. But if the knowledge available on conditions in the area shows that the occurrence of a decomposed granite is out of the question, such a value will correspond to an entirely satisfactory safety.

The considerations just presented clearly show that to compare, as is sometimes done, values of the factor of safety of a dam proper and of its foundation is meaningless. Thus, although failure tests on arch dams usually yield values above 10,[4] this does not by any means imply that the safety of the dam necessarily exceeds that of foundations with the usual factors of safety 2 or 3, as we are comparing entirely different magnitudes.

V. Values to be Ascribed to the Factors of Safety

A basic question, that affects the value to be adopted for the factor of safety, is the criterion to be followed in ascribing properties to a rock mass from test results. The problem is very delicate, given the complex properties of rock masses, the small number of tests usually carried out and their sometimes controversial meaning, and the scatter of the results. It seems advisable nevertheless to try to define a general criterion. Two methods of approach are possible: either to adopt mean values or values corresponding to small probabilities of failure, different factors of safety resulting from each alternative. In limit design of reinforced concrete structures, a problem which has been much studied and discussed, a trend is observed towards choosing for ultimate compressive strength of concrete the stress corresponding to a 95 % probability of rupture. If the number of tests usually available for determination of the shearing properties were not so small, we believe that a similar criterion should be adopted in the analysis of rock masses, due above all to the marked scatter of properties. Nevertheless for conditions usually occuring in practice, it seems preferable to characterize the shearing properties of a rock mass by the cohesion and angle of friction corresponding to the average Coulomb straight-line.

Ascribing values to the factor of safety cannot be done automatically. On the contrary it requires a careful assessment of the dispersion of the results available and the extent to which the test samples are representative, in the light of the knowledge available on the rock mass in reference. Another major question to be considered is the type of heterogeneity, i. e. the extent of the areas that can be regarded as homogeneous. In fact, if the foundation surface is large in comparison with that extent the rock mass will behave as if its properties approached those corresponding to the average Coulomb straigt-line and then the factor of safety can be lower.

The capacity of the structure for adjustment to local anomalous behaviour in the foundation by redistribution of the stresses acting on the latter must also be taken into consideration. Thus, for example the overall behaviour of arch dams is very insensitive to zones of weakness, even when these extend over comparatively large volumes of the foundation, which permits us to ignore certain lower values of the properties determined.

As for the factor of safety to be ascribed to the coefficient of friction it is important to note that, according to our experience, angles of friction in a given formation decrease slowly with the degree of alteration. We thus consider it sufficient as a rule to adopt a factor of safety from 1.5 to 2, according to the dispersion (scatter) and dependability of the results. Thus, to a coefficient of friction of 1.4, which is exceeded by practically all the values we have determined, there will correspond, as design values, 0.9 and 0.7. It is usual practice to assume values between 0.6 and 0.8, without performing tests on the rock. It follows, therefore, that these values can in fact be used with satisfactory degree of safety, even in rock masses which are considerably weathered. If tests are carried out, it will be possible, as a rule, to adopt higher values.

As cohesion in the rock mass derives from cohesion in the blocks, the corresponding factor of safety must be fixed taking into account the dispersion of the cohesion in the rock. If the correlation between cohesion and angle of friction of the rock is known, once the factor of safety is chosen for the latter, the one for cohesion will be implicitly determined. Since, however, this correlation, as a rule, is not available, it is advisable independently to obtain a factor of safety for cohesion. Its value depends on the knowledge available on the shear strength of the rock under very small or even zero normal stresses, and consequently on the way the tests were carried out. Contrary to what happens when interest centers especially on obtaining a fairly accurate value for the angle of friction, it is preferable, in the case of cohesion, to concentrate on determinations in the range of low values of the normal stress. As this is a question hardly dealt with by us, given the nature of the rocks tested — with a low cohesion — we possess but very little data on the dispersion of cohesion values. We believe, nevertheless, that, given the marked influence of fracturing on cohesion, the coefficients adopted should be at least the current values, 4 or 5.

It is noteworthy that in the case of dams, even neglecting cohesion, values of the coefficient of friction of an order of magnitude of 0.7 can, as a rule, give the foundation rock mass the required bearing strength. The delicate problem in safety lies in the possible occurrence of joints or faults filled, to a large extent, with materials having poor mechanical properties, oriented so as to make slidings possible. That is why no efforts should be spared to make exhaustive explorations of the geologic features of the rock mass, down to depths below the foundation level, so as to secure detection of every surface of weakness which may influence the behaviour of the structure.

Bibliography

[1] Rocha, M.: Mechanical Behaviour of Rock Foundations in Concrete Dams. Question No. 28, 8th Congress on Large Dams, Edinburgh, 1964.

[2] Creager, Justin and Hinds: Engineering for Dams. Vol. II, p. 298, 1954.

[3] Hamrol, A.: A Quantitative Classification of the Weathering and Weatherability of Rocks. Proc. 5th Int. Conf. Soil Mech. Found. Eng., vol. II, Paris, 1961.

[4] Rocha, M. and J. L. Serafim: Rupture Tests on Arch Dams by Means of Models. Water Power, London, March, April 1959.

Bemerkungen über eine Begriffsgliederung und Klassifikation der Gebirgsstrukturen im Hinblick auf theoretische Untersuchungen gebirgsmechanischer Probleme

Von

Tilo Döring*

Mit 6 Textabbildungen

Zusammenfassung — Summary — Résumé

Bemerkungen über eine Begriffsgliederung und Klassifikation der Gebirgsstrukturen im Hinblick auf theoretische Untersuchungen gebirgsmechanischer Probleme. Die Realstruktur des Gebirgskörpers ist durch geologische Faktoren sowie die spezifischen Verformungseigenschaften der Gesteine gekennzeichnet. Während das Verformungsverhalten der Gesteine in der Regel von der Zeit abhängig ist und darum die Berücksichtigung zeitabhängiger Spannungs-Dehnungs-Beziehungen erfordert, sind die Eigenschaften des Gebirgsverbandes noch zusätzlich durch die Schichtung sowie tektonische Einflüsse bestimmt.

Bei der theoretischen Behandlung gebirgsmechanischer Aufgaben sind die genannten Merkmale zu berücksichtigen. An Hand einer schematischen Gliederung der Gebirgsstrukturen läßt sich zeigen, daß die Kontinuumsmechanik ein brauchbares Hilfsmittel zur Untersuchung dieser Probleme darstellt und in einfachen Grenzfällen bereits die Lösung geomechanischer Aufgaben erlaubt. Diese Betrachtungsweise erscheint darüber hinaus auch geeignet, die oft noch unterschiedlichen Anschauungen über „Gesteins- und Gebirgsverhalten" zu vereinen, indem die Eigenschaften des Gesteins nicht formal auf den Gebirgskörper übertragen werden.

Notes on the Definition and Classification of Rock Mass Fabrics with Regard to Theoretical Investigations of Problems of Rock Mechanics. The real structure of the rock body is characterized by geological factors as well as the specific deformation properties of the rocks. Whereas, as a rule, the deformation behaviour of the rock depends on time and therefore requires taking into account stress-strain relations dependent on time, the properties of the rock mass are additionally determined by stratification and tectonic influences.

In theoretically solving problems of rock mechanics the above mentioned characteristics should be taken into account. It can be shown, in the light of a systematic classification of rock mass fabrics, that continuum mechanics proves to be a useful means for investigating these problems, and in some simple borderline cases even yields the solution of geomechanical problems. Moreover, this approach seems to be suited to combine the still widely varying views on "the behaviour of rocks and rock masses" because the properties of the rock are not formally related to the rock body.

Remarques sur un tableau récapitulatif et une classification des structures des roches en vue d'études théoriques de problèmes de mécanique des roches. La structure réelle du massif rocheux est caractérisée par des facteurs géologiques et par des propriétés de déformation spécifiques de la roche. La déformabilité de la roche est en général fonction du temps, ce qui se traduit par des relations entre les contraintes et les déformations où le temps intervient. En outre, les propriétés du massif rocheux dépendent de la stratification et des influences tectoniques.

---

* Dr.-Ing. Tilo D ö r i n g, Deutsche Akademie der Wissenschaften zu Berlin, Arbeitsstelle für Geomechanik, Freiberg/Sa., DDR.

Pour traiter la mécanique théorique des massifs rocheux, il faut tenir compte des caractéristiques indiquées cidessus. On peut montrer par un tableau schématique des structures rocheuses que la mécanique du milieu continu représente un moyen utilisable pour étudier ces problèmes et qu'il permet déjà de résoudre les problèmes géoméchaniques dans des cas limites. En outre, ce point de départ permet d'unifier les conceptions souvent encore diverdentes sur le comportement de la roche et du massif rocheux puisque les propriétés de la roche ne peuvent s'appliquer telles quelles au massif.

I. Allgemeines

Die theoretische Behandlung geomechanischer Aufgaben führt im allgemeinen zu Oberflächenproblemen in der Praxis des *Ingenieurbaus* und zu *Hohlraumproblemen* in der Praxis des Bergbaus. In beiden Fällen ist Gegenstand der Untersuchungen der *Gebirgskörper,* dessen komplizierte Realstruktur zu *berücksichtigen* ist. Die Frage nach der Verteilung von Spannungen und Verformungen zielt dabei auf die Grundbegriffe der Kontinuumsmechanik, deren Anwendung bei der Lösung dieser Aufgaben große Bedeutung besitzt[11]. Die Anschauung, daß der Gebirgskörper mit einem elastischen, homogenen und isotropen Kontinuum vergleichbar sei, bildete in der Vergangenheit häufig den Ausgangspunkt vieler Untersuchungen und führte natürlich zu der Vereinfachung, die Eigenschaften dieses Gebirgskörpers auch nur durch zwei Parameter zu beschreiben[9].

Im folgenden sollen, ausgehend von einer Klassifikation der Gebirgsstruktur, kurz einige neuere Untersuchungen in ihrer Bedeutung für die Gebirgsmechanik dargelegt werden, die unter Beachtung gewisser Merkmale der Realstruktur des Gebirges u. a. zeigen, daß die Annahmen der Isotropie und des elastischen Verformungsverhaltens im allgemeinen unzureichend sind.

II. Merkmale der Gebirgsstrukturen und ihre Gliederung

A. Geologische Faktoren

Die spezifischen Verformungseigenschaften der Gesteine bilden gemeinsam mit geologisch-tektonischen Merkmalen die Realstruktur des Gebirgskörpers. Während diese Verformungseigenschaften aber wichtiges Merkmal *aller* deformierbaren Körper sind, ergeben die geologisch-tektonischen Merkmale gerade die Besonderheiten der Struktur des Gebirgskörpers und müssen darum Ausgangspunkt einer Klassifikation sein. Wie C l a r[2] hervorhebt, sind die „Trennungsflächen" der Schichtung und Klüftung ein allgemeines Strukturmerkmal von Fels und Gebirge. Es erscheint darum naheliegend zu untersuchen, inwieweit eine Variation dieses Merkmals eine Einteilung der Gebirgsstrukturen erlaubt.

Zur Gliederung der geologisch-tektonischen Eigenschaften des Gebirgskörpers kann ein einfaches Schema nach Abb. 1 dienen. Im Falle der tektonisch ungestörten Lagerung lassen sich die Grenzfälle des ungeschichteten und des geschichteten Gebirges unterscheiden. Weitere Gliederungsmerkmale sind dann nach den spezifischen Verformungseigenschaften und nach den Haft- bzw. Berührungsbedingungen auf den Schichtflächen denkbar.

Im Falle tektonischer Einflüsse lassen sich die zwei Grundtypen derartiger Störungen: Bruch und Faltungserscheinungen, angeben[3].

Im ungeschichteten Gebirge sind ebenso wie im geschichteten Brucherscheinungen möglich und auch häufig. Man unterscheidet hier bekanntlich Klüfte und Spalten sowie Verschiebungsflächen[3]. In diesem Zusammenhang sind auch die geologischen Störungen (z. B. Sprünge) zu nennen.

Wendet man den Begriff der „Trennungsfläche" auf die dargestellte einfache schematische Gliederung der Gebirgsstrukturen an, so zeigt Abb. 2, daß sich alle

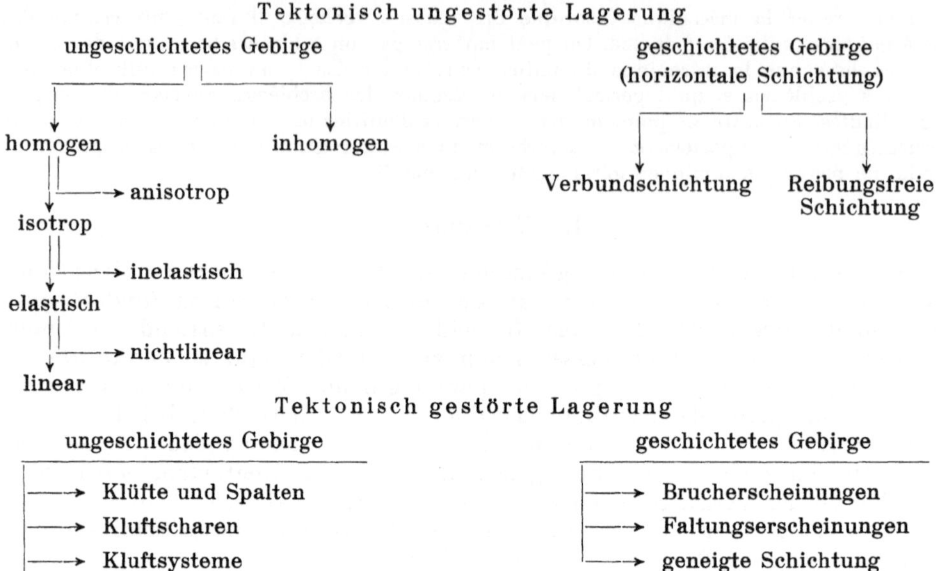

Abb. 1. Schematische Gliederung der Gebirgsstrukturen
Schematic arrangement of rock mass fabrics

genannten Strukturen einordnen lassen, wenn man die geometrischen Eigenschaften der Trennflächen als variables Klassifikationsmerkmal betrachtet. Neben diesen

A) Tektonisch ungestörte Formen:

1. Ohne Trennflächen
      homogen — inhomogen
      isotrop — anisotrop
      elastisch — inelastisch

2. Horizontale Trennflächen (Schichtung)
      Verbundwirkung
      Reibung
      Reibungsfrei

B) Tektonisch gestörte Formen:

3. Geneigte Trennflächen
      Kluftscharen
      Kluftsysteme
      Einfallende Schichtung

4. Gekrümmte Trennflächen (Faltungen)
      Fließformen
      Scherformen
      Biegeformen

5. Unstetige Trennflächen (Störungen)
      Verwerfungen
      Horizontalverschiebung
      Überschiebungen

Abb. 2. Klassifikation der Gebirgsstrukturen
Classification of rock mass fabrics

geometrischen Eigenschaften ist noch die Spannungsübertragung auf den Trennflächen wichtig. Drei Grenzfälle lassen sich hervorheben*:

a) Verbundwirkung $(\sigma_n \gtrless 0,\ \tau_n \gtrless 0)$,

b) Reibung $(\sigma_n > 0,\ \tau_n \gtrless 0)$,

c) Reibungsfrei $(\sigma_n > 0,\ \tau_n = 0)$.

* $(\sigma_n =$ Spannung normal zur Trennfläche, wobei Druckspannungen positiv sein sollen; $\tau_n =$ Schubspannung.)

B. Verformungsverhalten und Verformungseigenschaften

Beim Vorhandensein von Trennflächen wird der Gebirgskörper in Teilkörper zerlegt. Diese Teilkörper zeigen ebenso wie das Gebirge ohne Trennflächen in der Regel eine Abhängigkeit des Verformungsverhaltens von der Belastungs- bzw. Verformungsgeschwindigkeit sowie eine nichtlineare Spannungs-Dehnungs-Kennlinie. Diese Eigenschaften lassen sich im Labor an Gesteinsproben untersuchen. Als Beispiel demonstriert Abb. 3 das Verformungsverhalten eines Tonschiefers bei unterschiedlichen Belastungsgeschwindigkeiten [9].

Für theoretische Untersuchungen ist wichtig, daß die verwendeten Beziehungen und Differentialgleichungen möglichst linear sind, so daß für allgemeine Betrachtungen vor allem die „klassischen Kontinua" nach Hooke, Maxwell und Kelvin in Betracht kommen. Inwieweit lassen nun diese einfachen Spannungs-Dehnungs-Beziehungen bereits eine Interpretation des geschilderten Verformungsverhaltens zu?

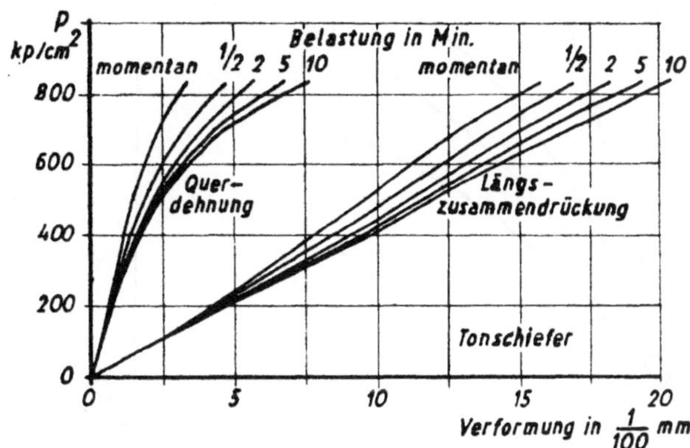

Abb. 3. Spannungs-Dehnungs-Linien für Tonschiefer bei verschieden schneller Belastung (nach Phillips) Stress-strain-curves of slate for constant rates of load

Für einen Maxwell-Körper lautet diese Beziehung zum Beispiel (der Punkt bedeutet Ableitung nach der Zeit)[13]:

$$\dot{\gamma}_{ik} = \frac{1}{2\,G}\,\dot{\tau}_{ik} + \frac{1}{2\,\eta}\,\tau_{ik} \tag{1}$$

mit der allgemeinen Lösung:

$$\tau_{ik} = e^{-\frac{G}{\eta}t}\left\{ \overset{\circ}{\tau}_{ik} + 2\,G \int\limits_{0}^{t} \dot{\gamma}_{ik}\, e^{\frac{G}{\eta}t}\, d\,t \right\}. \tag{2}$$

$$\overset{\circ}{\tau}_{ik} = \tau_{ik} \quad \text{für} \quad t = 0,$$

wobei gilt (über doppelt auftretende Indizes ist von 1 bis 3 zu summieren):

$$\tau_{ik} = \sigma_{ik} - \frac{1}{3}\,\delta_{ik}\,\sigma_{pp}, \tag{3}$$

$$\gamma_{ik} = \varepsilon_{ik} - \frac{1}{3}\,\delta_{ik}\,\varepsilon_{pp},$$

mit σ_{ik} und ε_{ik} als Spannungs- bzw. Verzerrungstensoren.

Eliminiert man für konstante Spannungs- bzw. Dehnungsgeschwindigkeiten in der zweiten Gleichung die Zeit t, so zeigen dafür die Abb. 4 und 5 die bekannten Spannungs-Dehnungs-Kurven, deren Verlauf zum Beispiel für verschiedene Werte der Spannungsgeschwindigkeiten mit den Versuchsergebnissen nach Abb. 3 qualitativ gut übereinstimmt. Die Spannungsabhängigkeit des Verformungsmoduls ergibt sich also aus dem rheologischen Verhalten dieses Materials. In erster Näherung erweist sich damit bereits der **Maxwell-Körper** zur Beschreibung des inelastischen Verformungsverhaltens der Teilkörper des Gebirges als gut brauchbar.

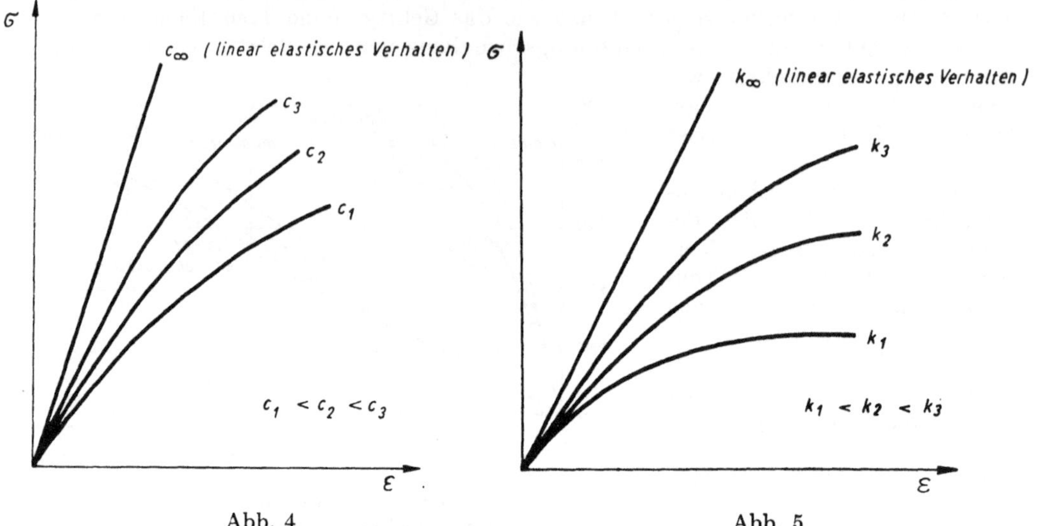

Abb. 4

Abb. 5

Abb. 4. Viskoses Verhalten für $\dot{\sigma} = c_\nu =$ const. (**Maxwell-Körper**)
Stress-strain-curves of a Maxwell-body for constant rates of stress

Abb. 5. Viskoses Verhalten für $\dot{\varepsilon} = k_\nu =$ const. (**Maxwell-Körper**)
Stress-strain-curves of a Maxwell-body for constant rates of strain

III. Über die theoretische Untersuchung geomechanischer Probleme auf der Grundlage der Kontinuumsmechanik

A. Aufgabenstellung

Sind Trennflächen (Schicht- oder Kluftflächen) vorhanden, so ist das Gebirge ein „zusammengesetzter Körper", der aus Teilkörpern, z. B. den einzelnen Schichten, besteht. Die Betrachtung der Spannungsverteilung in einem derartigen Medium erscheint auf der Grundlage der Kontinuumsmechanik wie folgt möglich:

a) an die Stelle des realen Gebirgskörpers tritt ein „Ersatzmedium";

b) aus der Betrachtung der Eigenschaften und der Beanspruchung der Teilkörper wird unter Berücksichtigung ihrer Randbedingungen die Spannungsverteilung im gesamten Gebirgskörper ermittelt.

Der erste Weg liefert je nach dem Grad der Annäherung des Ersatzmediums an die Realstruktur „statistische Mittelwerte", d. h. durchschnittliche Werte für den Spannungstensor in den einzelnen Punkten, während der zweite Weg einer strengen analytischen Lösung entspricht. Im ersten Fall liegen die Schwierigkeiten bei der Lösung der Aufgabe u. a. darin, die Parameter des Ersatzmediums eindeutig aus den gegebenen Kennziffern der Teilkörper zu ermitteln, während beim zweiten Fall vor allem die allgemeine Lösung des Spannungsproblems für die Teilkörper schwierig zu ermitteln ist.

B. Horizontale Trennflächen (Schichtung)

Als typische Beispiele der genannten zwei Betrachtungsweisen können bei elastisch-isotropem Verhalten der Teilkörper und ebenen Verformungszustand die Arbeiten von **Kafka**[8] und **Bufler**[1] dienen.

Kafka berücksichtigt den Einfluß der Schichtung durch die Einführung eines orthotropen Ersatzmediums (mit einer Isotropie-Ebene), wobei die Konstanten dieses Mediums aus den Parametern der Teilkörper (Schichten) berechnet werden. Mithin ist dieses Ersatzmedium nicht willkürlich vorgegeben, sondern seine Eigenschaften widerspiegeln eindeutig die Eigenschaften der Schichtung. Unabhängig von dieser Arbeit wandte auch Mincev[10] dieses Verfahren bei der Lösung von Hohlraumproblemen an.

Der analytisch strenge Weg Buflers geht von der allgemeinen Lösung für die streifenförmige Scheibe mit beliebiger Randbelastung durch Anwendung der Fourier-Transformation aus und berücksichtigt dann die Berühr- und Haftbedingungen auf den Schichtflächen des zusammengesetzten Gebirgskörpers für die Fälle der Reibungsfreiheit und der Verbundwirkung auf diesen Trennflächen.

Während der zweite Weg bisher nur bei oberflächenparalleler Schichtung und Oberflächenproblemen anwendbar ist, lassen sich die Ergebnisse von Kafka auch auf beliebig geneigte Schichten mit Verbundwirkung anwenden. Im allgemeinen muß dabei aber vorausgesetzt werden, daß die Mächtigkeit der Schichten relativ klein ist im Verhältnis zu den betrachteten Abmessungen des Untersuchungsgebietes bzw. zur Größe des Hohlraumes bei Hohlraumproblemen.

Für den Grenzfall horizontaler Trennflächen ohne Reibung hat Sonntag[15] auf der Grundlage der Biegetheorie vereinfachte Betrachtungen eingeführt. Die Anwendung der Biegetheorie empfiehlt sich aber auch dann, wenn kompliziertere Hohlraumprobleme zu untersuchen sind, insbesondere wenn diese Hohlräume parallel zur Schichtung bedeutend größere Abmessungen besitzen als in der Vertikalrichtung, wie dies beispielsweise auf den Strebbau zutrifft. Die dann gewöhnlich im Hangenden eintretenden Schichtablösungen und Risse erlauben faktisch nur noch eine modellartige Behandlung dieser Aufgaben[4].

Eine Verallgemeinerung aller genannten Untersuchungen ergibt sich zum Beispiel, wenn die Teilkörper selbst anisotrop sind oder einem zeitabhängigen Verformungsgesetz folgen, wie auch aus der beschriebenen Klassifikation (Abb. 2) hervorgeht. Die von Sonntag[15] angegebenen Beziehungen über die dünnbankig und reibungsfrei geschichtete Halbebene lassen beispielsweise die Berücksichtigung viskoser Verformungseigenschaften, die ja als brauchbar erkannt wurden (Abschnitt 2.2), leicht zu*.

Die Berücksichtigung anisotroper Verformungseigenschaften der Schichten ist andererseits nach der Methode von Kafka unschwer möglich.

Der von Bufler eingeschlagene Weg läßt ebenfalls die Lösung des Spannungsproblems, z. B. bei ortotropem oder inelastischem Verformungsverhalten der Einzelschichten, zu, wenn es gelingt, die entsprechenden allgemeinen Lösungen für die Teilkörper (Schichten) anzugeben; d. h. bei ortotropem Verhalten ist anstelle der bekannten Bipotentialgleichung $\Delta \Delta F = 0$ die Gleichung

$$a_{22}\,\frac{\partial^4 F}{\partial x^4} + (2\,a_{12} + a_{66})\,\frac{\partial^4 F}{\partial x^2\,\partial y^2} + a_{11}\,\frac{\partial^4 F}{\partial y^4} = 0 \tag{5}$$

---

* Unter den genannten Voraussetzungen läßt sich für die Spannungsverteilung folgende Differentialgleichung angeben (die y-Richtung zeigt senkrecht zu den Schichtflächen[15]):

$$t_0\left[\frac{h^2}{12}\,\frac{\partial^4 \dot{\sigma}_y}{\partial x^4} - \frac{d^2 \dot{\sigma}_y}{\partial y^2}\right] + \left[\frac{h^2}{12}\,\frac{\partial^4 \sigma_y}{\partial x^4} - \frac{\partial^2 \sigma_y}{\partial y^2}\right] = 0, \tag{4}$$

mit t_0 als Relaxationszeit. Bei zeitunabhängigem Verformungsverhalten geht diese Gleichung in die von Sonntag abgeleitete Beziehung über (Punkte bedeuten wieder Ableitungen nach der Zeit).

und bei inelastischem Verhalten z. B. nach der Art des erwähnten Maxwell-Körpers die Gleichung

$$t_0\, \Delta \Delta \dot F + \Delta \Delta F = 0 \tag{6}$$

für den unendlichen Streifen zu lösen, wobei hier die a_{ik} den elastischen Konstanten des orthotropen Mediums entsprechen und t_0 die Relationszeit darstellt.

C. Geneigte Trennflächen, insbesondere Klüftung des Gebirgskörpers

Die meisten Klüfte sind nach Cloos[3] „mathematisch eben". Parallele Klüfte bzw. Kluftscharen treten ebenfalls sehr häufig auf und sind beispielsweise auch typisch beim Strebbau. Kluftsysteme, d. h. sich kreuzende Kluftscharen, zerlegen den Gebirgskörper in mehr oder minder regelmäßige Teilkörper; sie gaben besonderen Anlaß, den Gebirgskörper als „geregeltes Teilkörpersystem" zu betrachten[12].

Es ist plausibel, daß in einem Kluftkörper die Frage nach dem „Grenzgleichgewicht", d. h. nach dem gerade noch möglichen Gleichgewicht, bei dessen Überschreiten ein Gleiten der Teilkörper auf den Kluftflächen einsetzt, eine besondere Rolle spielt. Über Abbauhohlräumen hat diese Frage unter Umständen große Bedeutung, weil mit wachsenden Hohlraumabmessungen die Gefahr einer eintretenden Instabilität und somit eines plötzlichen Einsturzes besteht. Als Ersatzmedium könnte für einen solchen Gebirgskörper beim Strebbau beispielsweise ein *zähes Kontinuum* dienen*.

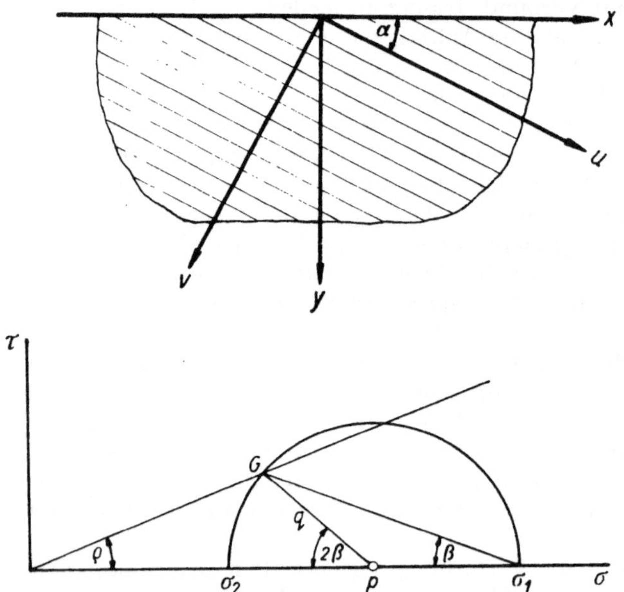

Abb. 6. Darstellung der Koordinaten und des Spannungszustandes in einem Punkt für einen Gebirgskörper mit einer parallelen Kluftschar

Coordination and state of stress at a point for a rock mass with an inclined joint set

Die Behandlung dieser Aufgaben stößt jedoch schon bei einfachen Annahmen (Biegetheorie) auf erhebliche mathematische Schwierigkeiten, weil sich im allgemeinen auch die Randbedingungen mit der Zeit verändern.

Die Untersuchung der statischen Seite des Grenzgleichgewichtes ist für den ebenen Verformungszustand unabhängig von den Verformungen in Anlehnung an das ebene Plastizitätsproblem möglich.

Zusammen mit der Gleitbedingung

$$g\,(\sigma_x,\, \sigma_y,\, \tau) = 0 \tag{7}$$

* Anregungen zu dieser Auffassung enthalten z. B. Arbeiten von Jacobi[7] und Znanski[15].

auf den Kluftflächen liefern die Gleichgewichtsbedingungen ($\tau_{xy} = \tau_{yx} = \tau$):

$$\frac{\partial \sigma_x}{\partial x} + \frac{\partial \tau}{\partial y} = 0,$$
$$\frac{\partial \tau}{\partial x} + \frac{\partial \sigma_y}{\partial y} = 0 \tag{8}$$

drei Beziehungen zur Bestimmung der drei Unbekannten σ_x, σ_y, τ.

Liegt beispielsweise nur eine parallele Kluftschar vor, deren Trennflächen mit der x-Achse den Winkel α einschließen (Abb. 6) und für die das Reibungsgesetz

$$\tau_{uv} = \mu \,.\, \sigma_v \tag{9}$$

auf diesen Flächen gelten soll, so folgt aus der Mohrschen Darstellung des Spannungszustandes in einem Punkte der Trennfläche wegen (9) die Beziehung (Abb. 6)

$$\mu \,.\, q \,.\, \cos 2\,\beta + q \,.\, \sin 2\,\beta = \mu\, p \tag{10}$$

mit $\mu = \mathrm{tg}\,\varrho$, wobei β der Winkel zwischen der Richtung der Kluftfläche und der Richtung der größeren Hauptspannung ist. Wird mit φ der Winkel zwischen der x-Achse und der Richtung der größeren Hauptspannung bezeichnet, so gilt:

$$\beta = \varphi - \alpha$$

und folglich:

$$q \,.\, \cos 2\,\varphi\,(\mu \,.\, \cos 2\,\alpha - \sin 2\,\alpha) + q \,.\, \sin 2\,\varphi\,(\mu \,.\, \sin 2\,\alpha + \cos 2\,\alpha) = \mu\, p \tag{11}$$

mit

$$\sigma_x + \sigma_y = 2\, p.$$
$$\sigma_x - \sigma_y = 2\, q \,.\, \cos 2\,\varphi, \tag{12}$$
$$\tau = q \,.\, \sin 2\,\varphi.$$

Für $\alpha = 0$ folgt aus (11) wieder (9). Das angegebene Gleichungssystem für die Unbekannten σ_x, σ_y und τ ist in allgemeiner Form lösbar*.

Sind Kluftsysteme, d. h. mehrere sich kreuzende Kluftscharen vorhanden, und ist die Größe der Teilkörper zwischen diesen Trennflächen genügend klein, so kann man diesen Kluftkörper in erster Näherung mit einem lockeren Körper im Sinne der Bodenmechanik vergleichen, um das Grenzgleichgewicht in ihm zu bestimmen**. Die Gleitbedingung (7) wird zur Hüllkurve der Spannungskreise, so daß keine bestimmte Gleitrichtung von vornherein ausgezeichnet ist, was für ein stark zerklüftetes Gebirge sicher zutrifft.

D. Über gekrümmte und unstetige Trennflächen

Faltungen des Gebirgskörpers sind in erster Linie bei der Lösung von Hohlraumaufgaben zu beachten, weil der Grundspannungszustand in einem derartigen Gebiet sehr kompliziert sein kann. Die Ursachen der Faltungsbildung sind sehr vielfältig und können horizontale und vertikale Kräfte, Scherbeanspruchung sowie Fließerscheinungen sein. Erschwerend tritt hinzu, daß die wirksamen tektonischen Kräfte im allgemeinen auch zeitabhängig sind.

---

* Vgl. eine demnächst in dieser Zeitschrift erscheinende Arbeit des Verfassers: „Das Gleichgewicht zerklüfteter Gebirgskörper mit einer parallelen Kluftschar."
** Die Berücksichtigung einer gewissen Kohäsion ist natürlich möglich.

Für ein isotropes elastisches Kontinuum hat Golecki[5] interessante Ergebnisse über die Spannungsverteilung in der Umgebung einfacher Faltungsformen, d. h. also in der Nachbarschaft einer gekrümmten Trennfläche, veröffentlicht, indem er diese Aufgabe auf ein Verschiebungsproblem für die Halbebene zurückführt.

Nadai[13] hat unter Berücksichtigung der Schichtstruktur des Gebirges und inelastischer Verformungseigenschaften der Schichten Faltungserscheinungen auf der Grundlage der Biegung auf nachgiebiger Unterlage interpretiert und dabei auch eine Zeitabhängigkeit der äußeren Kräfte berücksichtigt.

Über die Spannungsverteilung in der Umgebung vertikaler Verwerfungen, d. h. bei unstetigen Trennflächen, hat schließlich ebenfalls Golecki[6] für das elastisch-isotrope Kontinuum Untersuchungen vorgenommen.

E. Schlußfolgerungen

Die dargestellten Bemerkungen zeigen, daß infolge vorhandener Trennflächen das Gebirge aus Teilkörpern besteht. Die Formen dieser Teilkörper sind durch die Verteilung und die Geometrie der Trennflächen gegeben, d. h. der gesamte Gebirgskörper stellt ein „geregeltes Teilkörpersystem" dar, dessen „Regelung" eben eine Funktion der Trennflächengestaltung ist.

Auf der Grundlage der Kontinuumsmechanik erscheint die theoretische Behandlung derartiger „Teilkörpersysteme" (zusammengesetzte Körper) prinzipiell möglich. Dabei konnte an Hand der dargelegten Klassifikation der Gebirgsstrukturen kurz gezeigt werden, daß für einfache Formen dieser Teilkörpersysteme bereits allgemeine Lösungsverfahren entwickelt worden sind.

Die kontinuumsmechanische Betrachtung des Gebirgskörpers erscheint darüber hinaus auch geeignet, die oft noch unterschiedlichen Anschauungen über Gesteins- und Gebirgsverhalten zu vereinen, indem die Eigenschaften des Gesteins nicht formal auf den Gebirgskörper übertragen werden.

Literatur

[1] Bufler, H.: Die Bestimmung des Spannungs- und Verschiebungszustandes eines geschichteten Körpers mit Hilfe von Übertragungsmatrizen. Ingenieur-Archiv, *31*, (1962), 4, S. 229—240.

[2] Clar, E.: Gefüge und Verhalten von Felskörpern in geologischer Sicht. Felsmechanik und Ingenieurgeologie, *1* (1963), S. 4—15, Springer-Verlag, Wien

[3] Cloos, H.: Einführung in die Geologie. Verlag von Gebr. Borntraeger, Berlin 1963.

[4] Döring, T.: Modellvorstellungen zur Deutung von Senkungs- und Druckerscheinungen beim Strebbau in flacher Lagerung. Freiberger Forschungshefte, A 111 (1959), S. 35—101.

[5] Golecki, J. C.: Eine Methode der angenährten Bestimmung der Spannungsverteilung in der Umgebung von Faltungen. Arch. Gornictwa, Warszawa, *6* (1961), 4, S. 275 bis 282.

[6] Golecki, J. C.: Die Verschiebungs- und Spannungsverteilung im Gebirge in der Nachbarschaft zweier vertikaler Verwerfungen. Arch. Gornictwa, Warszawa, 7 (1962), 1, S. 27—48.

[7] Jacobi, O.: Der Druck auf Flöz und Versatz. Glückauf, *96* (1960), 7, S. 409—418.

[8] Kafka, V.: Der Spannungszustand in einem geschichteten Medium. Acta technica Nr. 3, 1957, S. 262—289.

[9] Link, H.: Zur Querdehnungszahl von Gestein und Gebirge. Geologie und Bauwesen, *27* (1962), 2, S. 89—100.

[10] Mincev, J. T.: Über den Spannungszustand in einem geschichteten Gebirge um Grubenbaue mit elliptischem, kreisförmigem und trapezförmigem Querschnitt. Bericht über das 3. Ländertreffen des IBG, Akademie Verlag, Berlin 1962, S. 52—63.

[11] Müller, L.: Die Geomechanik in der Praxis des Ingenieur- und Bergbaus. Geologie und Bauwesen, *25* (1960), 213, S. 203—214.

[12] Müller, L.: Die Standfestigkeit von Felsböschungen als spezifisch geomechanische Aufgabe. Felsmechanik und Ingenieurgeologie, *1* (1963), H. 1, S. 50—71.

[13] Nadai, A.: Therry of Flow and Fracture of Solids. Mc. Graw-Hill Book Company, New York 1963 (Bd. 2).

[14] Sonntag, G.: Die in Schichten gleicher Dicke reibungsfrei geschichtete Halbebene mit periodisch verteilter Randbelastung. Forsch. Ing. Wes., *23* (1957), S. 3—8.

[15] Znanski, J.: Kennzeichen eines allmählichen oder plötzlichen Lockerungsfließens des Gebirges. Int. Kongreß f. Gebirgsdruckforschung, Paris 1960, S. 223—233 (deutsch).

Large Scale Laboratory Tests
of the Shear Strength of Rocky Material

By

D. Krsmanović* and Z. Langof**

With 7 Figures

Summary — Zusammenfassung — Résumé

Large Scale Laboratory Tests of Shear Strength of Rocky Material. Large scale laboratory tests of shear strength in calcareous rock mass were recently made with the purpose of examining the stability of an arch (concrete) dam. To achieve this purpose the tests were carried out by means of a special large shear apparatus with a capacity of 60 (and 120) tons in the vertical and horizontal directions. The samples were 40×40 cm, with a height of about 20 cm.

The rock mass tested was a cleanly stratified limestone the layers of which were intersected by joint systems. The geological columnar profile showed here and there intersected with solid limestone thin plastic layers of clay, clayey marl and peat ranging in thickness from 1—40 cm.

A classification of stratification planes and joints was devised during the tests, giving particular consideration to their roughness and the nature of the filling. In addition, the thin plastic layers were also classified by thickness and composition of material.

Altogether about 70 samples were tested for shear strength, and the thin plastic layers were tested with respect to compressibility. The results of this kind of tests are of great value when the samples contain planes of different shear characteristics.

On the basis of these tests it was possible to gain additional knowledge both in regard to qualitative and quantitative parameters and also to determine their interrelation, obtaining a better knowledge of the nature of shear-lines in the "Shear Stress-Strain" diagram.

Together with other geological data the results of the tests can be of great use in the analysis of model tests and determination of potential sliding planes. On the basis of these investigations approximate values of the safety factor of the dam in relation to its abutments can be obtained.

Großmaßstäbliche Laboratoriumsuntersuchungen über die Scherfestigkeit des Gebirges. Im Rahmen der Untersuchungen über die Fundamentsstabilität einer Bogenmauer sind umfangreiche Prüfungen der Scherfestigkeit des dort anstehenden Kalkgesteines durchgeführt worden. Diese wurden mit Hilfe eines speziellen Schergerätes, das die Aufprägung einer Höchstlast von 60 Tonnen in senkrechter und 120 Tonnen in waagrechter Richtung ermöglicht, ausgeführt. Die Versuchskörper hatten 40×40 cm Grundfläche und eine Höhe von zirka 20 cm.

* Professor Dr.-Ing. Dusan K r s m a n o v i ć, Geotechnical and Foundation Engineering Institute, University of Sarajevo, Sarajevo, Yugoslavia.

** Ing. Z. L a n g o f, Collaborator of Institute of Geotechnics and Foundation Engineering, College of Civil Engineering, Sarajevo University, Sarajevo, Yugoslavia.

Die untersuchte Felsmasse bestand aus Kalkstein mit deutlich ausgebildeter Schichtung, der von Rißsystemen durchschnitten war. Das geologische Profil zeigt stellenweise teils dickere, teils dünnere plastische, tonige, mergelige und kohlehaltige Schichten. Die Dicke der Schichten bewegt sich zwischen 1 und 20 cm.

Anläßlich der Untersuchungen wurde eine Klassifikaton der Sedimentationsfugen und der Klüfte sowohl hinsichtlich der Rauhigkeit als auch des zertrümmerten Kluftfüllungsmaterials vorgenommen. Weiters wurden die plastischen Schichten nach der Art der Materialien und nach ihrer Dicke klassifiziert. Auf Grund dieser Einteilung wurden gegen 70 Versuchskörper unter Berücksichtigung der Zusammendrückbarkeit, insbesondere auch der plastischen Schichten, auf Scherfestigkeit untersucht. Die Versuchsergebnisse zeigten, daß bei Kluftflächen mit verschiedener Charakteristik der Scherfestigkei solche Untersuchungen von großem Nutzen sind. Auf Grund dieser Untersuchungen war es möglich, einzelne Parameter qualitativ und quantitativ eingehender kennenzulernen, ihre gegenseitige Abhängigkeit zu bestimmen und Scherspannungs-Deformations-Diagramme aufzutragen. Die Zuordnung der Spannungen σ und τ zu den Deformationen Δs wurde ebenfalls dargestellt.

Die Ergebnisse der durchgeführten Untersuchungen, zusammen mit anderen geologischen Angaben, dienen als Grundlage sowohl für die Berechnung als auch für die Modellversuche zur Bestimmung der potentiellen Bruchflächen. Daraus kann man die Sicherheit der Staumauerwiderlager abschätzen und in Beziehung zum Sicherheitsgrad der Staumauer bringen.

Essais de laboratoire sur la résistance au cisaillement des roches à grande échelle. Dans le cadre des recherches sur la stabilité des fondations des barrages voûtes, on a effec·tué des essais d'une grande portée sur la résistance au cisaillement des calcaires. Ces essais ont été exécutés à l'aide d'un appareil de cisaillement spécial capable d'efforts verticaux de 60 tonnes et d'efforts horizontaux de 120 tonnes. La taille de l'éprouvette est 40 sur 40 cm, avec une hauteur de 20 cm.

Le massif rocheux calcaire étudié avait une stratification très marquée, et les bancs étaient entièrement traversés par un réseau de joints. La coupe géologique comportait par endroits des couches plastiques, argileuses, marneuses ou charbonneuses, d'une épaisseur comprise entre 1 et 20 cm.

A l'occasion de ces essais, on a enterpris une classification des joints de sédimentation et de diaclases, tant d'après leur rugosité que d'après la nature du matériau détritique remblissant les joints. On a classé aussi les couches plastiques d'après la nature et l'épaisseur de leur matériau. Conformément à ces classifications on a examiné la résistance au cisaillement de 70 éprouvettes environ.

Les résultats obtenus ont montré l'utilité de ces essais lorsqu'il existe diverses surfaces à propriétés de cisaillement variées. On a pu connaître plus en détail certains paramètres, à la fois qualitativement et quantitativement, déterminer leurs relations, et se familiariser avec les graphiques donnant la contrainte de cisaillement en fonction de la déformation. On a représenté aussi la relation des contraintes normales et tangentielles avec les déformations.

Les résultats des essais effectués peuvent être utilisés avec d'autres données géologiques, tant pour le calcul que pour des essais sur modèle, pour déterminer les surfaces de rupture potentielles et pour obtenir des valeurs approchées du coefficient de sécurité d'une construction en ce qui concerne la stabilité de sa fondation.

Problems of foundations for heavy structures built on rocky material are mainly concerned with the question of determining the mechanical characteristics of the rock mass (shearing resistance, compressibility, permeability, etc.). In this regard, the rock mass should be considered as discontinua of very variable mechanical properties, dependent on a number of factors (kind and nature of intact rock, stratification and other surfaces of discontinuity, etc.).

In order to determine the mechanical characteristics of the mountain mass, it is general practice to undertake various kinds of mechanical tests, usually performed in-situ and categorized according to the size of test blocks or surfaces on the site.

Thus, among the tests for shearing resistance there are the *large scale* tests that have found wide application in recent years — on the recommendation of Dr. Müller[1] — at various damsites, where test specimens of several m³, or even several tens of m³ are used, the test surfaces measuring several m², up to approx. 15 m².

Next, there are the earlier *medium-scale* tests, made with considerably smaller test specimens and surfaces (of the order of approx. 1 m³ and 1 m² respectively).

Fig. 1. Laboratory direct shear press. Capacity 60/120 t

Laboratoriums-Scherpresse. Leistung 60/120 t

Presse de cisaillement 60/120 tonnes

Such *in situ* tests, particularly the former, and the resultant data are undoubtedly the most reliable and should therefore be made in any case.

However, owing to practical difficulties, the length of time needed and high costs, the number of tests is usually limited, so that it consequently becomes practically impossible to satisfy the other requirement, namely that the tests should produce data over a *statistical range,* and not merely *isolated* results.

To this end, it is expedient that *large-scale laboratory tests be combined with those in situ* to facilitate a more detailed classification of shear conditions and the shearing resistance of individual discontinous surfaces and, at the same time, to obtain relevant data on the correlation between stresses and deformations.

Tests of this kind were made in the building of the "Grančarevo" dam now under construction. The following is an account of the *methods used* and the experience gained.

The whole dam cross-section lies in liassic limestone with clearly defined stratification, revealing, at intervals, two systems of fissures of considerable depth. In addition, thin layers of plastic materials, approximately 1—20 cm thick, partly clayey and partly marlous-ligneous occur sporadically between the limestone layers.

The fissures are either closed or filled with detrital material. The nature of roughness of both the stratification planes and the fissures differ considerably; here and there between the stratification planes there are finely laminated rocky or quasi-rocky thin layers, which influence the shearing resistance.

It was obvious that in the presence of such a discontinuum complex satisfactory results could not be obtained by in situ tests alone. It was therefore necessary to make a series of tests that would take into account the values of shearing resistance of all varieties along the discontinuous surfaces.

Large-scale laboratory tests were carried out using a 60/120 ton capacity shear jeck. The size of samples was $40 \times 40 \times 20$ cm, so that the shear surfaces were approximately 1600 cm². Normal pressures ranged from 2 kg/cm² up to approximately 40 kg/cm² (see Fig. 1).

As shown in Fig. 3, there are three typical $\tau - \Delta s$ lines (obtained at approximately the same values of normal stress σ).

The first type occurs in the solid rock where there is cohesion (the line for the Series "A"). Its characteristics are very high shear resistance with slight deformations necessary to develop maximum shear resistance: Δs approx. 0.1 to

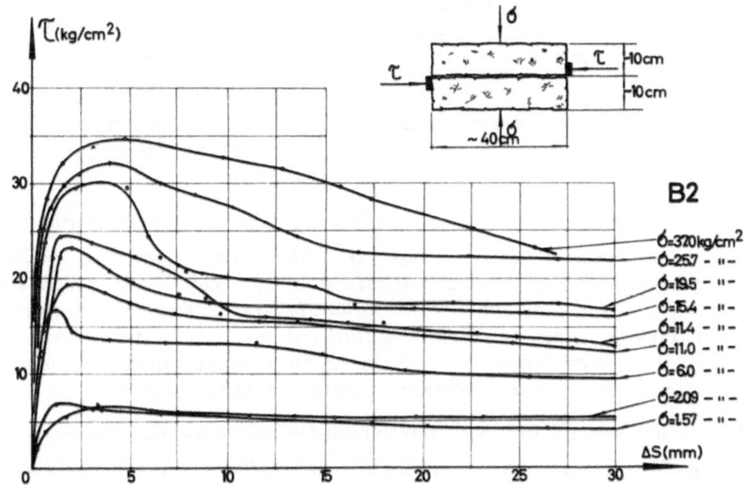

Fig. 2. Diagrams $\tau - \Delta s$ for samples of Series B 2. Illogical shapes of some lines evidence momentary influences of roughness

Diagramme $\tau - \Delta s$ für Versuchskörper der Serie B 2. Unlogische Formen einzelner Kurven sind auf augenblickliche Einflüsse der Rauhigkeit zurückzuführen

Graphique τ, Δs pour éprouvettes de la série B 2. Les formes anormales de certaines courbes témoignent de l'influence momentanée de la rugosité

0.5 mm. The high value of the relation τ_{max}/τ_{ult} is also to be noted, which in the cases tested was always greater than 2.0.

The second typical $\tau - \Delta s$ diagram, is observed for stratification surfaces of different degrees of roughness (lines for the Series B 1—B 3) and fissures of great roughness (line of Series D 2 and D 5). Their characteristics are (1) the deformations necessary to develop shear resistance are considerably greater (2—5 mm), and, (2) the value of the relation τ_{max}/τ_{ult} ranges between 1.0 and 2.0.

The third $\tau - \Delta s$ line is characterized by slight or great deformations necessary for activisation of shear resistance, but in this case the relation τ_{max}/τ_{ult} is approx. equal to 1.0 (Series $C\,1$ to $C\,6$, $D\,1$, $D\,3$, $D\,4$).

The influence of the amount of deformation on the decrease in shear resistance is shown in Fig. 4, where for different series ($B\,1 - B\,3$, $D\,1$, $D\,2$ and $D\,5$) the lines of shear for slight deformations (up to 2 mm, see Fig. 4, to the left) are compared

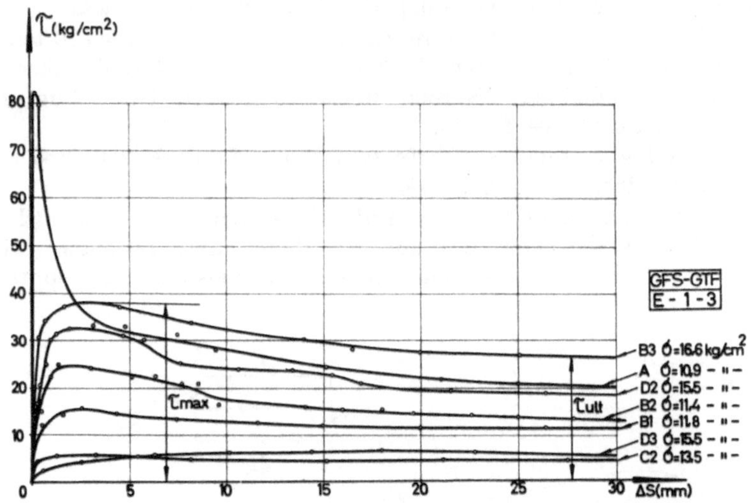

Fig. 3. Diagrams $\tau - \Delta s$ for typical samples of various series

Series "A": *Limestone, intact rock.* — Series "B": *Stratification surfaces (bedding joints).* *1* thin calcareous foliated layer; *2* rough stratification surface; *3* very rough stratification surface. — Series "C": *Thin plastic sedimentation layers.* *1* thick (5—20 cm); *2* medium (2—5 cm); *3* thin clayay layers (to 2 cm); *4* thick; *5* medium; *6* thin, marly and lignite layers. — Series "D": *Clean fissures.* *1* rough; *2* very rough surface of joints. *Fissures with detrital material.* *3* rough surface of joints; *4* very rough surface of joints

Diagramme $\tau - \Delta s$ für typische Versuchskörper verschiedener Serien

Serie „A": *Kompaktes Kalkgestein.* — Serie „B": *Schichtflächen (Bankungsklüfte).* *1* dünnblättrige Kalksteinschichten; *2* rauhe Schichtflächen; *3* sehr rauhe Sedimentationsflächen. — Serie „C": *Dünne plastische Sedimentationsschichten.* *1* dicke (5—20 cm), *2* mittlere (2—5 cm), *3* dünne tonige Schichten (bis 2 cm); *4* dicke, *5* mittlere, *6* dünne mergelig-lignitische Schichten. — Serie „D": *Offene Klüfte ohne Füllungsmaterialien.* *1* rauhe, *2* sehr rauhe *gefüllte Klüfte; 3* rauhe, *4* sehr rauhe Kluftflächen

Graphique τ, Δs pour éprouvettes typiques de diverses séries

Série "A": *Calcaire intact.* — Série "B": *Joint de stratification.* *1* banc de calcaire feuilleté; *2* joint de stratification non rugueux; *3* joint de stratification rugueux. — Série "C": *Couches sédimentaires plastiques.* *1* minces, *2* moyennes, *3* couches minces argileuses; *4* épaisses, *5* moyennes, *6* couches minces marno-charbonneuses. — Série "D": *diaclases.* *1* non rugueuses; *2* rugueuses sans matériau de remplissage; *3* non rugueuses avec matériau de remplissage; *4* rugueuses avec matériau de remplissage

with those for deformations up to 25—30 mm (see Fig. 4, to the right). It has been found that this influence is considerable in the case of shearing of rock surfaces (in clean fissures and stratification planes), and also that the influence is greater when rough rather than smooth surfaces are involved. As is well known, the fall off of shear resistance increases with increasing normal stresses; the fall off in our case amounted to over 30 per cent for greater deformations.

For comparison, the same Fig. also shows the dependence for plastic materials (clayey thin layers and detrital filling material of the Series $C\,1{-}C\,6$, $D\,3$ and $D\,4$). Within the mentioned deformation limits no fall off of shear resistance was observed, the materials behaving according to Coulomb's law (see Fig. 4, to the left).

The diminution of shear resistance with increase in normal stresses is shown in the diagram of τ_{max}/σ versus τ in the manner of Grišin and Evdokímov, 1961[2]. The line $\tau_{max}/\sigma = 1$ in the diagram corresponds to a shearing resistance angle of 45^0. The lines given are those for limestone without fissures (Series „A") and other shear surfaces. The curved lines correspond to shearing of rock surfaces, and the horizontal lines to surfaces of thin plastic layers (clayey or rocky with foliation) and layers with various filling material. The line "$D\,5$" is characteristic, and corresponds to samples of solid rock that have sheared and whose shear path has been lengthened subsequent to overcoming the rock cohesion. Shearing resistance in this sort of fissure proved to be considerably greater than in any

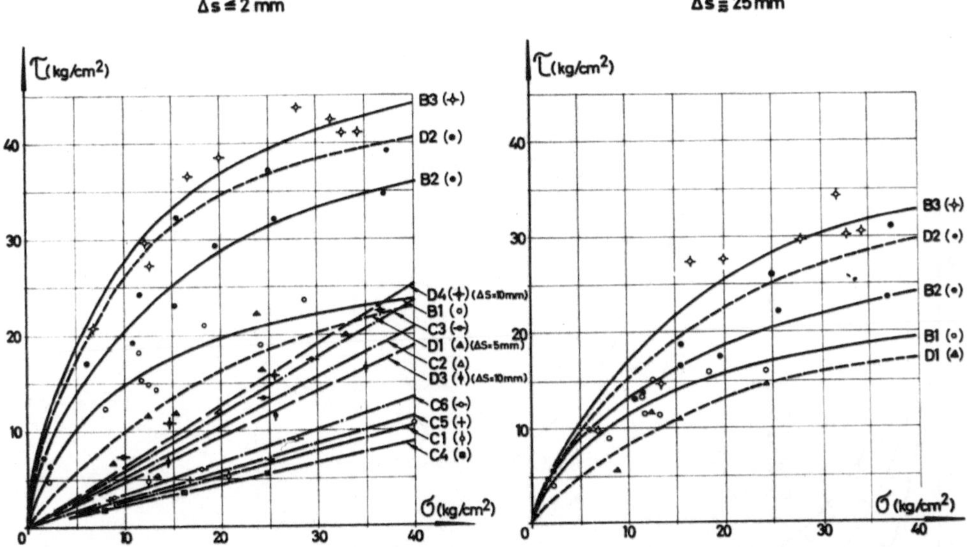

Fig. 4. Diagrams $\sigma - \tau$ for several different series. To the left, lines $\sigma - \tau$ for deformations 2 mm; to the right, for deformations of up to 25 mm

Diagramme $\tau - \sigma$ für einige verschiedene Serien. Links Kurven $\tau - \sigma$ für Deformationen bis 2 mm; rechts für Deformationen bis zirka 25 mm

Graphique τ, σ pour diverses séries: à gauche pour déformations jusqu'à 2 mm; à droite pour déformations jusqu'à 25 mm environ

fissure with filling. Indeed, this resistance is greater than that encountered in a clean smooth limestone fissure ("$D\,1$") while at the same time being considerably less marked in rough fissures.

In addition, the influence of roughness proved to be particularly strong in fissures, as can be seen by comparing the line „$D\,1$" with „$D\,2$". Rough surfaces without fillings and rough stratification surfaces show approximately equal shear resistance.

 D. Krsmanović and Z. Langof:

Correlation of Stresses and Deformations

In order to use the results of the shear resistance tests correctly it is necessary to understand the correlation between the normal stresses, σ, the shear stresses, τ, and the shear deformations, over the region in which the results are to be applied. The diagram of Fig. 6 should help to illustrate the significance of these mutual relationships and the extent of their influence.

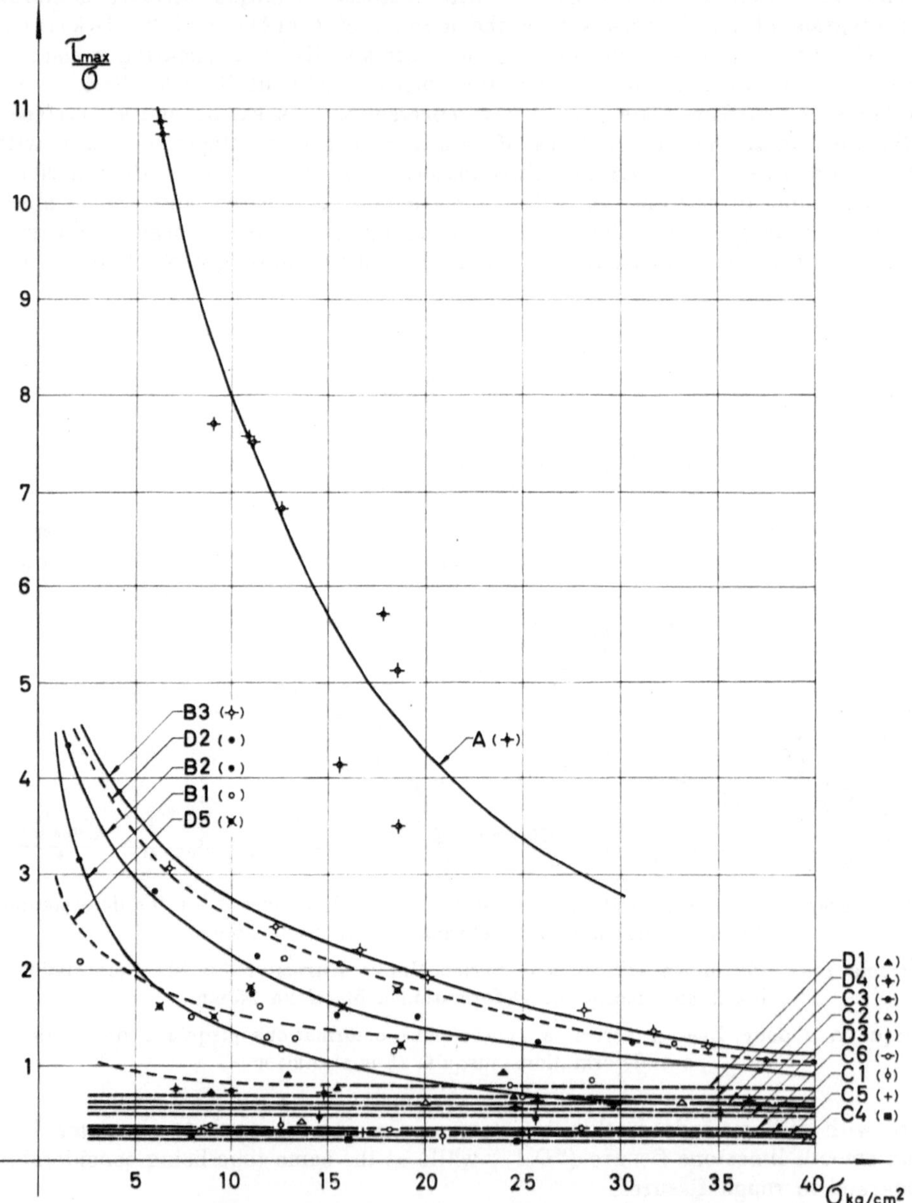

Fig. 5. Diagrams τ max $-\sigma$ for all tested varieties of shearing surfaces

Diagramme τ max $-\sigma$ für alle untersuchten Scherflächen

Graphique τ max, σ pour toutes les surfaces cisaillées

In addition to the relation τ_{max}/τ_{ult}, another characteristic relation proved to be $\tau_{max}/\Delta s$, where Δs represents the value of shear deformations necessary to develop maximum shear resistance. The change of these relations for various series can be seen in the diagram $\tau_{max}/\Delta s$ versus Δs, on the right of Fig. 6.

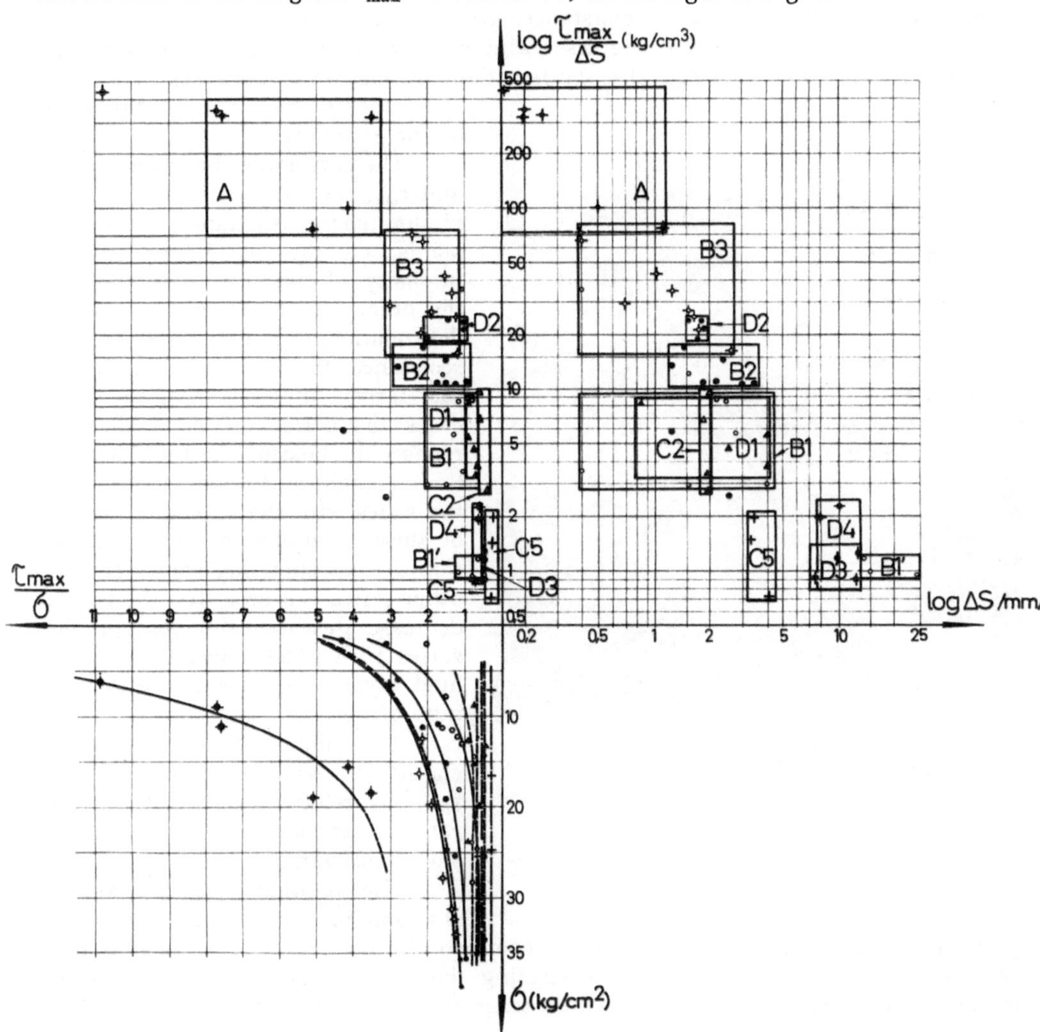

Fig. 6. Correlation of stresses σ and τ with deformations. The most characteristic relations are shown in quadrant IV

Zuordnung der Spannungen σ, τ und der Deformationen. Im IV. Quadranten sind die charakteristischsten Beziehungen angegeben

Corrélation des contraintes σ et τ et des déformations. Dans le 4ème quadrant on donne les rapports caractéristiques

The diagram also shows the correlation of $\tau_{max}/\Delta s$ versus τ_{max}/σ, and, next, the correspondence of these relations with the deformations Δs at τ_{max} and the normal stress σ. The relation $\tau_{max}/\Delta s$ and the deformations Δs are given, for the sake of convenience, in logarithic form and the other values in linear proportion.

The limits within the values range for individual series are marked in rectangles.

By analogy with the concept of brittleness of an intact rock at rupture, which can be represented in shear by the relation $\tau_{max}/\Delta s$, the concept of the rate of development of maximum shear resistance, given by the relation $\tau_{max}/\Delta s$, should also be inserted for the case of the discontinuum with jointed, hard (unweathered) rock. The greater the value $\tau_{max}/\Delta s$, the less the deformation of the mass of jointed unweathered rock due to the load of the structure.

The correlation between τ_{max}/σ, and $\tau_{max}/\Delta s$ aptly characterizes maximum shear resistance which depends, on the one hand, on the intensity of normal stress and, on the other, on the magnitude of the deformation, Δs, necessary for the resistance to be developed. Thus, the larger these values, in both cases the better the material with respect to shearing. In the diagram, the rectangle for Series "A" occupies the space with the maximum values of these relations. After the insertion of corresponding values for the other series it is seen that the values decrease steadily being least in the series with plastic and thinly foliated layers and fissures with fillings of detrital material.

The discontinuum may be classified, on the basis of the $\tau_{max}/\Delta s$ relation, according to the rate of development of maximum shear resistance.

The relations $\tau_{max}/\Delta s$ versus Δs, and τ_{max}/σ versus σ have been given in quadrants I and III in order to complete the picture of rock behaviour in shearing and for the purpose of illustrating the influence of the normal stress intensity on shear resistance and indicating the actual amount of deformation necessary to produce maximum resistance.

Fig. 7. Surface of sample at the end of shear test

Fläche eines Versuchskörpers nach erfolgtem Scherversuch

Surface d'une éprouvette après l'essai de cisaillement

It can be seen from Fig. 6 that the shear quality of hard rock, stratification surfaces and fissures (or of the material therein) is the poorer the lower the values of τ_{max}/σ and $\tau_{max}/\Delta s$, and the greater the deformations necessary to develop maximum shear resistance.

Given sufficient data on rocks of various kinds, it would be possible to classify a mountain mass with regard to shear characteristics according to the above parameters.

Nature of Rupture in Shearing of Rock Surfaces

The results of tests involving shearing along surfaces of stratification planes and limestone fissures of various degrees of roughness (smooth, free from roughness surfaces) indicate that at normal pressures, σ, direct contact between the surfaces is made only at a certain number of points, or where a number of small surfaces are involved. It is because the specific pressures at these points are 20, 50, or even more, times greater than the average pressures, σ, that crumbling or crushing of hard rock occurs at the points of contact.

In the process of shearing these planes tend to extend in a direction parallel to themselves, hence the beginnings of harnage.

In surfaces of considerable roughness, shearing can, of necessity, take place only after the cohesion of the hard rock has been overcome at certain points. This is the reason why in such cases considerably greater values of shear resistance and ratios τ_{max}/σ and $\tau_{max}/\Delta s$ are obtained. Consequently, particular attention should be paid to the nature of fissure surfaces and stratification planes; care should also be taken that the surfaces are thoroughly cleaned and consolidated.

It is common knowledge that the angles of friction on shearing surfaces decrease considerably with increase in normal stress, σ. This could probably be explained by the fact that the surfaces of crushed material, a material which in any case dimishes the shear resistance along these planes, are enlarged due to increased normal pressures (Terzaghi[3]).

Conclusions

(1) In a complex of sedimentary, jointed, unweathered limestone rock mass with marked stratification planes and various systems of fissures, the values of shear angles along stratification planes and fissures differ markedly. The differences are even greater if intercalary layers of plastic materials occur.

(2) The nature of the τ versus Δs diagram differs for various stratification planes and fissures, as shown in Fig. 3. Where deformations are large the shear resistance may decrease considerably; it is therefore advisable to introduce as a characteristic the ratio τ_{max}/τ_{ult} the values of which range from 1.2—2.0 for stratification planes and fissures with clean surfaces. If fissures are full of detritus or some softer material, in stratification planes, the value of the ratio is 1.0. This ratio exceeds 2.0 for an intact limestone rock.

(3) The dependence of the fall off in shear resistance with increase in normal stresses is characteristic of every kind of rock and the nature of stratification planes and fissures. This fall off in shear resistance along surfaces of stratification planes and fissures can be very considerable even within the limits of normal stresses generally permitted in practice. Tests for shear resistance (and slidings as well) should therefore cover the maximum stresses that are likely to occur. A most convenient way of representing these relations is shown in the diagram τ_{max}/σ versus σ, after Grišin and Evdokimov (Fig. 5).

(4) There is one further relation of utmost importance that characterizes the nature of the shearing process, viz. the ratio $\tau_{max}/\Delta s$ which in fact characterizes the rate of development of maximum shear resistance. The larger this ratio the better the quality of the rock mass with respect to shear. On the basis of this ratio, the value of which varies within wide limits, a classification of the shear qualities of a rock mass can be made.

(5) The ratios referred to under (3) and (4) completely characterize the behaviour of a rock mass with respect to shear considering both stresses, σ and τ, and deformation Δs, corresponding to the values of maximum shear stresses. Their correlation has been given by the ratios τ_{max}/σ and $\tau_{max}/\Delta \sigma$ in Fig. 6. It is this correlation that should be taken as a basis in analysing the stability and rupture of any rock mass.

Reference

[1] Müller, L.: Grundsätzliches über gebirgstechnologische Großversuche. Geologie und Bauwesen, H. 1, 1961.

[2] Grišin, M. M. et P. Evdokimov: Shear Strength of Structures Built on Rock. Proc. 5th International Conference of Soil Mechanics and Foundation Engineering, Paris, T. I, 1961.

[3] Terzaghi, K.: Stability of steep slopes on hard unweathered rock. Norwegian Geotechnical Institute, Publication No. 50, 1963.

Trends in Engineering Geology in the United States

By

George A. Kiersch*

With 8 Figures

Summary — Zusammenfassung — Résumé

Trends in Engineering Geology in the United States. The scope of geological sciences has shifted during the past generation in the United States. Today, geology is largely valued in proportion to its quantitative performance with applied science and engineering professions. Nearly 85 per cent of geologists are somehow involved, full-or part-time, with such applications. Although academic emphasis has been oriented towards the purely classical, a re-orientation is imminent within this decade.

Engineering geology is distinctive by its predominant confinement to the physical aspects of geoscience, i. e., physico-geology. An orderly grouping of the significant phenomena, principles, unique techniques and intangible factors constitutes the nucleus. Formerly practiced as an artful blending of general geology and judgment, today engineering geology is an applied science with a strong emphasis on quantification and solid geophysics.

The other branches of applied geology — petroleum, mining, and ground water — are distinctive by confinement to the dominantly mineral aspects of geoscience and demand for specific resources.

Tomorrow's undergraduate geology curricula will consist largely of preparation in the basic sciences, humanities, and engineering science, integrated with a few strong geological courses. Consistent with the improved caliber of pre-college training in the sciences, geology courses will start at a higher introductory level, be more rigorous and inclusive in content, and combine principles from overlapping courses into one strong presentation in geoscience. Graduate-level training in selected facets of geoscience is taken for granted for students planning a career in engineering geology. The integrated geoscience courses of the future will emphasize the environmental approach and "why" of processes and features.

Geology is changing from a predominately mineral-exploration science to one of diversification and acceptance by many other professions. All indications are for more physical data and proportionately less mineral. Henceforth, many petroleum and mining activities will fall within the broad category of heavy construction, i. e., open pits, secondary recovery, nuclear blasting for fragmentation. This trend, along with the expansive needs for physico-geology in applied science and engineering offers a growing diversification to engineering geology practice in such fields as: heavy construction of every type; municipal engineering and city and regional planning; water development; the construction materials and chemical industries; stream pollution and abatement, waste disposal, and public health; land evaluation, investment counseling, and legal aspects of materials and geologic conditions; oceanography and coastal environments; space exploration; nuclear explosions for industrial uses; and many military phases (*Appendix,* II — B).

The sophisticated and diverse problems destined for engineering geologists are fascinating. In addition to the many well-known ones of today, the future will require attention to such problems as: determining the stress retained in a rock mass by refining test

* George A. Kiersch, Professor of Engineering Geology and Structural Geology, Cornell University, Ithaca, New York, USA.

data with such geologic factors as age and deformational history, inherent properties, and
the factor of time; measuring strain-stress buildup along active faults; delineating near-
surface magma migrations; predicting volcanic eruptions and earthquakes; use of aeria'
sensoring to map and interpret terrain conditions; and adapting nuclear explosions t(
industrial uses.

Tendenzen der Ingenieurgeologie in den Vereinigten Staaten. Der Aufgabenbereich
geologischer Wissenschaften in den Vereinigten Staaten hat sich während der letzten Generation verschoben. Der Wert, welcher der Geologie heute beigemessen wird, richtet sich
hauptsächlich nach dem Umfang ihrer Anwendungsmöglichkeit in Verbindung mit ange-
wandter Naturwissenschaft und Ingenieurberufen. Annähernd 85 % der Geologen befassen
sich, als Haupt- oder Nebentätigkeit, mit solchen Anwendungen der Geologie. Obgleich die
akademische Betonung auf den rein klassischen Bereich gerichtet war, steht auch hier das
letzte Jahrzehnt im Zeichen einer Umorientierung.

Die Ingenieurgeologie zeichnet sich vor allem durch ihre Beschränkung auf die physikalischen Aspekte der Wissenschaften von der Erde aus, nämlich auf die Physikogeologie.
Den Kern bildet eine methodische Klassifikation der wesentlichen Erscheinungen, der Grundsätze, der besonderen technischen Einzelheiten und der unberechenbaren Faktoren. Die
Ingenieurgeologie, die früher als eine geschickte Mischung aus allgemeiner Geologe und
persönlicher Einschätzung betrieben wurde, ist heute eine angewandte Wissenschaft, die
großes Gewicht auf das Quantitative und auf solide geophysikalische Grundlagen legt. Die
anderen Zweige der angewandten Geologie — Erdöl, Bergbau und Grundwasser — zeichnen
sich aus durch ihre Beschränkung auf vorwiegend mineralische Aspekte der Wissenschaften
von der Erde und durch den Bedarf an spezifischen Hilfsmitteln.

Die Geologielehrpläne der Zukunft werden für den Studenten in der ersten Hälfte
seines Studiums (bis zum ersten Diplom, B. Sc., Anm.) vor allem aus Übungen in den
grundlegenden Naturwissenschaften, der klassischen Philologie und in Ingenieurwissenschaft
bestehen, vervollständigt durch einige straffe geologische Kurse. Im Einklang mit dem erhöhten Maß an naturwissenschaftlichem Unterricht vor dem Eintritt in die Hochschule werden
Geologiekurse auf einem höheren Eingangsniveau beginnen; sie werden strenger und umfassender sein. Grundsätze und Ergebnisse aus verschiedenen Kursen, die sich heute noch über-
schneiden, werden in eine konzentrierte Darstellung der Wissenchaften von der Erde zu-
sammengefaßt werden.

Nach der Graduation gilt eine Ausbildung in ausgewählten Abschnitten der Wissenschaften von der Erde als selbstverständlich für Studenten, die eine Laufbahn innerhalb
der Ingenieurgeologie ergreifen wollen. Der Unterschied zwischen der Ausbildung vor der
Graduierung und nach derselben wird in einem Wechsel von allgemeinem vorbereitendem
Unterricht — für diejenigen, welche ihr Studium nicht fortsetzen — zur Ausbildung auf
tatsächlich berufsmäßigem Niveau bestehen. Integrierte „geowissenschaftliche" Kurse dieser
Art werden die umfassende Annäherung verdeutlichen; die Fragestellung nach dem „Warum"
wird die gemeinsamen Bemühungen zweier oder mehrerer Spezialisten der Fakultät erfordern.

Die Geologie wandelt sich von einer vorwiegend der Erforschung von Mineralen
gewidmeten Wissenschaft zu einer vielgestaltigen, von vielen anderen Berufen mit Nutzen
aufgenommenen Wissenschaft. Die wirtschaftlichen Anzeichen sind auf eine Vermehrung
der physikalischen Daten und eine dementsprechende Verringerung der mineralischen gerichtet. So werden viele Gebiete von Erdöl und Bergbau von nun an in die breite Kategorie
der Großbauten fallen, z. B. offene Gruben, „secondary recovery" und nukleare Sprengungen mit Splitterwirkung. Diese Tendenz eröffnet, verbunden mit dem ausgedehnten Bedarf
an Physikogeologie in der angewandten Naturwissenschaft und im Ingenieurwesen, der
Praxis der Ingenieurgeologie wachsende und vielfältige Anwendungsmöglichkeiten in Gebie-
ten, wie: Großbauten aller Art; städtisches Ingenieurwesen, Stadt- und Bezirksplanung;
Entwicklung der Wasserwirtschaft; Baustoffindustrie, chemische Industrie; Wasserverschmut-
zung, Wasserabnahme, Abfallverwertung, öffentliches Gesundheitswesen; Bodenbewertung,
Investitionsberatung, juristische Aspekte von Stoffen und geologischen Bedingungen; Ozea-
nographie und küstennahe Gebiete; Raumforschung; nukleare Explosionen für industrielle
Zwecke; und viele militärische Gebiete.

Die verfeinerten und vielgestaltigen Aufgaben, die sich dem Ingenieurgeologen anbie-
ten, sind faszinierend. Zu der Vielzahl an heute bekannten Problemen werden in Zukunft

weitere Probleme hinzukommen, wie z. B. die Entspannung von großen Gesteinsmassen, die aus der geologischen Vergangenheit tektonische Spannungen zurückbehalten haben; Beschreibung von oberflächennahen Magmawanderungen; Vorhersage von Vulkanausbrüchen und Erdbeben; Messung von Spannungs-Beanspruchungs-Zuständen entlang aktiver Störungen; Anwendung von „aerial sensoring" zur Interpretation von geothermalen und Terrainbedingungen; Anwendung nuklearer Explosionen auf industrielle Aufgaben.

Evolution de la Géologie de l'Ingenieur aux Etats-Unis. Aux Etats-Unis, l'aspect des Sciences géologiques vient de se modifier en une génération. Aujourd'hui la géologie est largement valorisée par son important apport aux sciences appliquées et aux métiers de l'ingénieur. Près de 85 % des géologues sont quelque peu concernés, à plein temps ou partiellement, par ces applications de la géologie. Comme l'essentiel de l'enseignement avait gardé une orientation purement classique, un changement d'orientation est imminent dans les dix années à venir.

La géologie de l'ingenieur se caractérise par l'intérêt prédominant qu'elle porte aux aspects physiques des Sciences de la Terre. Le noyau de cette géologie physique est un groupement ordonné de phénomènes significatifs, de principes, de constantes et de techniques exclusives. Pratiquée autrefois comme un mélange expérimental de géologie générale et de jugement, la géologie de l'ingénieur est aujourd'hui une science appliquée particulièrement axée sur la mesurable et sur la physique du solide.

Les autres branches de la géologie appliquée — pétrole, mines, eau souterraine — se caractérisent par l'intérêt qu'elles portent aux aspects purements minéraux des sciences de la terre et elles exigent des moyens particuliers.

Les programmes d'études géologiques pré-universitaires de demain comporteront surtout une préparation dans les sciences fondamentales, l'enseignement général, et les sciences de l'ingénieur, associée à quelques cours de géologie très spécialisés. Grâce à cette amélioration des études scientifiques pré-universitaires, l'enseignement de la géologie pourra commencer à un niveau plus élevé et leur contenu sera plus rigoureux et plus sélectif. Les principes actuellement dispersés dans plusieurs cours qui se chevauchent seront rassemblés dans une vaste introduction aux sciences de la terre.

La formation universitaire dans des secteurs choisis des sciences de la terre est considérée comme souhaitable pour les étudiants qui se destinent à la géologie de l'ingénieur. L'entrée à l'université marquera le passage d'une formation préparatoire générale — pour les élèves qui ne continuent pas leurs études — à une véritable formation professionelle. Ces cours unifiés sur les sciences de la terre insisteront sur l'étude des conditions du milieu et sur le pourquoi des phénomènes. Ils nécessiteront les efforts réunis de deux professeurs spécialisés ou davantage.

La géologie qui était surtout la science permettant l'exploration des ressources minérales est en train de se diversifier et d'intéresser de nombreux métiers. La tendance économique demande davantage de données physiques, et comparativement moins de données sur les minerais. Ainsi beaucoup d'activités pétrolières et minières, se rattacheront au domaine de la construction lourde: exploitation à ciel ouvert, récupération secondaire, utilisation d'explosions nucléaires. Cette évolution et les besoins accrus de géologie physique dans la science appliquée et la construction apportent une diversification grandissante à la géologie de l'ingénieur dans les domaines suivants: constructions lourdes quel qu'en soit le genre; travaux publics y compris l'aménagement urbain ou régional; mise en valeur des ressources hydrauliques; matériaux de construction et matières premières pour l'industrie chimique; polution des rivières et contrôle de leur débit; évacuation des déchets et hygiène publique; valorisation des terres, conseils financiers, problèmes juridiques posés par les matières premières et les conditions géologiques; océanographie et correction des rivages; utilisation industrielle des explosions nucléaires; et plusieurs aspects de la défense nationale.

Les problèmes complexes et variés que rencontrera l'ingénieur-géologue sont passionnants. Outre les nombreux problèmes déjà bien connus, il faudra s'occuper des problèmes suivants: décompression des massifs rocheux affectés de contraintes géologiques résiduelles; tracé des déplacements du magma sous la surface du globe, prédiction des éruptions volcaniques et des tremblements de terre; mesure des contraintes et des déformations le long des failles actives; prospection aérienne géothermique et géophysique; adaptation à l'industrie des explosions nucléaires.

Introduction

The field of engineering geology in United States denotes a somewhat different technical background and professional practice than implied by many in Europe. Engineering geology is now embraced by the International Society for Rock Mechanics with inclusion in the title of the Journal. Consequently, comments on academic training, subdivisions of professional practice, and some expanding sub-fields may be of interest.

Today, geology is largely valued in proportion to its quantitative performance in the applied-science and engineering professions. Although nearly 85 per cent of geologists are somehow involved, full-time or part-time, with applications of geology, academic emphasis has been oriented toward the purely classical. A re-orientation in the academic approach is imminent, and within the decade it will be reflected by major changes in the preparation for and practice in all branches of applied geology.

Geology is changing from a predominantly mineral-exploration science to one of diversification and of acceptance by many professions; this expanded utilization is largely due to a strong emphasis on the physical aspect of geoscience. An examination of applied geology and the branch of engineering geology in the United States, illustrates the causes of some changes during the past decade, the needs of the future, and their influence on academic training and professional practice.

Applied Geology

Four Branches

A clear-cut basis exists for distinguishing four branches of applied geology. Three branches — petroleum, mining, and ground water — are distinctive by confinement to the dominantly mineral aspects of natural materials and the demand for resources (Fig. 1). In contrast, engineering geology is distinctive by its dominant confinement to the physical aspects of geoscience (Kiersch 1958).

Engineering Geology

A confused image of engineering geology currently exists among many students, engineers and geologists alike. Unfortunately this has been compounded by arguments over semantics e. g., geology for engineers, engineering geology, engineer-geologist, and geological engineering. Too often these arguments disregard the underlying scientific reasons for distinction, such as the ones set forth by the Division of Engineering Geology, Geological Society of America (Kiersch, et al. 1957).

Engineering geology is a distinctive branch of applied geology primarily restricted to the physical aspects of geoscience (Fig. 1). An orderly grouping of significant phenomena, principles, and techniques, plus intangible factors, constitutes the nucleus of this field. Professional practice, primarily based on a thorough knowledge of physical geology and natural materials, could be conveniently termed physico-geology, thus eliminating certain misconceptions produced by the adjective "engineering".

The theory and practice of engineering geology are outlined in the *Appendix* where a tabulation is given under the major groupings of: geologic phenomena of chief concern; geologic materials of special importance; principles of mechanics, analysis and properties of materials; techniques and methods of analysis, investigation and remedial treatment; and geology applied to engineering and construction projects. The outline summarizes an advanced-level geology course for students interested in the professional practice of engineering geology. The integrated presentation presupposes a strong background in the basic sciences, humanities, and engineering science, with a sound prerequisite training in geological science. This systematic approach is not a borderland between geology and engineering, as some

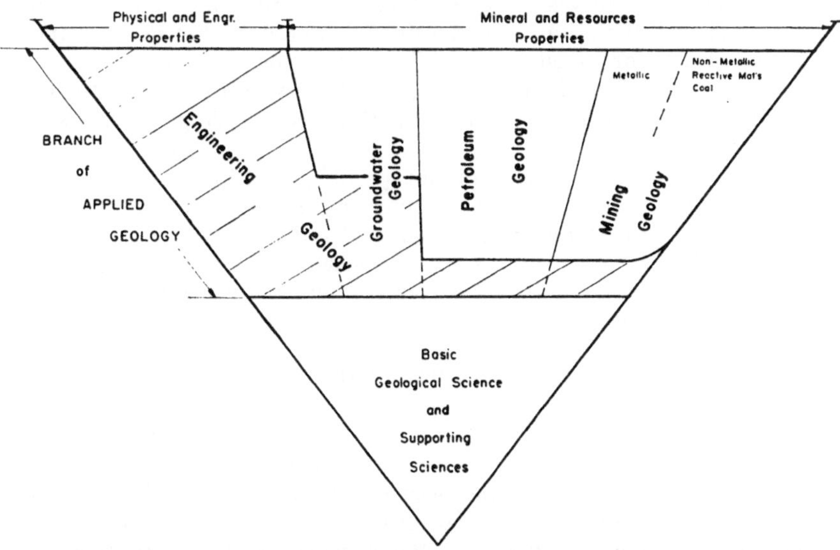

Fig. 1. Classification — four branches of applied geology based on physical or mineral aspects dominant. Note that groundwater geology is about one-half developing a mineral resource and the other half as a component of earth materials and concerned with the physical aspects of engineering geology. Likewise, the physical aspects are of prime importance to some projects in petroleum and mining such as: pipelines, areal subsidence from de-pressuring of reservoirs, underground storage chambers, stabilizing mine workings and open cuts, and dewatering for mining operations (Schematic-scale does not represent relative proportion of geologists in each branch)

Einteilung — vier Zweige der Angewandten Geologie, dargestellt auf Grund von vorherrschend physikalischen oder mineralogischen Aspekten. Bemerkenswert ist, daß sich die Grundwassergeologie teilweise nach mineralogischen Methoden entwickelt, teils aus Komponenten der Bodenkunde, und sich als solche mit den physikalischen Aspekten der Ingenieurgeologie befaßt. Die physikalischen Aspekte sind von höchster Bedeutung bei Erdöl- und Bergbauprojekten, wie Pipelines, Gebietssenkungen unter dem Einfluß der Auflast von Staubecken, unterirdischen Vorratsräumen, Sicherungsarbeiten in Gruben und Tagebauen sowie Entwässerungen im Bergbau (Die schematische Darstellung gibt nicht das Verhältnis der Anzahl von Geologen in jedem Bereich wieder)

imply. All four branches of applied geology follow this "overview" line-of-attack, as do most professions today.

Geology for Engineers. This introduction to the general principles of geology is given as a 1 or 2 semester course and constitutes a survey of the science (Kiersch, et al. 1957). Geology for civil engineering students is taught in a similar manner at colleges throughout the United States. This limited training in no way qualifies the student for practice as an applied geologist. Rather, the intent is to train him to recognize when a geologist is needed.

Engineer-Geologist

Training to practice in both fields simultaneously is not encouraged — as is common in Europe. Professional demands render this two-fold practice impossible, except in the small technical organization concerned with the more routine, simpler design and construction or as an administrator.

Geological Engineer

Geological engineering is the training of an engineer with a general knowledge of geological science. He serves as an exploration or production technician, or administrator in any one of the four engineering fields allied with geology (petroleum, mining, ground water, or construction). This is a sound undergraduate training for subsequent advanced-level study in geoscience with practice ultimately as a geologist, in one of the four applied branches. Many engineering geologists have been trained in this manner.

Geotechnical engineering is used in Europe for practitioners in the inter-related fields of soil mechanics and foundation engineering, along with some rock mechanics and a few sub-fields of engineering geology. In the United States, the applied science and/or engineering categories of geoscience are practiced by the specific professional subdivisions mentioned, and geotechnical engineering is not appropriate for any professional group.

Professional Practice

Historical Development

Engineering geology has made important strides in the past generation. This growth and wide acceptance today has been possible because a group of geologists shifted from the long-standing classical traditions. Instead of purely descriptive information, they initiated attention to the physical and quantitative aspects of geology whether for:

a) the works of civil engineering and the heavy construction industry;

b) the conservation of land and development of water resources by agricultural and irrigation specialists;

c) industrial or radioactive waste disposal by the sanitary engineer;

d) specialized requirements for physical data in such fields as coastal engineering, mineral engineering, and military operations (see *Appendix* for complete list);

e) and now, exploration of basing on the moon and outer space, and harnessing nuclear explosions.

However, the acceptance engineering geology enjoys today was not without help from Mother Nature and a long list of famous engineering dilemmas and catastrophies (Kiersch 1955, p. 13—34). For example, the San Francisco earthquake of 1906 focused attention to the influence of geologic conditions on effects of seismic shock intensity and foundation conditions; landslides at Culebra Cut of Panama Canal between 1910 and 1915 developed a recognition of slide characteristics and their importance in open-cuts; World War I dramatically demonstrated the advantages of utilizing geology and terrain conditions as practiced by the German Army; the St. Francis Dam failure in 1928 alerted engineers to the significance of an inferior-quality foundation; with the advent of many high concrete dams in the 1930's, super-highways, and airports the seriousness of the cement-aggregate reaction was exaggerated to proportions requiring geologic attention; furthermore the large reservoirs increased sediment-filling to the status of a serious geological problem; and then ironically the cleansed reservoir water created leakage problems downstream in the very canals that were formerly self-sealed by the natural silt; and so — each difficulty ultimately provided an improved state of knowledge and progress in engineering geology.

Since World War II, highway and airport programs created demands for larger sites with greater bearing capacities and enormously expanded construction material needs. Of a similar nature has been the expansion of state and Federal conservation measures, urban developments with necessity to reclaim marginal lands, military planning both on land and beneath the water and then the urgency for Arctic construction with its problems of frozen ground. In the 1960's, scientific demands require the active participation of engineering geologists for the needs of lunar and planetary exploration, and the utilization of nuclear explosions for industrial purposes.

Common to all these geologic advances has been the underlying needs of an expanding population for "social engineering projects", economic development of virgin lands, and the proportional decrease of acceptable natural sites for engineering purposes.

From this seemingly disconnected and diversified growth, to meet the day to day needs of engineering and applied science, has evolved the orderly grouping of the principles and techniques that constitute engineering geology today (listed *Appendix*).

Specialist-Team

The specialist-team approach (Arbeitsgemeinschaft), so successful in practice, is overdue for full academic recognition. From it has evolved the concept of physical environmental study and evaluation, a realization of paramount importance to engineering geology of the future. The team must reconstruct the geological history and determine the causes for features and/or conditions by means of exploration and studies prior to forecasting the effect of a proposed project on the physical site. This approach is particularly effective when the individual members develop mutual respect for the other specialities.

Regional-Areal Investigations

The importance of regional-areal investigations for critical geologic data pertaining to a natural site or material is frequently overlooked. However, today this approach is being increasingly accepted as the initial step in any major exploration program. Features controlling a site rarely conform to a type interpretation; and an understanding of the causes thereof can prove more advantageous to an engineerning objective than the detailed knowledge of a restricted site.

The most successful method of utilizing regional-areal studies is the "convergence-of—evidence" plan that integrates four to six interdependent stages as shown on Figure 2.

1. *Initial* — regional-areal reconnaissance, compilations, and maps incorporating direct and indirect data.

2. *Interim summation* — interpret existing data, extrapolate at proposed sites.

3. *Site* — detailed scale mapping supplemented by specialized exploration and research techniques as justified.

4. *Interpretation* — synthesis of all physical-engineering data that includes; origin, changes after deposition or emplacement, and chemical petrographic factors. Recommendations conform to the engineering objectives.

Some common desiderata determined primarily from regional-areal investigations and the manner of utilizing this approach in highway geology have been described elsewhere (Kiersch 1962).

The initial investigation for any engineering works should begin with an assessment of the regional-areal geologic setting of the proposed site. This study provides: recognition of the geologic origin; a systematic geologic history and the important sequence-of-events responsible for site conditions; and the causes responsible for physical properties of the natural materials. The appraisal will invariably disclose some inherent problems of a site which are fundamental to the geologic origin, primarily regarding formational units, structural features, erosional features, and the surficial deposits. This investigation is on a limited basis with duration and extent a function of the complexity of geology and scope of the project. The preliminary geological data is synthesized with findings of the other specialist team members for the interim summation and decisions.

The geologist is ideally qualified to correlate the natural processes in the historic and geologic past with the future action of such processes and land uses proposed. Every effort should be made to evaluate processes in terms of their rate

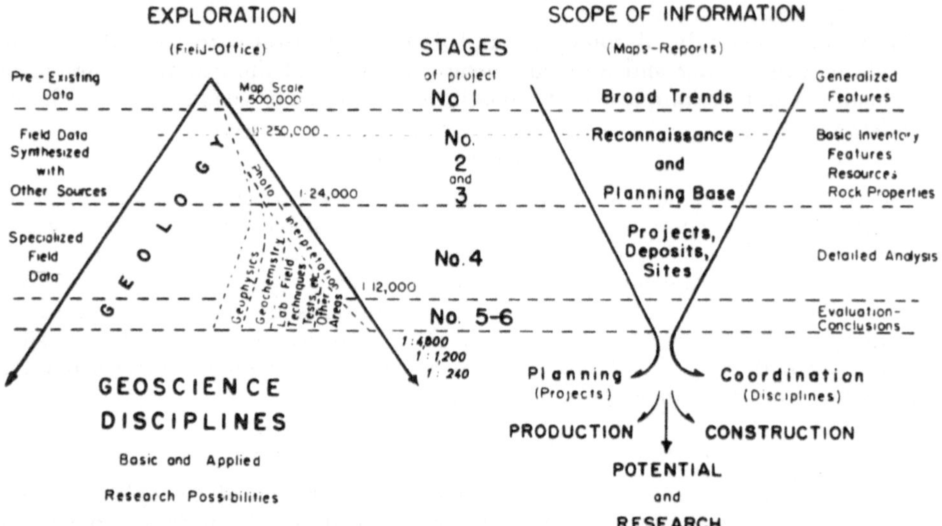

Fig. 2. Regionalized approach for geoscience exploration and research investigations from the initial reconnaissance of broad trends to the detailed study of site conditions

Weg der erdwissenschaftlichen Erkundung und der Forschungsuntersuchungen für ein bestimmtes Gebiet, vom einleitenden Erkennen umfassender Grundtendenzen bis zum detaillierten Studium der Gegebenheiten der Baustelle

— specifically, to distinguish between geologic time and time in the engineering or human sense. Quantitative estimates are needed concerning such features as natural compaction, bearing characteristics, transmissibility rate of groundwater flow, gross stability of earth and rock materials, and the expected yearly rates and magnitudes of erosional processes and ground movements. In essence — how will the site react to the proposed engineering works.

Paramount, then, is the ability of the engineering geologist to analyze the evidence of a problem from more than one point of view. Many geologic situations initially require the creative, fragmentary evidence approach, with the remaining parts of the puzzle available only after exposure by subsurface excavation or laboratory test data (detailed exploration of Stage IV, Fig. 2).

The site conditions as shown on Fig. 3 are an excellent example of the necessity to understand and utilize areal geologic information to unravel a complex site in glacial terrain. These unsuspected conditions of unstable, highly solutioned, deformed, and areally-subsided and locally-collapsed beds were encountered during an excavation for building sites and proposed highway alignment in Syracuse, New York. To reconstruct the circumstances briefly — initially, thick evaporite beds were laid down throughout the region, followed by alternating sandstone, siltstone, and limestone strata, with frequent interbeds of salt and gypsum (in Silurian time).

Since deposition, at least five separate geologic events have contributed to the complex subsurface rock conditions as follows:

1. The groundwater level changed greatly at least twice and each time affected evaporite beds.

2. Extensive natural solutioning removed at least 200 feet of salt beds and interbedded gypsum layers at depth. Regional subsidence resulted, with locally strong collapse.

3. Following salt removal, original anhydrite absorbed water with a volume expansion to gypsum that created upheaval and deformation. When gypsum was subsequently removed by solution action, a further local subsidence and collapse. Black muds accumulate in voids as a residue from evaporite removal.

4. Glacial scouring (Pleistocene) modified surface. Moving ice abutted against the ridge (site) with some squeezing and further distortion of site rocks.

5. Since removal of continental ice sheet. Rebound phenomenon active, causing further deformation and sliding.

As demonstrated by circumstances of Fig. 3, the most successful practice of engineering geology comprises two major parts:

1. Large-scale or background geology (regional-areal); and

2. Small-scale or site geology (ASTM 1951).

The same geologist may or may not perform both types of investigations, according to needs and circumstances. In either case, data from both sources is essential for the maximum utilization of geology in the planning-design, construction, maintenance and/or operation of engineering works.

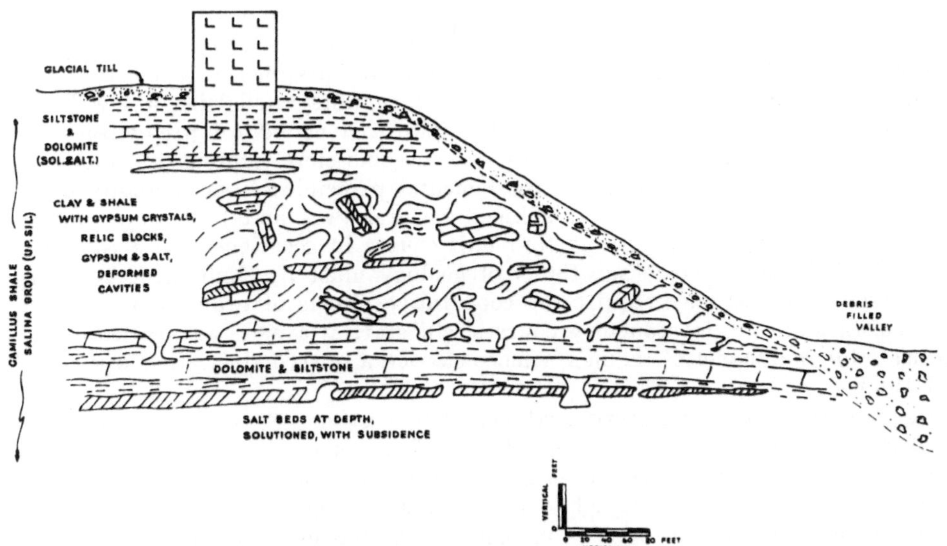

Fig. 3. Schematic geologic cross section in glacial terrain, Syracuse, New York. The subsurface conditions of unstable and deformed rocks, beneath the thin caprock, is due to a complex regional-areal history and at least five separate geologic events (After 1960 report by Paul H. B i r d, Dept. Public Works, Albany, N. Y.

Schematischer geologischer Querschnitt in einem eiszeitlich geformten Gelände in Syracuse, New York. Unter der dünnen Gesteinsdecke befindet sich lockeres und deformiertes Gestein. Dies ist zurückzuführen auf eine komplexe örtliche Geschichte und mindestens fünf verschiedene geologische Ereignisse (Nach einem 1960 veröffentlichten Bericht von Paul H. B i r d, Dept. Public Works, Albany, N. Y.)

Contributions of Other Disciplines

Engineers and applied scientists have provided geologists with an ever-increasing amount of otherwise unobtainable scientific information, and are themselves beginning to make their own contribution to man's knowledge of the earth. In some instances, these investigators have advanced well beyond the understanding of geologists themselves. An extensive list of ingenious technical equipment has been contributing to geological advances; such as x-ray diffractometer and numerous mineral-chemical techniques; infra-red and radiowave aerial sensoring; TV cameras for borehole and underwater observation; computers and analytical procedures for statistical analysis; stress-strain measurement apparatus; and many others.

This growing knowledge of geological science supplied by allied disciplines is clearly demonstrated in the field of geomechanics. The excellent contributions of engineers and other scientists have advanced certain aspects of structural geology far beyond the state of knowledge supplied by geologists only.

Academic Training[1]

To meet professional demands of the future, geological course content must be strongly quantitative and less descriptive. To accomplish this shift will require a truly interdisciplinary approach to the study of geologic events and processes, thereby providing insight into physical environmental studies. Facts and the quantitative, separated from inference, conjecture, and hypothesis, are the needs and demands not only of applied science and engineering, but of theoretical research as well.

Curricula

Tomorrow's undergraduate geology curricula are destined to consist of the basic sciences, humanities, and engineering science, integrated with a few high-caliber geology courses. Consistent with the improved caliber of pre-college and undergraduate training, geological courses must become more rigorous and inclusive in content; this means combining the fundamental principles from currently overlapping courses into one integrated subject. Several two- and three-unit courses should be integrated into one strong four-unit course in geoscience. To accomplish this, artificial fences in some sub-fields of geology must be eliminated. Teaching such integrated courses may require the cooperative efforts of two or more faculty specialists. In addition, introductory courses will start at a more advanced level to compensate for the impact of the new science curricula in secondary schools (Gatewood 1963).

Many of the major engineering schools (USA) offer a 4- or 5-year program for a Bachelor of Science degree, of which 2 or 3 years are liberal arts and basic sciences, and 1 or 2 years are the engineering sciences with allied courses. Professional-type courses are restricted to the graduate school.

This diversified basic program for engineering students is ideally suited, with certain modifications, for undergraduate geology training. The principal difference is that a few selected geoscience courses are taken at the upper-division level in lieu of some engineering science courses. This approach is particularly suitable for the student who plans on further study for subsequent practice in engineering geology. It has long been realized that the serious, mature student — with a sound

---

[1] Condensed from G. A. Kiersch, 1964: Academic trends in engineering geology. Jour. Geol. Educ., vol. 12, p. 14—21. Reader is referred to original paper for full explanation on academic curricula and planning.

background in the basic sciences and engineering science — grasps the fundamentals of undergraduate geology courses in a fraction of the time required by the poorly prepared student.

Graduate-level training in selected facets of geoscience is taken for granted for students planning a career in engineering geology, or any other branch of applied geology. Duration and scope is dependent on the individual. The break between undergraduate and graduate school should be a change from the general preparatory training — for those not continuing — to the truly professional-level school. The graduate student then concentrates on a specified field of geoscience interest and professional-type courses.

Academic Planning

Academic policies concerning engineering geology must take into account such critical professional factors as:

1. Tomorrow, engineering geology will require practitioners of diverse and strong scientific backgrounds.

2. Students should be trained for a growth potential to deal with the more complex problems of the coming generation. Geology and its applications are becoming more complex — not simpler.

3. Technical advances require the quantitative approach. The artful approach only cannot meet contemporary demands. Although, today's student commonly has strong preparatory courses in mathematics, most geology courses ignore its application.

4. An increased amount of instruction by the experienced practitioner is one manner of introducing a greater emphasis on the quantitative.

5. Courses should emphasize the environmental approach to geology, in which the inter-related phases of geoscience are utilized — as per professional practice. For example, the effect of man-made changes, the geologic processes active in modifying surface features, and the stress distribution in rock masses in the sense of human time.

6. One of the most retarding situations is the schism between geology and solid geophysics. This should be corrected by truly interdisciplinery courses where a geologist teaches the processes and why for field conditions, and a geophysicist teaches the measurement of physical parameters. Interpretation then becomes inter-related and emphasizes the environmental control.

Trends in Utilizing Applied Geology[2]

Utilization of mineral-geology by the major industries of petroleum and mining reached new heights within the past decade. During the next generation its future seems geared largely to a status quo based on refinements and reworking of the *known*. Large, low-grade deposits ("rock") are fast becoming the principal source of many major mineral commodities. To these sources is being added the largest mineral storehouse of all — the ocean.

Instead of the petroleum and mining industries as we know them today utilizing over two-thirds of the geological talent, a major use will develop for physico-geology in other fields of applied science and engineering, particularly for heavy construction of every type and the list of sub-fields given in *Appendix*. III — A, B.

---

[2] Modified from G. A. Kiersch: Engineering geology — today and the expansive tomorrow. American Geological Institute Lecture Series, 1961 and 1962.

Many mining and petroleum activities will themselves fall within the broad scope of heavy construction, with engineering geology important for such problems as: excavation and slope stability in open pits; secondary recovery techniques for petroleum; application of nuclear blasting for fragmentation; *et al.* The economic indication are for more physical geologic data and proportionately less mineral.

Sub-Division of Engineering Geology

Today, urbanization, industrialization, and population growth daily exert pressures on our needs for natural sites and social engineering projects. Consequently, engineering geologists are involved in greater numbers with increasing types of projects. A single practitioner can not expect to qualify for practice in all sub-fields of engineering geology (concurrently). Therefore, after acquiring general field experience and serving an "internship", he will practice in only one, two or possibly three of the following major sub-divisions within this broad field (*Appendix,* III — A, B).

1. *Foundation* — Surface Excavations and Structures.
   Includes: Geomechanics of open cuts.
2. *Hydraulic Structures* — Water Retention and Recharge.
   Includes: Coastal environments, Control works.
3. *Water Supply and Irrigation Works.* Includes: Conservation of land.
4. *Underground* — Excavations and Structures. Includes: Geomechanics.
5. *Construction Materials.*
6. *Military Functions.*
7. *Astrogeology* — Lunar and Planetary Operations.
8. *Nuclear Explosions* — Industrial Purposes.
9. *Public Health* — Disposal of Wastes.
10. *Regional Economic Development* — Partly overlaps no. 1—5.
    Includes: Urban Geology.
11. *Legal Aspects of Geology in Engineering Practice.*
    Includes: Land evaluation, Investment counseling.

Some Expanding Sub-Fields of Engineering Geology

The sophisicated and diverse problems destined for engineering geologists tomorrow in the United States are fascinating. Many of them are not very different from those in other parts of the world. It has been estimated that in the year 2025 A. D. our earth will be saturated population-wise. If this is true, consider the greatly increased needs and demands for construction sites, social engineering projects, water, and natural materials that will fall within the provenance of the engineering geologists.

Continuing then, will be an expansion of the well-known needs for geology as related to such engineering projects as: dams, tunnels, powerhouses, reservoirs, highways and railroads, canals, buildings and bridges. Each type of works has inherent stability and foundation problems and needs for construction materials (Sub-Divisions no. 1—5). In addition to these and other sub-fields active today (*Appendix,* III — A, B). demands of the future will require that engineering geologists give increasing attention to a growing list of new problems such as: the de-stressing of large rock masses that retain tectonic stress; delineating near-surface magma migration; predicting volcanic eruptions and earthquakes; measuring stress-strain buildup along active faults; use of aerial sensoring to interpret geothermal and terrain conditions; and adapting nuclear explosions to industrial uses. Such demands on engineering geology will obviously require a different academic training than most schools are offering today. They will require competent practitioners who possess a well-rounded training in geoscience, the necessary attributes, and an extra-broad understanding of applied science and engineering. A brief discussion of several emerging sub-fields follows, that illustrates the expansion.

Geomechanics. As this is a Colloquium of the Society for Rock Mechanics, comments on trends in this sub-field in United States are likely to be expected. This distinguished group of engineers and scientists is indeed aware of the progress and excellent outlook for an expanded growth and application of rock testing and geomechanics — due in large part to your pioneering work in Europe of the past two generations. The conference on "State of Stress in Earth's Crust" at Santa Monica, California, in June 1963 was an up-to-date synthesis of the state of knowledge in geomechanics for the diversified aspects of practice and focused attention on research needs (Judd 1963).

Many disciplines of engineering and science utilize certain principles of geomechanics and vise versa. The following summarizes some current thinking on geomechanical principles that will strongly influence the practice of engineering geology in the future.

1. A correlation exists bewteen the regional-areal geologic features (megageologic) and geomechanical stress distribution. The rate of rebound of retained tectonic stress is more or less proportional to the horizontal load of a region (Birch 1963) and a function of the properties of a mass. In a viscous material, hundred to thousands of years would dissipate stress, while in a non-viscous rock the stress dissipates very slowly (to tens of millions of years).

2. A correlation exists between the — rate of surface erosion and unloading, the amount of de-stressing or rebound accomplished, and the inherent physical properties of a rock mass. For example, the work of A. Kieslinger (1958, and 1960) for numerous sites in Europe has established a maximum depth of some 50 metres for the zone-of-rebound in plutonic rocks, exposured since mid-Tertiary time. Below this level conditions are related to the regional stresses and effects.

3. Weakly cemented rocks and sediments rebound very quickly when unloaded. Stress relief can occur within the brief interval of days, weeks, or months, a function of the type and degree of bonding. Consequently, the typical surficial conditions investigated by the principles of soils mechanics involve only weak or no retained stress. In contrast, the average rock mass possesses varying degress of retained stress which must be considered in applying the principles of geomechanics.

4. The realization is growing that ancient mountain systems — such as the Appalachians — retain a portion of their active tectonic stress at reasonable depths today. This explains why large-scale uplifts in Devonian time, as the Green Mt. anticline, Vermont, continue to stand above the surrounding countryside due to an imbalance of stresses in the crustal rocks, even though the surrounding ancient mountain system has long since been reduced (Birch 1963). If a truly hydrostatic stress distribution existed at depth, then the Green Mt. surface today, excluding differential erosion, would be more or less level.

5. The hydrostatic theory, generally accepted since 1880's, must be modified on basis of recent field and laboratory data, such as:

a) A very large stress imbalance may be encountered; up to 6 times greater in the deep underground excavation at Poatina Power Station, Tasmania (Judd 1963).

b) Within a single rock mass, the stress magnitude can be many times greater between adjoining rock types, beds, layers, and laminae or bands of minerals (Emery 1963).

c) Experience from deep oil wells confirms strong differences in stress distribution at depths to 25,000 feet, especially in areas of Tertiary rocks. The tectonic origin of this stress seems confirmed.

d) Evidence that stresses are not truely hydrostatic within the core of the earth has been gained from recent space satellite surveys. Differences of stress from 10 to 100 bars are estimated within the interior of earth (Birch 1963).

6. The "elasticity theory" must be modified for the majority of rock masses. Although low-stressed rocks confirm the elastic theory, high-stressed rocks react in-elastically and the theory is inappropriate (*Appendix*, II — B, 3). Invariably the latter create the major problems at an engineering site.

7. The statistical approach to understanding the properties of a rock mass is limited in value, because each one is strongly influenced by the geologic events at the affected area. (See Approach below.)

8. Geological defects in a rock mass are a primary consideration in the determination of the deformation modulus of the rock (Müller 1962). The axis of maximum deformation is invariably parallel to the strike of the major fracture system.

A technique that offers promise for improving geomechanical properties of a site rock is artificial de-stressing; such as by use of nuclear blasts and shock waves (Press and Archambeau 1962) on a rock mass. Preliminary experiments are encouraging. Small-scale blasting has been utilized for several years in deep mines of South Africa and Canada. Shock waves apparently traverse grain boundaries of a rock and temporarily reduce the stress concentrated there, but are not strong enough to fracture the rock mass.

The long-standing question being studied continually is — how to make full use of the basic theories of geomechanics to forecast the strength of a rock mass in-situ. One technique that offers excellent possibilities according to D. Wantland (1963) is based on putting an artificial stress in a rock mass, measuring the damping effect (geophysical data), and thereby determine the viscosity of rock mass.

Approach — A theoretical analysis of rock masses only, in a manner paralleling the soil mechanics discipline, is wholly unrealistic and inadequate due to: 1. the inherent nature of rocks with greater inhomogenity; and 2. the occurrence of strain energy in a non-uniform pattern. The soundest approach to unravelling the geomechanical conditions at a complex site incorporates a comprehensive geologic knowledge of conditions, the tests and measurements of engineering specialities, and a theoretical assessment in the following sequence:

Step I — A geological assessment, progressing from the regional to site scale, with appropriate quantitative studies.

a) A detailed mapping of the structural and compositional features of site mass; incorporate interplay of areal features.

b) Ascertain the three-dimensional composition of rock mass and physical features.

c) Determine the origin and any associated conditions that influence physical properties (hydrothermal alteration, abnormally high geothermal gradient, et. al.).

d) Determine the changes that have taken place in mass since origin, according to historical events. Date each event, if possible, to ascertain time factor active.

e) Determine the spacing and pattern of structural features, as discontinuum by percentage.

f) Analyse the potential for active stresses in mass; forecast the amount of retained strain energy and degree of unrelieved stress at depth of the proposed engineering works. Together, these data are utilized in *Step II* to determine the apparent and true strength of rock (safety margin for engineering load).

Step II — The specialized testing, measurements and techniques of engineering specialities, both laboratory and field (in-situ).

a) Determine inherent strain conditions of site rocks, ascertain whether design needs can be met.

b) Determine the failure type for rock or mass: e. g., prone to fracture, creep, or flow and whether latter two occur before or after failure.

Step III — A theoretical analysis of site conditions and problems of the engineering works.

Step IV — An assimulation of all data from *Steps I, II* and *III* with evaluation to disclose their interdependence. Conclusions incorporate the in-situ conditions and are consistent with field circumstances.

Step V — Provide a forecast of the geologic reactions to the physical changes (to be) imposed by engineering works on rock mass(es) at a site. Outline methods of observing and measuring the magnitude and physical characteristics of changes and reactions forecast.

Step VI — During operation of the engineering works, geologic observations are complimented by appropriate engineering techniques to continually assess any changes that may occur in mass, e. g., increased stresses, reduced strength, creep. The effects of time on geologic conditions are thereby closely charted in case changes induce a weakening to impending failure.

Concerning *Steps I—IV,* the geologic approach is outlined on Fig. 1. Potential problems are initially considered as influenced by the large-scale features with progressive attention to smaller scales and more details. Depending then on the complexity of the engineering problems, the investigator must analyze smaller and smaller details of the rock mass. The more detailed the investigations, the more techniques are required for analysis, including the special strength and strain measurements of rock mechanics and other disciplines.

Groundwater Basin Management. The increasing demands for groundwater conservation to fulfill the ever expanding water consumption needs in United States is of national importance. To meet this number *1* dilemma, the engineer-geologist-economist team is confronted with a multitude of projects for dams, tunnels, canals, aqueducts, and storage facilities.

A new approach to the complex water problems is basin management — in which resources and supply, conservation, pollution, and industrial waste disposal are integrated. This trend offers great promise for both an improved production of water and more efficient use of the total surface and underground water available within any geologic province. Basin management requires a thorough physical and quantitative knowledge of the inherent geologic conditions. Founded on a sound understanding of the three-dimensional geologic setting, the ground water and surface reservoirs can be operated as an integral unit — whether for storage, supply, replenishment, or the disposal of polluted waters.

As an example, Fig. 4 is a block-diagram reconstruction of the subsurface conditions within three adjoining intermontane basins, typical of the Basin and Range Province in western United States. Extensive surface and subsurface studies of these basins in recent years has determined that:

1. deposition of basin fill was not all due to alluviation from nearby mountain ranges;

2. the dominant sediment is not a coarse gravel and sand alluvium; and

3. the total reservoir potential for groundwater is only some one-third of what had been anticipated.

The critical features of these basins are:

Unit C — Oldest is largely coarse alluviation debris from adjacent mountains. Well-cemented and of low to moderate permeability.

Unit B — The major filling is largely fine-grained sediments of silt and clay of very low permeability with some lenses of gravels; laid down by ancient drainages and intermittent lakes. Includes intercalated lavas.

Unit A — The young, near-surface gravel and sand alluvium is partly deposited by present drainage, and partly by slope alluviation.

The unravelling of this sedimentation sequence of the intermontane basins, whose boundaries are usually fault controlled, is an excellent example of research trends. From this quantitative understanding of the subsurface, a more adequate

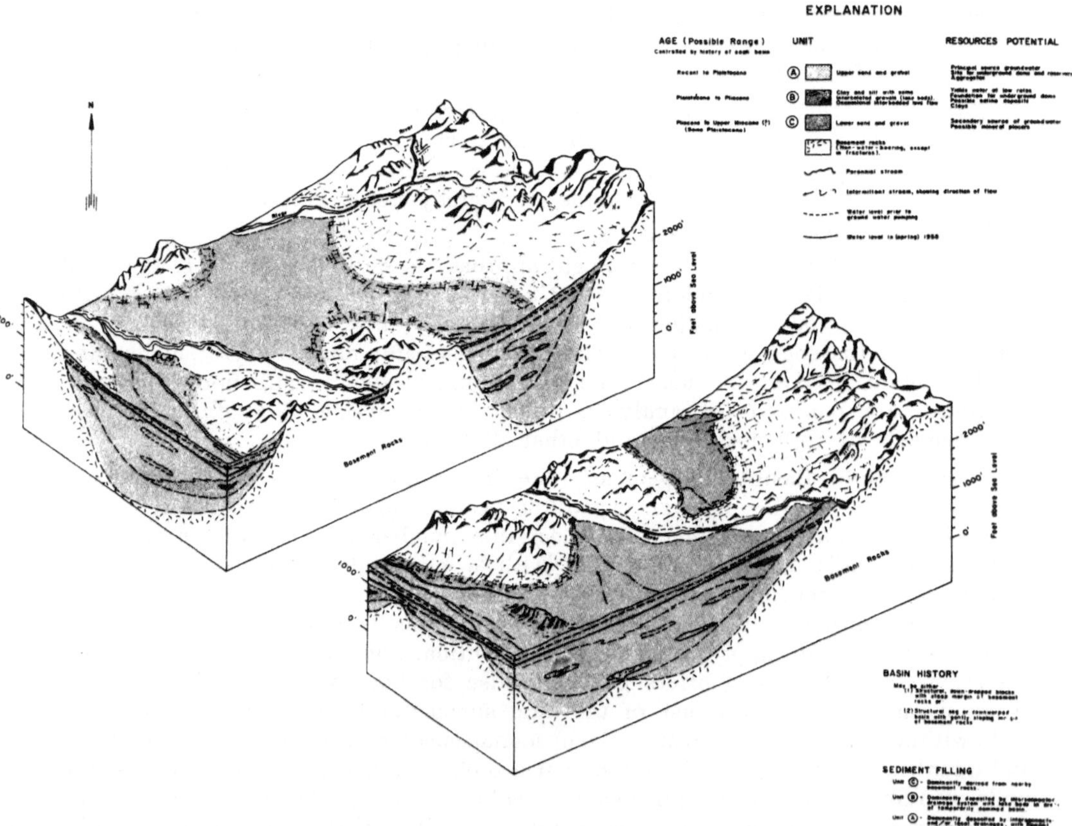

Fig. 4. A generalized three-dimensional diagram of subsurface geology within alluvial basins of the Basin and Range Province, western United States. The three major sedimentary units are delineated, along with possible age range, groundwater level conditions, and structural history of basins. Note excellent site for underground dam (Fig. 5) on upper block, western edge

Schematisiertes dreidimensionales Diagramm der Untergrundgeologie innerhalb alluvialer Becken der „Basin und Range"-Provinz im Westen der Vereinigten Staaten. Die drei größeren sedimentären Einheiten sind eingezeichnet; außerdem die wahrscheinliche Altersabstufung, die Lage des Grundwasserspiegels und die Baugeschichte der Becken. Zu beachten ist der hervorragende Platz für ein Untergrundstauwerk (Abb. 5) am westlichen Rand des oberen Blockes

water storage can be provided economically through ground water basins in conjunction with surface reservoirs. This practice imposes new engineering needs and likewise additional geologic problems relating to groundwater, such as: recharge, water quality, salt balance, sewage and industrial waste disposal, and in some coastal areas seawater intrusion.

Besides basin management of water resources, these regional-areal relationships are critical to the important phenomenon of areal ground subsidence, both shallow (above) and deep (below water table) and experiments to artificially induce subsidence for foundation consolidation or large-scale "excavations". Another emerging sub-field utilizing this knowledge is that of underground dams.

Subsidence. The current drawdown of groundwater basins, such as shown on Fig. 4, has created an additional problem for the engineering geologist in this setting — that of land subsidence. This phenomenon has long been recognized in places like Mexico City, Long Beach, California, and some Gulf Coastal cities[3]. In the past few years, communities in the intermontane setting are experiencing serious areal subsidence, such as Phoenix and Tucson in Arizona, and Santa Clara and San Joaquin valleys in California.

Recently it was recognized that subsidence (shallow) occurs above the water table in sedimentary units, when virgin lands of arid climates are put under irrigation. This constitutes a re-orientation of the individual grains of the sediments, into a closer-knit, compacted mass, which results in an areal subsidence of the land surface. Deep subsidence due to lowering of water table has long been recognized.

Underground Dams. The improved understanding of basin sedimentation is critical to another aspect in engineering geology, the grouting of foundation material for major dams and construction of underground dams and storage reservoirs. Many excellent hydrologic sites have been unused heretofore, because it was uneconomical to excavate and backfill the hundreds of feet of ancient alluvium occuring beneath the stream channels. Success by the French at sites on the Rhine and Rhone Rivers has opened up new vistas for dam construction and such sites are now receiving attention around the world (Cambefort 1963).

The setting of an underground dam is illustrated by a small, alluvial-filled valley of an intermontane basin as shown on Fig. 5. The channel is carved in essentially impermeable bedrock, the overlying alluvium consists of open-void gravel (principal aquifer) with a unit of fines and coalescing lenses of clay (unit B material of Fig. 4) above. This is capped by Recent, coarse alluvium (Unit A). Recharge is near the mountain front.

A heavy draw-down by well "B" located down-gradient is rapidly depleting the ground water of the up-valley section. It is not uncommon for wells such as "A" to go dry.

To insure against a complete draw-down in the upper-valley area, an underground dam offers a possible solution; a grout barrier could be placed from holes on a grid pattern, using clay and a clay-cement slurry to seal the open-void gravel in the lower section of the channel[4]. This, creates a small underground dam founded on bedrock perhaps 30 or 40 feet high (could be higher); design allows for movement of water to the downstream area by overtopping after the upstream reser-

---

[3] There are many other cases of this phenomenon throughout the world due to either oil and/or water production, or the withdrawal of solids by mining.

[4] Use varying mixes on any site according to the alluvial composition. Technique satisfactory if 90 per cent of alluvium is above 0.40 mm. in size (medium sand). Size range of 10 per cent fines determines ultimate permeability of grout; only one-half (5 per cent of gross) can be less than 0.25 mm. (fine sand).

voir is filled. By this design, potential ground-water losses to downstream land owners would be avoided, once the reservoir for well *"A"* was charged.

The clay lenses overlap one another at different elevations and grossly constitute a *Unit B* of low permeability. The clay bodies are conveniently utilized as both a "cap" on the grout barrier and as a keystone to stabilize the groundwater dam.

Other methods to construct a barrier might be feasible depending on the geologic conditions of a site. If the channel is not too deep a trench (limit about

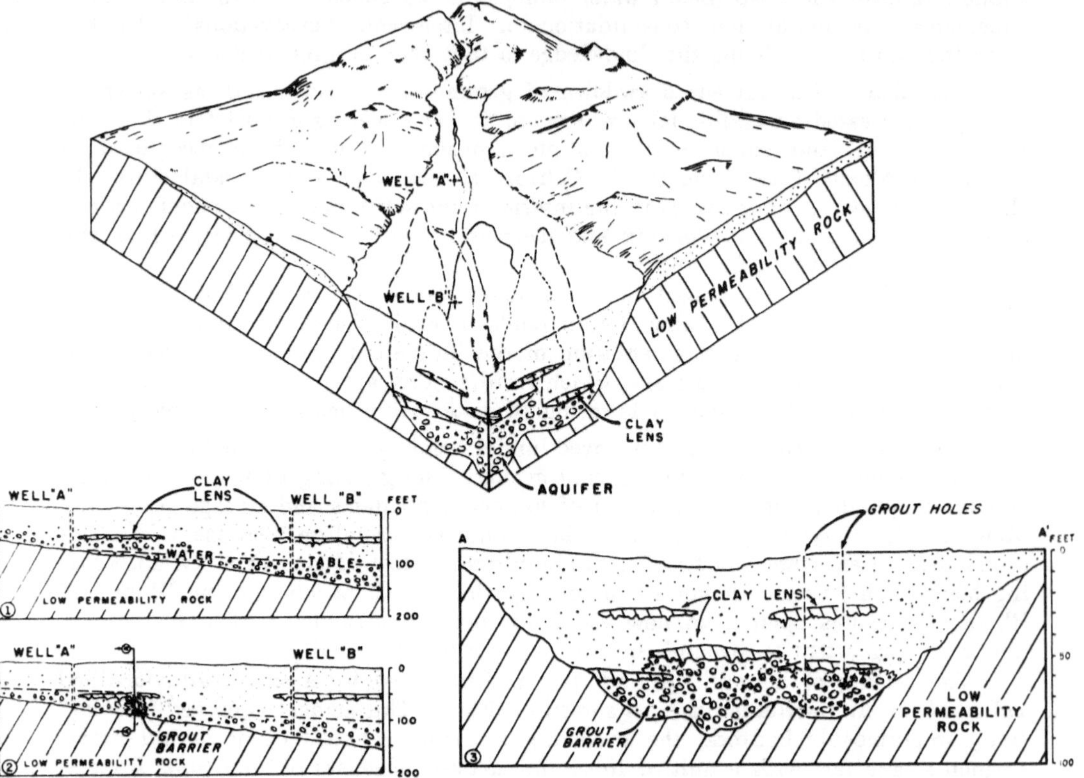

Fig. 5. Block diagram of alluvial-filled basin with an underground dam to create ground-water reservoir. Longitudinal subsurface section 1 shows conditions between wells *"A"* and *"B"* of natural gradient and water level. Section *2* is along same line after grout barrier constructed, showing underground reservoir for well *"A"*. Cross section *3* is along the line of grout barrier across valley (see *2*) and shows placement keyed to the inter-bedded clay lenses of alluvium (From Kiersch 1954)

Blockdiagramm eines von Alluvionen erfüllten Beckens mit einem Untergrundstau-werk zur Anlage eines Grundwasserbeckens. Der Untergrundlängsschnitt *1* zeigt die Gegebenheiten zwischen den Quellen „*A*" und „*B*" bei natürlichem Gefälle und Wasser-spiegel. Schnitt *2* verläuft entlang derselben Linie nach der Errichtung eines Dichtungs-schirmes; er zeigt das unterirdische Staubecken für die Quelle „*A*". Der Querschnitt *3* ver-läuft entlang dem Injektionsschirm durch das Tal (siehe *2*) und zeigt, daß die Einsatz-stellen auf die eingebetteten alluvialen Tonlinsen gerichtet sind (Nach Kiersch 1954)

70 feet) could be excavated to bedrock and backfilled to the desired height with a clay slurry or clay core; also sheet piling across narrow constructions may be feasible. Some legal aspects are apparent, however, a barrier as suggested would not violate the groundwater codes of many states. Such "reservoirs" could serve

as ideal storage basins to supply local needs that otherwise would require a more expensive surface storage structure, and so eliminate evapo-transporation loss be-

sides providing an improved moisture content of the soil for cultivation and no land costs for the site. Note the good site for an underground dam shown on Fig. 4 left frontblock, where Unit *C* is shallow and enclosed by impermeable Unit *B*.

Nuclear Explosions for Industrial Uses. Underground nuclear explosions test data have focused attention on many potential industrial uses. A summary of such uses and research aspects for geoscience are given by N. M. Short (1961). Of particular significance to Colloquium members and geomechanics is the use of underground test data for: 1. static and dynamic measurements of mechanical properties of natural materials; 2. improved techniques for in-situ measurement of physical properties; and 3. the forming of "sinks" by induced subsidence over a contained nuclear explosion.

Induced Subsidence — In a geologic setting of alluvium and sediments as shown on Fig. 4, relatively small yield underground explosions produced spectacular subsidence[5] of the alluvial material (Houser and Eckel 1962). Subsidence "sinks" up to several hundred feet in diameter and over 100 feet deep were formed (a function of charge, depth of burial, and material). The "sink" structures develop after the initial overpressure from the explosion dissipates and the fusion cavity

Fig. 6. Geothermal steam well, near Beowawe, Nevada, discovered in 1960. Flow is currently strongest well in United States at some 41,500 lbs./hr. of steam and 1,430,000 lbs./hr. of hot water at a well-head pressure of 116.5 psig. and temperature of 342⁰ F.
(Kiersch 1964) Photo by W. A. Oesterling

Geothermale Dampfquelle in der Nähe von Beowawe, Nevada, entdeckt 1960. Diese Quelle hat zur Zeit die stärkste Ausschüttung in den Vereinigten Staaten: etwa 18 824 kg/h Dampf und 648 636 kg/h heißes Wasser bei einem Druck von 8,19 atü an der Quellenöffnung und einer Temperatur von 172,2⁰ C
(Kiersch 1964) Photo von W. A. Oesterling

collapses. This reduction in volume and subsequent induced subsidence of the sediments is created by a combination of:

---

[5] This induced phenomenon has been understood since the early work of Lyman (1942) when high explosives were used to consolidate cohesionless foundation soils. However, the areal scale, magnitude of subsidence, and potential for utilization are quite different for underground nuclear explosions.

    1. the re-arrangement of sand-gravel grains of alluvium giving rise to a consolidation within area affected by strong shock waves; and

    2. the actual slumping when the underlying cavity collapses and a chimney is formed.

The possible engineering applications for such induced subsidence are quite promising such as for: preconsolidation of foundation alluvium before new construction, both small and large scale; and the "excavation" of harbors or large-scale depressions without removal of material. The advantages for preconsolidation in seismically active areas is obvious, however its use in previously underdeveloped regions of virgin lands may be even greater. This method of creating an "excavation" contains all the radioactive material and there is no detrimental waste and fallout, as per normal cratering.

Geothermal Energy. Although developed industrially in Italy and Iceland for over a hundred years, only since 1955 when new energy sources were needed has geothermal energy been given serious attention in the United States. Fig. 6 is the strongest steam well discovered to date and is located in central Nevada — an area with no surface water potential for hydro-electric power and its value is obvious.

Although potential for natural geothermal energy is largely limited to western United States, this study has established fundamental information of wide importance, such as the nature of geothermal gradients, flow pattern of heat and the con-

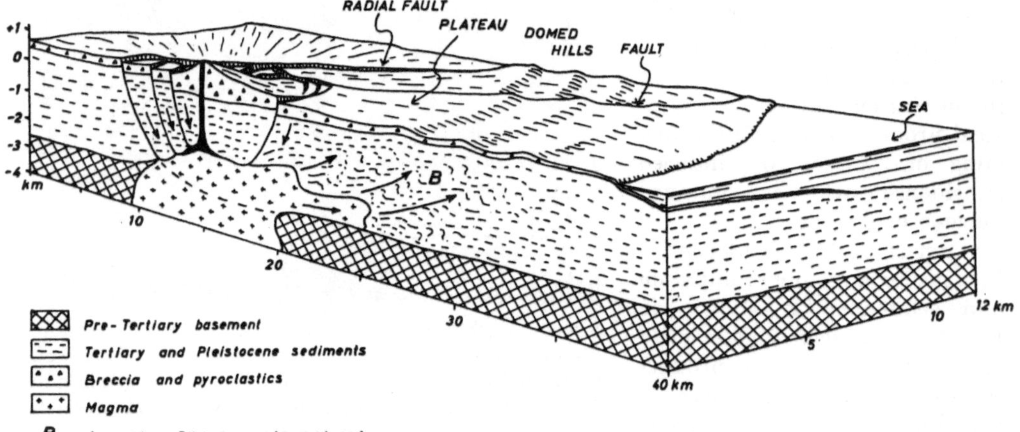

Fig. 7. Volcano-tectonic structures. The negative or subsidence type is in the central part with distinctive boundary faults and pattern. The positive protuberances type is shown in right central part with distinctive domal structures (hills) in overlying rocks and area of potential blisters

Vulkanotektonischer Bau. Der negative Typ (Senkungstyp) erscheint im mittleren Teil mit deutlichen Randstörungen und deutlichem Gefügemuster. Der positive Protuberanzentyp tritt im rechten mittleren Teil mit deutlicherem Kuppelbau (Domen) im darüberliegenden Gestein auf

duction and convection properties of certain rocks, and the distinctive features of the areal volcano-tectonic structures so commonly associated with geothermal districts (Kiersch 1964). The latter has given further insight into magma migrations, which is now aiding the development of techniques to forecast eruptions, and furthermore prognosticate conditions along active faults as to their strain build up and potential for a sudden release (seismic event). This study again demonstrates how the fundamental knowledge of a geologic phenomenon can be utilized by the interdependent geologist-engineer team. Some new applications of geology to engineering practice are emerging as a result. However, before mentioning these new developments, a brief description of volcano-tectonic structures to clarify.

Volcano-Tectonic Features — Structures are of two major types (Fig. 7):

1. *protuberance (positive),* in which the magma volume greatly exceeds the volume extruded and therefore doming of the enclosing rocks occur; and

2. *rifts or subsidence (negative),* in which the magma is withdrawn, creating a collapsed structure due to the weight of the overlying rock column (Van Bemmelen 1949).

Aerial Sensoring. The abnormally high ground temperatures within a geothermally active volcano-tectonic structure have been successfully recorded and contoured by aerial sensoring. The technique uses a combination of aerial photography and infrared data — thermal imagery is recorded by radiometers or scanners which is then reduced to temperature (MIST, 1962). Of particular interest to engineering geologists is the encouraging outlook for the utilization of this technique to delineate positive blisters of magma within the near-surface rocks. The blisters tilt and inflate the rock cover, and frequently continue to an eventual eruption. Such protuberances (as shown on Fig. 7) were accurately delineated on the slopes of Mt. Kilauea, Hawaii, in February 1963 using the infrared technique. The same survey also delineated concealed fault zones that were conducting excessive heat and themselves were potential conduits for eruption. The value for public safety and long-range engineering planning is obvious — both matters of concern for engineering geologists.

An adoption of the aerial sensoring infrared technique to active fault zones will be attempted soon. Periodic mapping of the associated geothermal temperatures (in units of one degree) may eventually lead to a correlation between: the geothermal temperature; the strain energy build up along zone; and actual conditions preceding fault movement. Obviously a systematic case history of the full cycle of an active fault zone is needed to determine any such correlation, and requires many years to verify.

Lunar Basing. Studies of the lunar surface (Green 1962) have lead to the belief that some of the structural features are volcanic in origin (as shown on Fig. 8). The presumed lunar volcanic fissure system would be similar in many ways to rift systems in active volcanic areas on the earth.

Alphonsus has been selected as an example of a potential lunar base that is located in a probable volcanic crater (negative volcano-tectonic structure) and adjacent to a Mare (Fig. 8). The crater shows peripheral fracturing and is considered to have

Fig. 8. Photo of Moon, region of Alphonsus, at last quarter. Note crater features, structural features of region, and central cone within the crater. Photograph from the Mount Wilson and Palomar Observatories, Hooker (100-inch) telescope (Courtesy E. Shoemaker 1959)

Lichtbild vom Mond, Gebiet Alphonsus, im letzten Viertel. Zu beachten sind Kratermerkmale, strukturelle Merkmale des Gebietes und ein zentraler Kegel innerhalb des Kraters. Die Photographie stammt von den Observatorien am Mount Wilson und Palomar, Hooker- (2500-mm-)Teleskop (mit freundlicher Genehmigung von E. Shoemaker 1959)

internal heat possibilities. The crater is 70 miles across, and has a maximum wall height of 6,900 feet above the crater floor. Evidence strongly suggests a volcanic-like history throughout *Alphonsus,* and the volcano-tectonic structure is believed similar to a caldera in Iceland (Green 1962).

If *Alphonsus* possesses the origin suggested, then of significance to engineering geologists is the utilization of such volcanic craters on the lunar surface as sites for building underground lunar bases. The subsurface heat possessed by the lunar material could be used for warming underground openings carved at a shallow depth. This would give a more or less constant temperature to the lunar quarters.

The quantitative understanding of a geologic phenomenon, volcano-tectonic structures, has thus given rise to new possibilities for engineering geology. Today, experienced engineering geologists are members of the space exploration team of the United States.

Research for Engineering Geology

Research activities in engineering geology fall within four basic categories, as they pertain to the geological and inter-disciplinary specialities. Results of the research can be grouped in two broad classes: 1. small-step, interval advances to knowledge; and 2. gap advances, where a major break through or discontinuity in progress pushes knowledge ahead to new vistas. The following summarizes (Kiersch 1958).

Step I — Conventional research on an individual scale. Investigations by geologists or any of allied disciplines on: laboratory or field study of a geologic material, process or phenomenon; a compilation and synthesis of hitherto scattered physical data; or case histories and performance records of materials or features in service. Advances are small-scale.

Step II — Group research in one of the geological sub-fields. Involves modest tools and a wide variety of research problems with team approach, such as for: rebound phenomenon; seismic intensity maps; ground water movement and permeability; causes of creep and gravity movement correlated with conditions inherent to the rock mass; and rock-soil properties that influence excavation characteristics. Advances are small-scale.

Step III — Group research of an interdisciplinary character. Involves complex tools and a synthesis of results by specialists from geoscience and related disciplines. Numerous fundamental problems exit in this category that possess the potential for a major gap advance in engineering geology, such as: groundwater recharge; sea water intrusion; areal subsidence; predicting movement by active faults, and magma eruptions; developing new mapping techniques using aerial, infrared sensoring and radiowave transmission; determining viscosity of a rock mass by measuring effect of an induced stress; and determining a means of analyzing the influence of time, both geologic and human, on the inherent properties of rocks and engineering sites.

Step IV — Group research of an interdisciplinary character involving major and expensive tools, or very large efforts. A number of speculative to very promising research areas exist in this category and offer potential for major gap advances. Currently, the most important with potential for major advances is underground nuclear explosions and their applications to industry. Another is space research and studies of the lunar surface for basing and potential development. Obviously, very few projects can be pursued at this scale, due to the high cost of research.

For an exhaustive list of research problems and projects concerned with dams and hydraulic structures, the reader is referred to the compilation prepared by P. D. Trask (1959).

Appendix

Engineering Geology[6]

An outline of the subject matter of advanced-level course in Engineering Geology. Presupposes student has a background in physical sciences, some engineering science and humanities and the basic geology courses[7].

Part I — Fundamentals

A. *Introduction.*
   1. Scope.
   2. Historical Development.
   3. Utilization.

B. *Geologic Phenomena of Special Importance or Concern.*
   Presupposes general knowledge of all geologic processes from prerequisite courses.
   1. Properties of rock, mass-vs-substance.
      Partly, solid geophysics and geomechanics.
   2. Seismology.
   3. Natural stresses and geomechanics.
      Surface excavations.
      Underground openings.
      Time factor, geologic-vs-human.
   4. Dynamic stresses (explosions).
      Surface excavations.
      Underground excavations.
   5. Weathering of rock, soil derivatives.
      Properties of surficial materials.
   6. Dynamic geomorphology.
      Erosion and mass wasting, surficial changes.
   7. Sedimentation-silting.
   8. Downslope-gravity movement.
      Falls, Slides, Creep.
   9. Geohydrology.
      Underground fluids.
   10. Bodies of water (ocean, estuary, bay, lake, reservoir).
       Currents, Erosion, Shorelines, Sedimentation.
   11. Frozen ground.
       Perennial, Seasonal.
   12. Subsidence.

---

[6] From G. A. Kiersch and V. Mencl: Engineering Geology, Theory and Practice. Prentice-Hall Publ. Co., New York (in prep.), and Kiersch 1955.

[7] These are: physical, historical, mineralogy, petrology-petrography, structure, sedimentation, geomorphology, mineral resources, and field geology.

Some advanced courses for further basis are solid geophysics, seismology, soil science and/or soil mechanics, and exploration geology.

C. *Geologic Materials of Special Importance.*
   1. Aggregate.
   2. Cement and pozzolan.
   3. Rock-fill.
   4. Earth-fill.
   5. Ice and snow.
   6. Miscellaneous construction materials, e. g., dimension stone, riprap, clays, sediment-lining, stabilizer, coral, water for construction.

Part II — Methods of Analysis and Investigations

A. *Investigations — Approach.*
   1. Stages.
      Planning; Design; Construction; Operation and/or Maintenance.
   2. Environmental Studies.
      Regional-areal (background geology); Small-scale (site geology); Specialist-team.

B. *Principles of Mechanics, Analysis and Properties of Materials.*
   1. Fundamentals, scope.
   2. Elastic state of rock-soil masses.
   3. Elasto-plastic state of rock-soil masses.
   4. Stability of rock-soil masses.
      Influence of water.
   5. Loose state of rock-soil masses.

C. *Techniques and Methods, Unique or Standard.*
   1. Field exploration (direct and indirect).
      Drill holes, man-sized openings; aerial photos; borehole photos; TV-observations; geophysical methods; geochemical methods; groundwater tests; mapping surface and underground; aerial sensoring by infrared, radiowave and radar.
   2. Field and laboratory testing.
      Sampling of holes, openings and cuts.
      Soil mechanics-tests (correlate IB - 1, 5, 7).
         Physical-engineering properties surficial materials.
      Rock mechanics-tests (correlate IB - 1, 3, 4, 8, 12).
         Physical-engineering properties rock masses.
         Measure inherent strain, analyze stresses active, time factor.
      Petrography.
         Physical-chemical properties of earth materials.
         Petro-fabrics of rocks, correlate testing data.
   3. Evaluation-interpretation of comprehensive data (geostatistis).
      Direct and indirect exploration.
      Field tests.
      Laboratory tests.
      Geologic models.
   4. Plans and specifications.
   5. Remedial treatment for construction and maintenance.
      Grouting; drainage; dewatering; tracing ground water flow, leakage; channel, shoreline stabilization; embankment, open-cut stabilization; underground stabilization; excavation, blasting.

Part III — Geology Applied to Engineering and Construction Projects

A. *Engineering Works.*

Engineering nomenclature; Geologic Problems Both Common and Exceptional; Remedial Procedures; Case Histories.

1. Surface excavations and structures (exclusive of hydraulic works).

   Open-cuts, construction and mining; pipelines; quarrying; buildings, all types; power plants; bridges, piers; highways; railroads; airports; launching pads; urban development, homesites; land reclamation.

2. Hydraulic structures.

   Dams and reservoirs, surface; underground dams and reservoirs; canals and irrigation works, sediment-lining; conservation works, water and soil; stream-channel control works; beach and harbour control works; tidal energy (harnessing), oceanographic projects.

3. Underground excavations and structures.

   Tunnels, conduits, aqueducts, and siphons; mine workings and shafts; power plants; subterranean installations, storage and factories; drainage works; storage reservoirs-chambers, petroleum, gas-liquid, and water; missile launching installations.

4. Construction materials.

   List given under Part I, C.

5. Military.

   Intelligence-terrain factors; general military construction, ground and air operations; naval operations.

   Dynamic stresses induced by explosions, protective construction.

6. Astrogeology.

   Lunar and planetary operations; sites and resources for basing, protection, natural sources of heat-energy.

7. Regional economic development (partly overlaps A, 1—4).

   Basis for interrelated development of natural resources, agriculture and industry.

   Land management; engineering works (categories A, 1—3); water supply projects; city planning for site selection.

8. Nuclear explosions for industrial uses.

   Excavations, surface; induced subsidence and depressions; underground fragmentation, reservoirs; earth- and rock-fill dams, et al. underground storage of energy, geothermal heat; mineral exploitation, especially oil in shale, "sand", secondary recovery.

9. Public health.

   Disposal of radioactive and industrial waste; stream pollution and abatement control.

10. Urban geology (for cities located and growing).

    Largely background geology for certain engineering works overlaps with category A-7.

B. *Legal Aspects of Geology in Engineering Practice.*

1. Responsibility of the project sponsor-vs-contractor.

   Changed conditions; limits of accuracy; factors of safety.

2. Responsibility of the insuree-vs-insured.

3. Mineral rights and claims; title ownership for construction materials, groundwater, surface water, access roads, disposal areas; adequate rent or royalty payment.

4. Land evaluation.

5. Accretion-vs-avulsion of sedimentation and channels.
Riparian land rights lost or gained.

6. Investment counseling.

C. *Failures of engineering Works.*

1. Failure due to geologic weakness or condition (single or combination of adverse features).
Surface excavations and structures (A - 1).
Hydraulic structures (A - 2).
Underground excavations and structures (A - 3).
Time factor, geologic-vs-human (life of works).

2. Failure due to the man-made changes emplaced on the geologic setting or site by operations of works.
Same sub-division as C - 1.

3. Failure due to a combination of adverse geologic features and man-made changes.
Same sub-divisions as C - 1.

References Cited

ASTM: Symposium on surface and subsurface reconnaissance. Amer. Soc. Test. Materials, Spec. Techn. Public. 122, 228 p., 1951.

Birch, F.: Megageological considerations in rock mechanics. In: Judd, W. R. (ed.), RAND Corp., Santa Monica, Calif., 1963.

Cambefort, H.: Entachement et consolidation des roches. Felsmechanik und Ingenieurgeologie, I, no. 2, 1963.

Emery, D. E.: Discussion in: Judd, W. R. (ed.), RAND Corp., Santa Monica, Calif., 1963.

Gatewood, C. E.: Impact ahead. Geo Times, 7, no. 5, p. 8—12, 1963.

Green, J.: Geology of the lunar base. North American Aviation Co., Los Angeles, Calif., 1962.

Houser, F. N. and E. B. Eckel: Induced subsidence by underground nuclear explosions. U. S. Geol. Survey Prof. Paper, 450-C, no. 66, p. 17—18, 1962.

Jaeger, J. C.: Discussion in: Judd, W. R. (ed.), RAND Corp. Santa Monica, Calif., 1963.

Judd, W. R. (ed.): State of stress in earth's crust. Symposium RAND Corp., Santa Monica, Calif., June 1963.

Kiersch, G. A.: Uses of clay in the development of water resources. Bull Geol. Soc. Amer., 65, p. 1273—74, 1954.

Kiersch, G. A.: Engineering geology. Colorado Sch. Mines Quart., 50, no. 3, 123 p., 1955.

Kiersch, G. A., McGill, J. T. and J. Mann: Teaching aids and allied materials in engineering geology. Geo. Soc. Amer. Spec. Publ., 36 p., 1957.

Kiersch, G. A.: Quantitative trends and research needs in engineering geology. Bull. Geol. Soc. Amer., 69, no. 12, p. 1597, 1958.

Kiersch, G. A. Regional-areal geologic investigations in highway geology. Proceedings 13th Annual Highway Symposium. Ariz. Hwy. Dept. Phoenix, p. 121—160, 1962.

Kiersch, G. A.: Academic trends and engineering geology. Jour. Geol. Educ., 12, no. 1, p. 14—21, 1964.

Kiersch, G. A.: The global distribution and physical characteristics of geothermal steam resources. Air Force Cambridge Research Laboratory (OFOAR), contract no. AF-19 (628)—293, 170 p., 10 figs, 3 tables, 1964.

Kieslinger, A.: Restspannung und Entspannung im Gestein. Geologie und Bauwesen, 24, no. 2, p. 97—112, 1958.

Kieslinger, A.: Residual stress and relaxation in rocks. Intern. Geol. Congress, Copenhagen (XXI), Part 18, p. 270—276, 1960.

Lyman, A. K. B.: Compaction of cohionless foundation soils by explosives. Amer. Soc. Civil Engr. Trans., 107, p. 1329—1348, 1942.

MIST: Second symposium on aerial sensoring techniques. Mich. Inst. Science and Technology, Ann Arbor, Michigan, 1962.

Müller, L.: Der Felsbau, Part I, Theoretical. F.-Enke-Verlag, Stuttgart, Germany, 1963.

Press, F. and C. Archambeau: Release of tectonic strain by underground explosions. Jour. Geophys. Review, 67, no. 1, p. 337—344, 1962.

Short, N. M.: Applications of geology to underground nuclear explosions. Geo Times, 6, no. 3, p. 18—23, 42—43, 1961.

Trask, P. D.: Barrages et bassins de retenue. Les Congres Et Colloques de L'Université de Liège, 14, 29—42, 1959.

Van Bemmelen, R. W.: Geology of Indonesia, Govt. Printing Office, The Hague, Netherlands, vol. I A, p. 207—213, 1949.

Wantland, D.: Personal communication, June 12, 1963.

The Determination of In Situ Stress and Strain Using Photoelastic Techniques

By

A. Roberts and I. Hawkes*

With 10 Figures

Summary — Zusammenfassung — Résumé

The Determination of In Situ Stress and Strain Using Photoelastic Techniques. This paper describes the technique of measuring in-situ stress and strain by means of photoelastic devices. The photoelastic coating technique uses a layer of plastic cemented on the surface of the object under examination. When the object is strained, the layer becomes birefringent and when observed through a reflection polariscope an optical pattern is seen which identifies the strains in magnitude and direction, by reference to the isochromatics and isoclinics. For ease and convenience of measurement, when body forces are under examination, a birefringent biaxial gauge is used which consists of a hole at its centre which acts as a stress concentrator in the gauge. The gauge integrates the strains within the area enclosed by its perimeter and the resultant produces a symmetrical isochromatic pattern when viewed under polarised light. The fringe order identifies the magnitude of strain and hence of stress. The axes of symmetry of the pattern identify the directions of principal stresses. By making a number of such measurements over an area, the character of the stress field in an exposed rock mass may be deduced. The technique is applied to the exposed surfaces of an excavation in solid rock, and also at the back of boreholes drilled into the rock mass, in which case it provides a simple method of determining the state of stress in the rock at the back of the borehole, by the overcoring "stress-relief" technique.

When the elastic properties of the rocks are unknown a direct method of determining in-situ stress is provided by the photoelastic stressmeter, which consists of a glass plug containing an integral polarised light source. Any increase in stress within the rock, after the plug is inserted, is measured both in magnitude and direction, by reference to the optical pattern displayed, which is calibrated under controlled biaxial conditions in the laboratory. The plug acts as a high modulus inclusion in most sedimentary rocks, one calibration serving for all materials whose elastic modulus is not more than half that of glass. For materials of higher modulus the plug is calibrated for a limited modulus range, to include that of the material concerned.

Die Bestimmung der in-situ-Spannung und -Dehnung durch die Photoelastizitätstechnik. Dieser Vortrag beschreibt die Technik, wie man die Messung der in-situ-Spannung und -Dehnung mittels photoelastischer Apparate durchführt. Bei der photoelastischen Technik verwendet man eine auf der Fläche des Probestückes angeklebte Schicht aus Kunstharz. Bei der Beanspruchung des Probestückes wird der Überzug optisch doppelbrechend; wird dies durch einen Spiegelpolarisator beobachtet, so sieht man ein optisches Muster, das in bezug auf Gleichfarbigkeit und Isokline die Größe und Richtungen der Dehnung bestimmt. Um die Körperkräfte bequemer und einfacher messen zu können, verwendet man ein doppelbrechendes, doppelachsiges Meßgerät, das aus einer nur am Rand aufzementierten Scheibe besteht. Die Scheibe hat im Mittelpunkt ein Loch, das eine Spannungskonzentration im Meßgerät bewirkt. Das Meßgerät integriert die Beanspruchung der in seinem Umkreis

---

* Dir. Dr. Albert Roberts, Postgraduate School in Mining, University of Sheffield, Mining Department, St. George's Square, Sheffield 1, England.

eingeschlossenen Fläche; wenn dies bei polarisiertem Licht untersucht wird, ergibt die Resultierende ein symmetrisches, gleichfarbiges Muster. Die Ordnung der Strahlenbrechung bestimmt die Größe der Dehnung und damit auch der Spannung. Die Symmetrieachsen des optischen Musters bestimmen die Richtungen der Hauptspannung, und die Form des optischen Musters zeigt das Größenverhältnis der Hauptspannungen auf. Nach der Durchführung zahlreicher Messungen auf einer Fläche kann auf den Charakter der Spannungsfelder in einer unter Beobachtung stehenden Felsmasse geschlossen werden. Die Technik wird auf vorhandenen Oberflächen einer in Abtrag befindlichen Felsmasse angewendet und auch bei Bohrlochwandungen von Bohrlöchern in der Felsmasse. In einem solchen Falle ist es bei Anwendung des Kernbohrens möglich, mittels des einfachen Verfahrens den Spannungszustand des Gebirges zu bestimmen.

Wenn die elastischen Eigenschaften des Gebirges unbekannt sind, gibt das photoelastische Spannungsmeßgerät durch ein direktes Verfahren die Möglichkeit, die in-situ-Spannung festzustellen. Das photoelastische Spannungsmeßgerät hat einen Glasstöpsel, der das aus einer Lichtquelle kommende Licht integriert und polarisiert. Nach Einführung des Glasstöpsels kann jede Spannungsänderung innerhalb des Felsens sowohl der Größe wie auch der Richtung nach mit Hilfe des optischen Musters bestimmt werden. Dieses Muster wird unter kontrollierten doppelachsigen Bedingungen im Laboratorium kalibriert. Der Stöpsel funktioniert auch bei den meisten sedimentären Felsen, und zwar auch bei Felsen mit hohem Modul. Eine Kalibrierung dient für alle Materialien, deren Modul nicht über der Hälfte des Glases liegt. Für Materialien mit hohem Modul ist der Stöpsel für einen begrenzten Modulspielraum kalibriert.

La détermination des contraintes et des déformations in situ par photoélasticité. Cet article décrit les techniques de mesure des contraintes et des déformations in situ au moyen de systèmes photoélastiques. La technique de la couche photoélastique emploie une feuille plastique collée sur la surface de l'objet à examiner. Lorsque l'objet est déformé, la couche devient biréfringente et lorsqu'on l'observe en lumière réfléchie polarisée, on voit un réseau de courbes qui caractérise les déformations en grandeur et direction grâce aux courbes isochromatiques et isoclines. Pour la facilité et la commodité des mesures, lorsqu'on étudie des forces de volume, on utilise comme jauge biréfringente à deux dimensions un disque collé seulement sur son périmètre. Un trou percé au centre assure la concentration des contraintes. Cette jauge intègre les déformations à l'intérieur de son périmètre et la résultante fournit un réseau symétrique d'isochromatiques lorsqu'on l'observe en lumière polarisée. Le nombre des franges donne la grandeur des déformations et par conséquent des contraintes. Les axes de symétrie du réseau donnent les directions des contraintes principales. En faisant un grand nombre de ces mesures dans une région, on peut analyser la distribution des contraintes dans un massif rocheux. Cette technique s'applique aux surfaces libres d'une excavation et aussi au fond des trous forés dans le massif rocheux. Dans ce cas elle fournit une méthode simple de détermination des contraintes dans le rocher par la technique de libération des contraintes par carottage autour de l'appareil.

Lorsqu'on ignore les propriétés élastiques de la roche, l'extensomètre photoélastique fournit une méthode directe de détermination des contraintes in situ, grâce à un bloc de verre contenant une source de lumière polarisée. Tout accroissement des contraintes à l'intérieur du massif après la mise en place de ce bloc est mesurée à la fois en grandeur et direction par référence au réseau de franges étalonné au laboratoire grâce à un module élevé dans la plupart des roches sédimentaires et un seul étalonnage convient pour tous les matériaux dont le module reste inférieur à la moitié de celui du verre. Pour des matériaux à module plus éléve, le bloc est étalonné pour une faible plage de modules autour de celui du matériau étudié.

Introduction

The stability of excavations in rock and the stability of rock as a foundation for structures built upon it is determined by the behaviour of the rock as it exists in situ in relation to the stresses acting within it. The stress that exists, for example in the rock comprising the foundations of an engineering structure, is the resultant of the original in-situ stress that existed before the excavation was made, the effect that the excavation has on this original stress field, and the effect of the loads superimposed by the engineering structure itself. It is possible, by various means, such as by the aid of models, and by the assumption of idealised conditions in the foundation rock, to estimate the magnitude of the concentration of stress that the excavation will produce about itself and the stresses imposed by the structure upon the foundations. The application of the results of such work to a field situation, however, must take into account the fact that the rock may differ substantially from the idealised material and that it may possess unpredictable properties due to inhomogeneities or anisotropy. The original state of stress in the rock is quite unpredictable by any method and can only be approximately determined by such techniques of in situ measurement as are practised at the present time.

The determination of the in-situ strength of rock in mass is one of the most urgent problems yet to be solved in rock mechanics. It is logical to approach the problem by observing the material properties of the rock concerned, in the laboratory, as a stage towards understanding the mechanism of deformation and fracture of the same rocks in mass. Again, we must modify the results of our laboratory tests to include the effects of anisotropy, inhomogeneity, and discontinuities, that can only be observed in the field. The strength of a rock is, in the end, evidenced by its deformation under load. Observation of in-situ strains are therefore essential to our understanding. Such observations may include measurement of strain at localised positions, using conventional strain gauge techniques, subsequently combined to give information as to the condition of the rock mass, or by direct measurement over a distance when mass deformation is of sufficient magnitude to permit this.

Photoelastic Strain Gauges

During recent years considerable advances have been made in the development of instruments for the measurement of stress and strain in rock in situ. Broadly speaking, in the past, the equipment has been such that the observed signal has originated from a mechanical, hydraulic, or electrical component, but a more recent introduction is that of the optical strain gauge or transducer which utilises the property of birefringence that certain materials display when strained. Observation of surface strains by the use of photoelastic coatings is now widely practised in many branches of engineering. The mode of application of the same technique to some of the problems presented by rock mechanics, both in the laboratory and in the field, have previously been described by the authors and their colleagues[1,2].

Since the first paper was presented further experience has been gained, the technique has been considerably extended, and some most encouraging results have been obtained. The attachment of a photoelastic coating cemented over the whole of its surface to a rock surface enables strains to be observed on a point by point basis, all over the area so covered, which is usually, for convenience, a circular patch of birefringent plastic some 50 mm diameter. On a granular material such as rock or concrete, the isochromatic pattern produced by strain when the patch is viewed through a reflection polariscope, is likely to be extremely variable, both in magnitude and direction, over the area viewed. While some rocks may show evidence of a directional trend giving a resultant "preferred" strain on the surface as a whole, the magnitude of that resultant may be difficult to determine, even if its direction is self evident. Quite often, neither the magnitude nor the direction of a resultant can be deduced without the aid of many measurements made on a point-by-point basis. Such techniques, while presenting little difficulty in the laboratory are generally unsuited to the conditions that exist underground in mines and on civil engineering sites.

However, if the circular patch is cemented to the rock only around its perimeter, and if in addition the patch has a hole through its centre, so that it now becomes an annulus, two important effects become apparent. 1. The presence of the hole produces a concentration of stress in the birefringent patch, so that any measurable strain in the rock is magnified and produced a higher fringe order on the patch than that which would exist if the hole were not present. 2. The patch exhibits an isochromatic pattern which is produced by the integration of the strains around its periphery. It becomes a biaxial strain gauge in which the fringe pattern identifies the directions of principal strains by its axes of symmetry, the magnitude of the strains by its fringe order, and relationship of the pricipal strains by the fringe shape. The sign of these strains, that is whether they are tensional or compressive, is also identified.

Only one measurement is now required to determine the magnitude and direction of the resultant strain on the rock surface to which the patch is cemented. By making a number of such measurements the characteristics of strain over a wide area, such as may be exposed in a mine stope or over a section of mine workings, now becomes a practicable operation, and it is possible to deduce stress trajectories, representing the directions of major and minor principal stress, over the area concerned. Figs. 1 and 2 show two examples, each being a plan of a section of mine workings where this procedure has been followed. The results of such observations are of value in that they provide the necessary data upon which rational decisions may be taken as to the optimum lay-out of the workings. They help the mining engineer decide on the best orientation of supporting pillars, lay out of rock-bolting systems etc., to resist possible movements of the strata along the directions of "preferred" strain. In a

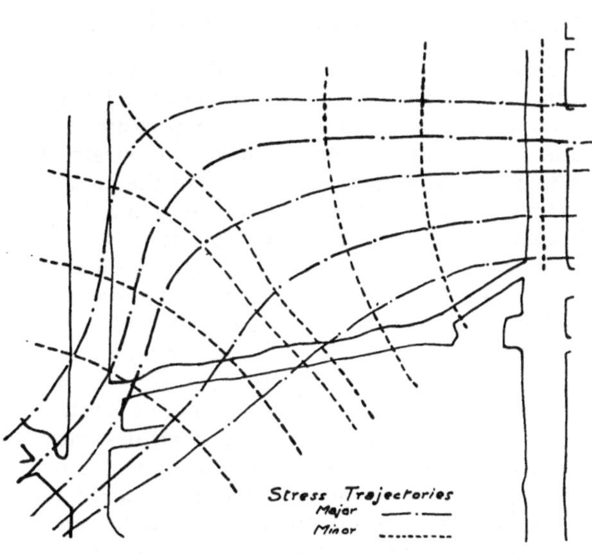

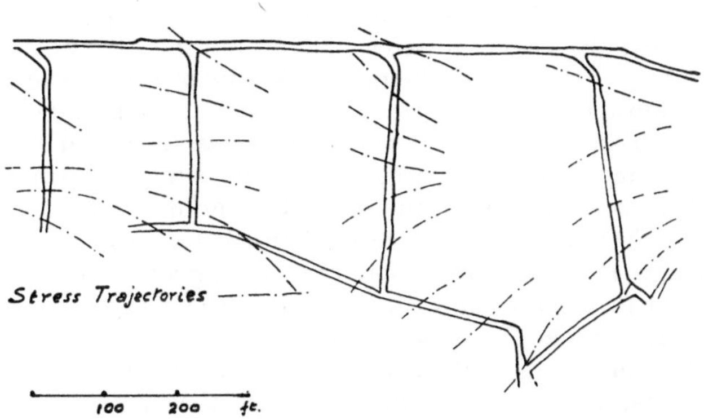

Fig. 1. Plan of section of mine workings for which major and minor stress trajectories are plotted

Plan eines Querschnitts durch ein Abbaugebiet, für den größere und kleinere Spannungstrajektorien eingezeichnet sind

Fig. 2. Major stress trajectories over a section of a mine working

Größere Spannungstrajektorien über einem Querschnitt eines Abbaugebiets

converse sense they also help him to decide on the orientation of the working faces so that the "preferred" strain directions may be utilised to promote fracture of the mineral and thus improve the efficiency of his drilling, blasting, and rock breaking techniques.

Determination of in-situ stress by observation of „elastic rebound"

Several investigators have observed the component of elastic recovery that occurs when a portion of rock is relieved from constraint, and have used it as an indication of the original confining pressure that existed when the rock was constrained. Tincelin[3] applied electrical strain gauges to the surface of the rock face in iron ore mines. Slots cut into the rock face adjacent to the gauges, caused the stress on the central rock to be relieved and hence to expand by an amount indicated by the output signal from the strain gauges. By repeating the operation with slots cut in other directions, some indication as to the direction of the in-situ stress was obtained, on the rock face. Similar techniques were subsequently applied by other investigators in various parts of the world. Serafim[4] has carried out a detailed study into the technique of applying electrical resistance strain gauges to a rock surface and then subsequently overcoring them.

Observation of surface strains are of limited value in that the rock immediately adjacent to the surface may have been de-stressed as the result of fractures produced during the processes of excavation. A more suitable place of measurement is at the end of a borehole drilled into the solid rock.

Authoritative descriptions of this technique are given by Slobadov[5], Hast[6], Leeman[7] and Obert[8]. Slobadov's method consisted of drilling a hole in the rock, grinding the back of this hole flat, and then bonding two electrical resistance strain gauges at right angles to each other on the end face. The strain gauges and the rock to which they were bonded were subsequently trepanned out in the form of a core, on which the elastic rebound resulting from the relief of constraint was measured. Hast, Leeman and Obert use borehole plugs embodying magneto-strictive, L. V. D. T.* transducers, and electrical resistance strain gauges respectively. Other borehole plug deformation meters include those of Potts[9] which embodies electrical resistance strain gauges, and the well known Maihak acoustic gauge. All these boreholes plugs must be pre-stressed when set, if they are to record the relief of stress on overcoring. They are also directional in their response, most of them being capable of measuring in one direction only although by mounting plugs in series one behind the other, two or possibly three signals may be obtained, oriented at known angles to one another. From the measured strains the magnitude and direction of the principal strains must then be calculated, or deduced from the results of calibration of the device under known conditions.

The whole procedure is simplified very considerably if the biaxial photoelastic gauge is used. Chakravarty in England[10] and Seelers in Canada[11] have both used this technique with success. Chakravarty measured the stress in an anhydrite mine at the back of holes drilled to a depth of 1.5 m, by a 140 mm diameter diamond core drill, any core left at the back of the hole being wedged out by a tapered rod. A grinding tool was then attached to the drill and the back of the hole ground flat.

A 50 mm diameter disc of photoelastic material was then bonded to the back of the hole and subsequently trepanned out. A study of the relaxation pattern in the disc, observed under normal and oblique incident polarised light, enable the initial state of stress in the rocks to be deduced. Fig. 3 shows typical cores extracted in this manner, each with the photoelastic disc attached.

---

* Linear, Variable, Differential, Transformer.

Sellers drilled holes 64 mm diameter to a depth of from 1 to 2 m in hard quartzite, flattened the back of each hole with a diamond drill bit and cemented a photoelastic disc 29 mm diameter to the back of each hole, using the apparatus

Fig. 3. Stress-relieved cores and rock fragment with photoelastic gauges attached
Spannungsentlastete Bohrkerne und Gesteinsbruchstück mit angefügten spannungsoptischen Meßinstrumenten

sketched in fig. 4. The patches were overcored some 10 to 14 days after being set. Strain began to appear in the discs when the overcoring had penetrated 6 mm into the rock, and increased to a maximum when the overcoring depth was 13 mm. Over-

Fig. 4. Arrangement for setting photoelastic disc at the back of a borehole
Einrichtung für das Anbringen eine spannungsoptischen Platte am Boden eines Bohrloches

coring beyond this depth did not alter the strain, but provided a core which could be removed without damaging the photoelastic disc. Control tests established that the process of overcoring did not in itself induce strains in the disc.

The Photoelastic Borehole Plug Stressmeter

The "low modulus" borehole plugs, and deformation meters in general, present certain difficulties as stress indicating devices, since stress must be deduced from the measured strains using equations involving the elastic moduli of the rock. Since the rocks generally do not show linear elastic behaviour and the moduli are not constant, this is an uncertain procedure. Also, since the response of the instruments is directional, multiple measurements are required. If the measuring instrument is uniaxial it is necessary to make 9 measurements, three each mutually disposed at 60 degrees, in three mutually perpendicular planes, in order to determine the triaxial stress field. Thirdly the deformations that must be measured are so small that the associated instrumentation is liable to become somewhat costly and complicated, and not always suited to the arduous conditions that are inevitable in mines and constructional sites.

The photoelastic stressmeter overcomes many of these difficulties.

Coutinho[12], Hiramatsu and others[13] have indicated that instruments which could be regarded as rigid inclusions would have many advantages as stress measuring tools. Basically the theory is that if a rigid inclusion is inserted into a body which has a modulus of elasticity less than one half that of the inclusion, the stress induced in the inclusion will only be slightly dependent on the elastic moduli of the surrounding body. Coutinho and Hiramatsu both applied this theory to a glass plug inserted in a borehole and proved that such a simple device could be used to form a stressmeter. We have taken this basic technique and developed it to a produce a sensitive biaxial stressmeter which will accurately measure any changes in the stress field and which, with ancillary techniques, can be used to measure the absolute stresses existing in a rock mass.

The stressmeter consists of a glass cylinder ($E = 7 \times 10^5$ kg/cm²) with an axial hole giving the device radial symmetry about the axis which is also the direction of observation. When the cylinder is inserted into a body subsequently subjected to load the glass is strained and exhibits an isochromatic pattern when observed under circularly polarised light, the fringe order of the pattern being proportional to the stress difference at the point observed. The glass cylinder thus shows increasing fringe orders with increasing stress in the body within which it is inserted, the order being proportional to body stress provided that the geometry of the stress field and the ratio between the principal stresses remain constant. Chan-

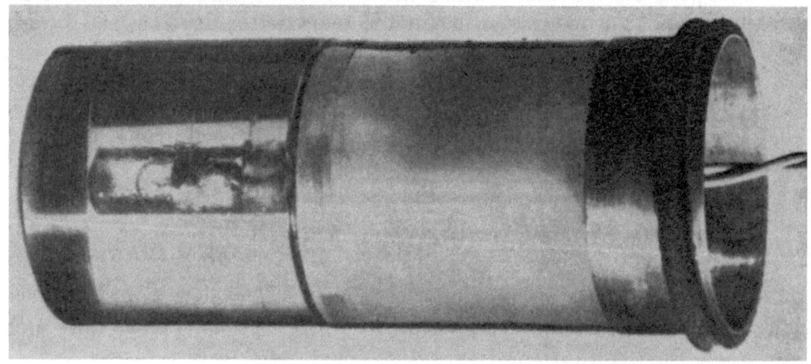

Fig. 5. The photoelastic stressmeter
Das spannungsoptische Meßinstrument

ges in the principal stress ratio produce corresponding changes in the shape of the fringe pattern, and thus the pattern can be used to identify the principal stress ratio. The axes of symmetry of the pattern identify the directions of stress.

Stresses which are aligned in a direction parallel with the light path produce no photoelastic effect in a birefringent material, so that, when the photoelastic stressmeter is inserted in a body which is subjected to a triaxial stress field of unknown geometry the observed fringe pattern identifies the stresses in a plane at right angles to the axis of view.

The range of sensitivity of the device depends upon the dimensions of the cylinder (in particular the length of the light path, the ratio between external diameter and the diameter of the axial hole), and the stress-optical properties of the glass. The example illustrated in fig. 6 has an external diameter of 32 mm, central hole 6 mm diameter, and length 38 mm, although the cylinder may range from 6 mm to 76 mm in length.

Hiramatsu used a reflection polariscope to observe glass stress meters set in concrete. The authors also use this technique on occasion, but when the stressmeter has to be observed from a distance of more than a few inches reflection polarimetry becomes difficult, for a variety of reasons, which have resulted in the tech-

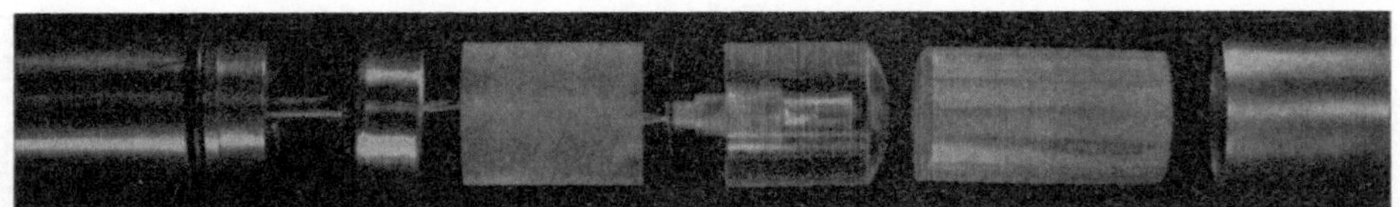

Fig. 6 a. Exploded view of photoelastic stressmeter and cement tube assembly
Auseinandergezogene Ansicht eines spannungsoptischen Meßinstrument und des Montierens eines Zementrohrs

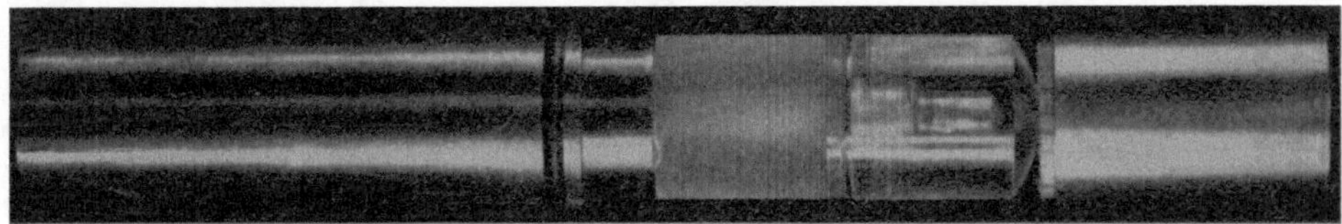

Fig. 6 b. Photoelastic stressmeter in its setting position before the rods are withdrawn
Spannungsoptisches Meßinstrument im Einsatzzustand bevor das Gestänge zurückgezogen ist

nique being applied only on a very limited scale. Similar difficulties do not arise if the stressmeter is observed by transmitted light and the device designed by the authors uses this principle.

As illustrated in fig. 6 the light source is carried in a transparent acrylic resin capsule, the back of which serves as a reflector to project a parallel beam of light through the meter. Leads for the lamp are carried through the axial hole of the meter and are powered by an external battery when an observation is to be made. A circularly polarising filter is fitted between the light capsule and the glass member, the whole being cemented together into a simple, compact, and robust unit.

Either of two methods of installation are applied. In the first a hole is drilled from the observation point into the rock or structure to be examined and the plug is cemented in this hole using the equipment illustrated in fig. 6. In essence the setting procedure consists of placing at the back of the hole a cylindrical tube containing the cement, sealed by an ejection plunger. This is inserted by the aid of an attachment to the drill rods, which are also used to insert the stressmeter, the leads from which are threaded through the hollow rods. Fig. 7 illustrates the appearance of the stressmeter when observed in a uniaxial stress field of progressively increasing magnitude, and fig. 8 shows examples of typical biaxial stress fields, the ratios of the principal stresses being indicated in each case.

The stressmeter is observed through a crossed circular polariscope, which enables the isochromatics to be seen without confusion from the isoclinics. The use of circularly polarised light also means that the stressmeter itself is nondirectional in its response and needs no specific radial alignment in the setting procedure. The isochromatic pattern is the same no matter what the radial position of the stressmeter may be, relative to the direction of view, and the position of the isochromatic pattern as shown by its axes of symmetry, is determined by the directions of principal stress in the body under examination. If those stresses change in direction after the stressmeter is inserted the observed pattern will change its

                    A. Roberts and I. Hawkes:

orientation accordingly, the axes of symmetry always identifying the direction of stress in a plane normal to the axis of view.

The fringe order identifies the magnitude of stress and for precision measurements it is necessary to measure fractional fringes. The method adopted by the

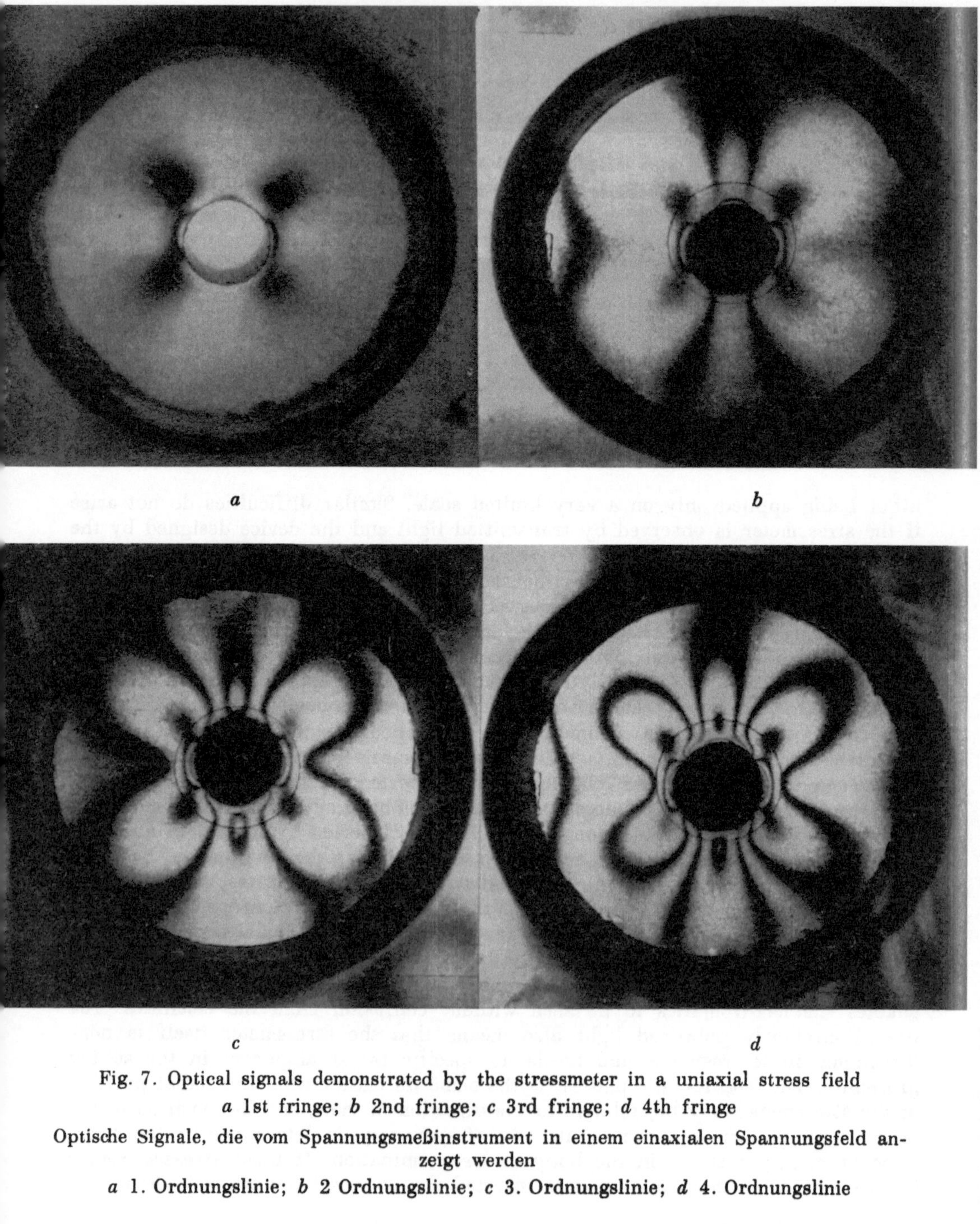

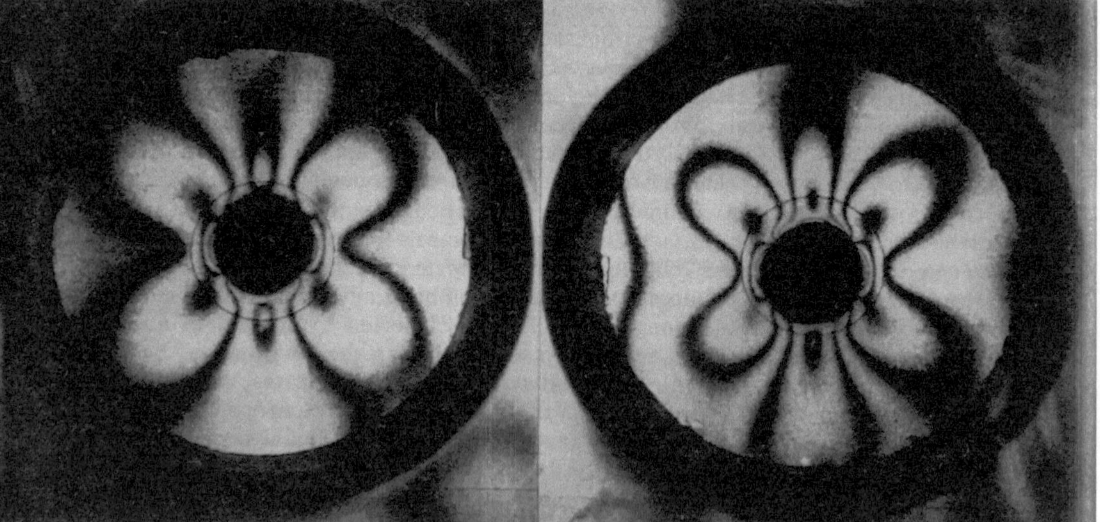

Fig. 7. Optical signals demonstrated by the stressmeter in a uniaxial stress field
a 1st fringe; *b* 2nd fringe; *c* 3rd fringe; *d* 4th fringe
Optische Signale, die vom Spannungsmeßinstrument in einem einaxialen Spannungsfeld anzeigt werden
a 1. Ordnungslinie; *b* 2 Ordnungslinie; *c* 3. Ordnungslinie; *d* 4. Ordnungslinie

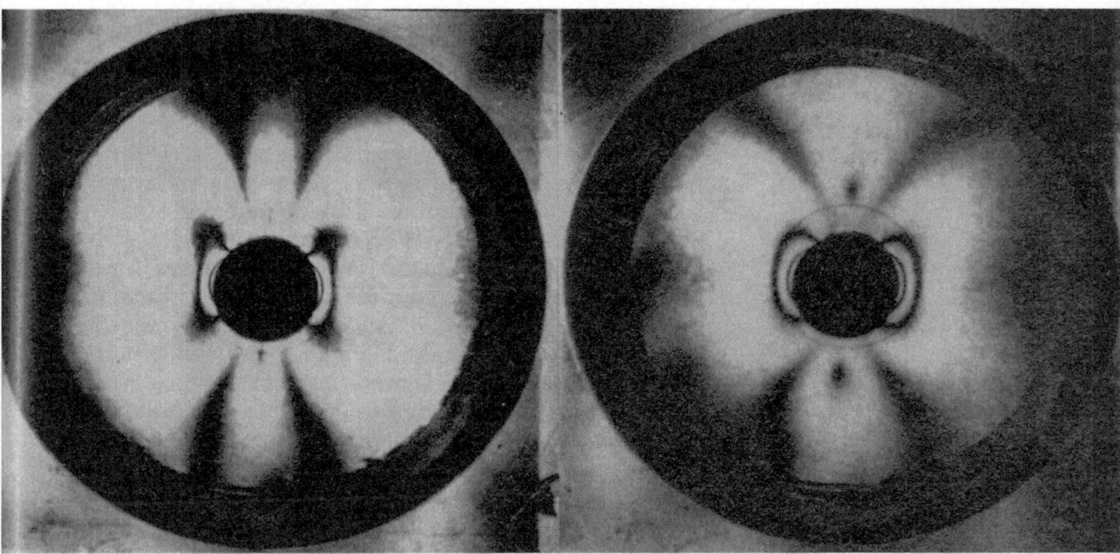

a   b

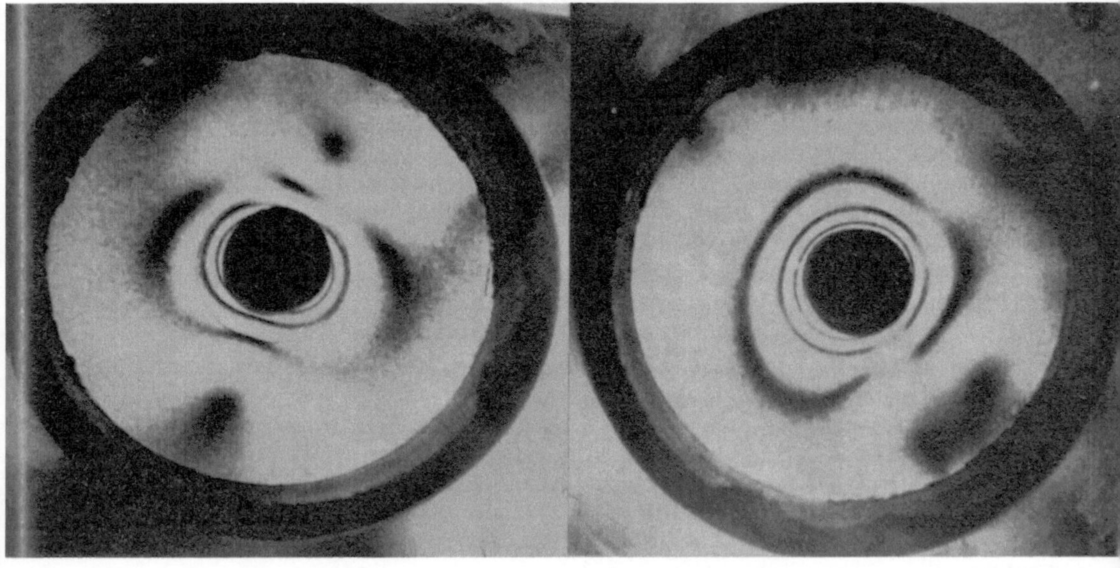

c   d

Fig. 8. Signals demonstrated by the stressmeter in biaxial stress fields (all at 2nd fringe)

a Ratio of principal stresses $1 : {}^{1}/_{4}$; *b* Ratio of principal stresses $1 : {}^{1}/_{3}$; *c* Ratio of principal stresses $1 : {}^{3}/_{4}$; *d* Ratio of principal stresses $1 : 1$

Signale, die vom Spannungsmeßinstrument in biaxialen Spannungsfeldern angegeben werden (alle bei der 2. Ordnungslinie)

a Verhältnis der Hauptspannungen $1 : {}^{1}/_{4}$; *b* Verhältnis der Hauptspannungen $1 : {}^{1}/_{3}$; *c* Verhältnis der Hauptspannungen $1 : {}^{3}/_{4}$; *d* Verhältnis der Hauptspannungen $1 : 1$

authors is to make use of goniometric compensation by the Tardy method. The
process of compensation consists of rotating the linear polariser relative to the
quarter-wave retardation plate in the analyser of the viewing polariscope. Such
a rotation alters the apparent fringe pattern by an amount such that 180 degrees
rotation corresponds to one fringe. The rotational movement is thus calibrated in
terms of percentage and observers have no difficulty in measuring to within
3 per cent of one fringe. A viewing polariscope designed by the authors is illustrated in fig. 9.

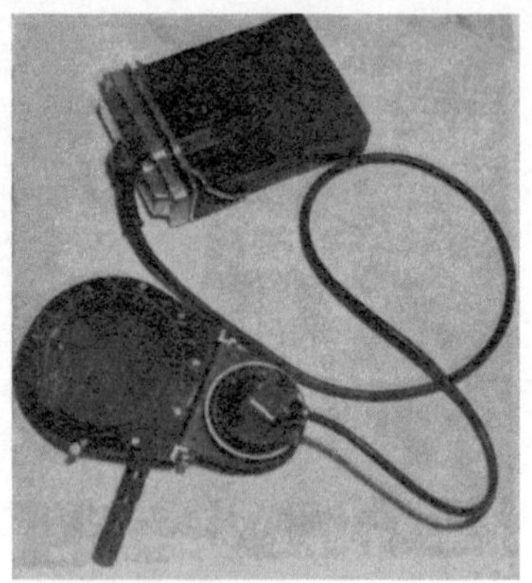

Fig. 9. Viewing polariscope with
a miner's cap lamp as the light
source

Beobachtungspolariskop mit Gru-
benlampe als Lichtquelle

Stressmeters viewed by transmitted light contain their own integral circularly polarised light source and provide an illuminated field which is observed through an analyser consisting only of quarter wave plate and linear polarising filter, mounted in a hand viewer. For observing stressmeters at a distance or at depth in a borehole the viewing filters are incorporated into a telescope. The use of a telescope is, in any case, desirable when maximum precision is required, for a graticule fitted into the eyepiece facilities observation of fringe orders at specific points precisely located on the fringe pattern.

Reading the stressmeter thus consists of the following steps:

1. With the viewing polariscope in the fully crossed position the handle of the instrument (if a hand-viewer), or the graticule (if a telescope), is aligned with the major stress direction indicated
by the symmetry of the observed pattern. (Identification of the major and minor
stress directions in a biaxial pattern is possible by observing whether the fringe
order decreases or increases in the process of goniometric compensation. Compensation when the viewer is aligned on the major principal stress produces a reduction
in the fringe order and vice versa.)

2. The fringe pattern is examined, and if it is not (as is probable) an exact
unit fringe order the analysing filter is rotated in the process of goniometric compensation until an exact fringe order is seen on the stressmeter.

The reading of the stressmeter is then that number of unit fringes plus the
percentage registered on the scale of the analyser.

Observation of the stressmeter in a biaxial stress field involves identification
of the fringe shape which gives the ratio between the principal stresses. This is
performed by comparison with photographs of patterns obtained under controlled
laboratory calibration. The identification of a unit fringe increment is performed
under white light by the recognition of the tint of passage as each new fringe
emerges.

Calibration of the Stressmeter

The meter is calibrated under controlled uniaxial and biaxial stresses on the
laboratory testing machine. See fig. 10. An extensive series of investigations was
conducted with the stressmeter in various materials, to test the validity of the
theory of the high modulus inclusion. The materials included a number of sedimentary rocks, mainly carboniferous sediments but including anhydrite, various
sandstones, and concrete. A number of tests were also made in transparent acrylic
resin which was itself birefringent under load, the purpose of these being to study

platen end effects and also to establish an optimum size of calibration block. All these materials have E less than 3.5×10^5 kg/cm², or less than half that of the stressmeter, and were found to give an identical linear calibration characteristic, such that a stressmeter of the size illustrated in fig. 5 gives a uniform increase of birefringence with stress at the rate of 1 fringe order for 21 kg/cm² increase of stress. Linear characteristics were also obtained under biaxial loading, also identical, for the various materials ($E < 3.5 \times \times 10^5$ kg/cm²) independent of changes in the elastic modulus of the material in which they are inserted. It is known, for example, that the elastic modulus of rocks changes with change in ambient stress level and that of concrete changes due to ageing effects and creep. All the evidence from tests carried out to date indicates that these factors have no appreciable influence on the reading of the glass stressmeter, and one calibration serves for all materials of low modulus.

Fig. 10. Calibrating the stressmeter in the laboratory

Eichung des Spannungsmeßinstruments im Laboratorium

In materials of higher modulus, approaching or exceeding that of glass, the stressmeter will no longer function as a high modulus inclusion and separate calibration will be necessary for different materials. These materials will include many igneous and metamorphic rocks as well as most metals.

Determination of Absolute Stress

The photoelastic stressmeter as described can only measure any increase of stress that occurs after the time it is inserted. This will not be the absolute stress level unless the meter was inserted at a zero condition of stress. The stressmeter can be suitably constructed, however to make possible the measurement of absolute stress, or to register the effects of decreases in load after the time of its insertion. In either case, the stressmeter must be pre-loaded, that is, placed under load at the time of its insertion, which is achieved either by mounting it in a tapered sleeve which is operated by a hydraulic ram, or incorporating wedges to place it under diametral loading to produce a recognisable fringe pattern at the time of setting.

Advantages of Photoelastic Gauges

There are several features of the photoelastic gauges which render them excellently suited for field applications in industry, and in mining and engineering generally. The gauges are robust and self contained. Simple in construction and in use, they need no associated costly or delicate mechanical or electrical equipment. The gauges that are observed by reflection polariscope need no ancillary equipment at all, other than the viewer and hand torch or lamp, the operation of which has been brought to a very simple procedure. They can therefore be used as control tools by which the men on the job can observe stress, strain and load, in mines and earthworks, in structures and in support systems, as the operations proceed.

At the same time, the gauges, when used in conjunction with the more refined measurement techniques and instrumentation, lend themselves very usefully as research tools for the purpose of stress and strain analysis.

The photoelastic stressmeter is unique in that it produces an optical pattern which gives directly, without mathematical computation, the following information:

1. The magnitude of the increase in stress subsequent to the time of setting the meter in place, and the magnitude of the two principal stresses perpendicular to the axis of the borehole in which the meter is set.

2. The direcion of the two principal stresses perpendicular to the axis of the borehole.

3. The direction of the two principal stresses perpendicular to the axis of the system of loading as to whether it is hydrostatic, uniaxial or biaxial, in the plane of measurement.

The biaxial bircfringent plastic gauge, when applied by the stress concentration technique, gives in addition to all the above information:

4. The magnitude of the load in that part of the structure on which the gauge is set, at the time the gauge is set. That is the gauge measures absolute load as well as change of load.

The optical gauges are also unique in that the validity of the signal given out, in relation to the calibration of the gauge, is self evident. If the gauge is functioning properly, the shape of the pattern indicates that this is so. Conversely a pattern shape that does not conform to the standards for that type of gauge indicates that the gauge is malfunctioning and should not be read. These properties are not possessed by electrical or mechanical gauges, from whose signals it is often impossible to tell whether or not the gauge is in order and properly set. When one also takes into consideration the simplicity and robustness of construction of the photoelastic gauges there can be no doubt that they promise to have a wide application in many fields of engineering, and particularly in rock mechanics.

Acknowledgements

Acknowledgement is made to Dr. J. B. Sellers and Denison Mines Ltd., for permission to use information on which fig. 1 and 3 are based, and to C.-I.-M. Consultants Ltd., for fig. 2. This work forms part of the rock mechanics research programme currently in progress at the University of Sheffield Postgraduate School in Mining. The instruments and equipment described are manufactured by Messrs. Horstman Ltd., Locksbrook Road, Bath, England.

References

[1] Roberts, A., Emery, C. L., Hawkes, I., Williams, F. T. and P. K. Chakravarty: The Photoelastic Coating Technique Applied to Research in Rock Mechanics. Trans. Inst. Min. Met., *71* (1962), 581.

[2] Roberts, A. and I. Hawkes: The Application of Photoelastic Devices for the Measurement of Strata Pressures and Support Loads. Mine and Quarry Engineering, July 1963.

[3] Tincelin, E.: Mesures des pressions de terrains dans les mines de fer de l'est. Methode de mesure. Ann. Inst. Tech. Bat. Trew. Pub. Supplement, Oct. 1952.

[4] Serafim, J. L.: Internal stresses in galleries. Septième Congres des Grands Barrages. Rome 1961.

[5] Slobadov, M. A.: The Method of Relief from Load in Studying Stresses in the Depth of Rock Mass. Ugol. Yr. 33, No. 7, July 1958, 30—35.

[6] Hast, N.: The Measurements of Rock Pressure in Mines. Arsbok 52, Novstedt and Smev, Stockholm (1958).

[7] Leeman, E. R.: The Measurements of Stress in Abutments at Depth. C. S. I. R., South Africa. Publication RN 122.

[8] O b e r t, L.: In situ measurement of stress in rock. Mining Engineering, New York. August 1962.

[9] P o t t s, E. L. J.: The Practical Application of Scientific Measurement to Problems in Strata Control. Iron Coal Trades Review (Nov. 1955), 1169—1179.

[10] C h a k r a v a r t y, P. K.: The Application of the Photoelastic Technique to the Problems of Rock Mechanics. Ph. D. Thesis 1962. University of Sheffield.

[11] S e l l e r s, J. B.: Private Communication.

[12] C o u t i n h o, A.: Theory of an Experimental Method for Determining Stresses, not Requiring an Accurate Knowledge of the Elasticity Modulus. Internat. Assn. Bridge and Structural Engrs. Congress No. 9 (1949), 83—103.

[13] H i r a m a t s u, Y., N i w a, Y. and Y. O k a: Measurement of Stress in the Field by Application of Photoelasticity. Technical Report Engineering Research Institute, Kyoto University, VII, 3, No. 37 (1957).

Model Experiments on Pressure Distribution in Some Cases of a Discontinuum

By

D. Krsmanović, S. Milić*

With 16 Figures

Summary — Zusammenfassung — Resumé

Model Tests on Stress Distribution in Some Cases of a Discontinuum. Only a very small number of tests concerning the behaviour of a rock mass under load have as yet been carried out systematically. In an effort to arrive at a better understanding of the behaviour of a rock mass under these conditions, model tests have been carried out to determine stress and strain distribution.

The programme of the proposed tests, now in progress, is one of wide scope. A few results, however, have already been made available and have proved most illuminating. To start with, the tests were made with models of simulated rock mass and stratification in relief intersected with vertically orientated systems of fissures.

The results obtained so far demonstrate that — the structure of the discontinuum being equal — stress distribution is influenced to a very great extent by the primary state of stress in which the rock mass happens to be. In addition, it has been shown that rigidity of structure or compressibility of rock mass also plays a certain role.

The nature of rupture is altogether different where granular material is involved. On this point there is no reasoning by analogy or drawing of comparisons. When progressing in a discontinuum, both rupture and its nature and the critical bearing capacity — all are governed by a large number of factors.

Hence the necessity for any mountain mass to be dealt with and examined as a separate problem together with all the factors affecting stress distribution and critical bearing capacity.

Modellversuche über die Spannungsverteilung für einige Fälle des Gebirgskontinuums. Es gibt augenblicklich nur sehr wenige systematisch durchgeführte Untersuchungen der Felsmasse hinsichtlich ihres Verhaltens unter der Wirkung der Belastung. Mit der Absicht, eine bessere Kenntnis ihres Verhaltens unter solchen Bedingungen zu erzielen, werden Untersuchungen hinsichtlich der Spannungsverteilung und der Deformationen in einer Felsmasse bestimmter Kluftstruktur und Schichtung durchgeführt. Die Art des Bruches und das Tragvermögen wurden ebenfalls untersucht.

Es ist ein ziemlich umfangreiches Untersuchungsprogramm vorgesehen. Die Untersuchungen sind im Gange. Schon jetzt ist es jedoch möglich, Teilergebnisse bekanntzugeben, die sehr aufschlußreich sind. Die Untersuchungen begannen an nachgebildeten Gebirgsmodellen ausgeprägter Schichtung mit senkrechten Kluftsystemen.

Die bisherigen Untersuchungen haben gezeigt, daß die primären Spannungszustände der Gebirgsmasse sehr großen Einfluß auf die Spannungsverteilung haben. Desgleichen spielen auch die Steifigkeit der Konstruktion und die Verformbarkeit der Gebirgsmasse, die Beschaffenheit der Klüfte und deren Füllungsmaterial eine sehr große Rolle.

* Dr.-Ing. D. K r s m a n o v i ć, Professor of Civil Engineering, University of Sarajevo, Yougoslavia. — Ing. S. M i l i ć, Lecturer of Civil Engineering University of Sarajevo, Yougoslavia.

Seiner Natur nach ist der Bruch im Gebirge ganz verschieden von dem in gekörnten Materialien. In diesem Sinne können keine Vergleiche und Analogien erfolgen. Der Bruch im Diskontinum ist progressiv, seine Art und die Größe der kritischen Kraft hängen von vielen Parametern ab.

Aus diesem Grund drängt sich die Notwendigkeit auf, jedes Gebirge als gesondertes Problem zu behandeln und zu untersuchen. Dabei sollen alle Faktoren berücksichtigt werden, die von Einfluß sowohl auf die Spannungsverteilung als auch auf die maßgebende kritische Bruchkraft sein können.

Etude sur modèle de la distribution des contraintes pour quelques cas de milieu discontinu. Il n'existe encore que très peu d'études systématiques sur le comportement des massifs rocheux sous l'action des charges. Pour en obtenir une meilleure connaissance, on a entrepris des recherches aussi bien sur la répartition des contraintes et des déformations dans un massif rocheux défini par sa structure (joints de stratification et diaclases), que sur la détermination du type de rupture et de la force portante.

Les essais en cours comportent un programme assez important. Dès maintenant, on peut en extraire quelques indications significatives. Les essais ont porté d'abord sur des modèles de massifs rocheux bien stratifiés horizontalement et recoupés par des réseaux de joints verticaux.

Les résultats actuels montrent que l'état initial des contraintes dans le massif a une très grande influence sur la distribution des contraintes nouvelles. De même la rigidité relative du barrage et du massif de fondation, la nature des joints et les propriétés de leurs matériaux de remplissage jouent un très grand rôle.

La rupture de ce matériau est tout à fait différente de celle des sols. Aucune comparaison ni analogie ne peut être retenue. La rupture du milieu discontinu est progressive, sa nature et la grandeur de la force critique dépendent de nombreux paramètres.

Il faut donc étudier chaque massif rocheux comme un problème particulier en tenant compte de tous les facteurs qui peuvent avoir une influence tant sur la distribution des contraintes que sur la force critique conduisant à la rupture.

Introduction

Model experiments concerned with pressure distribution in a discontinuum formed of parallelepiped shaped elements were carried out last year at the Institute of Geotechnics and Foundation Engineering, College of Civil Engineering, Sarajevo University. The models used were made from clay and gypsum mixed in proportion suitable to ensure values of strength and modulus of elasticity low enough for the tests to proceed under the laboratory conditions available and with a gradual increase in load up to the point of rupture.

All tests were made with models with dimensions of $100 \times 100 \times 16$ cm, and their elements were fashioned from parallelepipeds measuring $4,0 \times 4,0$ cm with a half-space (plane) 16 cm thick. The elements were arranged so as to form parallel layers perpendicular to the direction of force (models $M\,1$—$M\,3$), their joints being shifted in the direction of force and covering half the surface of the element. A graph of the model is shown in fig. 1.

The load was applied through a steel plate which served as a foundation slab. The loads acting on the discontinuum were exerted by means of hydraulic presses, which enabled the desired degrees of intensity to be reached in stages.

During each test normal stresses at a definite number of fixed points were measured in cross-sections perpendicular to the direction of the load. This was followed by measurement of the displacement of the points in the direction of the load and perpendicular to it.

Measurement of normal stresses and displacement was by means of specially constructed measuring pressure cells and comparators respectively.

In the following, the results of measurement on four different models are shown, the first three of which are relative to the semi-plane of the discontinuum

that was subjected to load applied through a slab three to four times the width
of one element of the model. The fourth model illustrates the case of a semi-plane
of discontinuum subjected to load applied through a slab by a force directed at

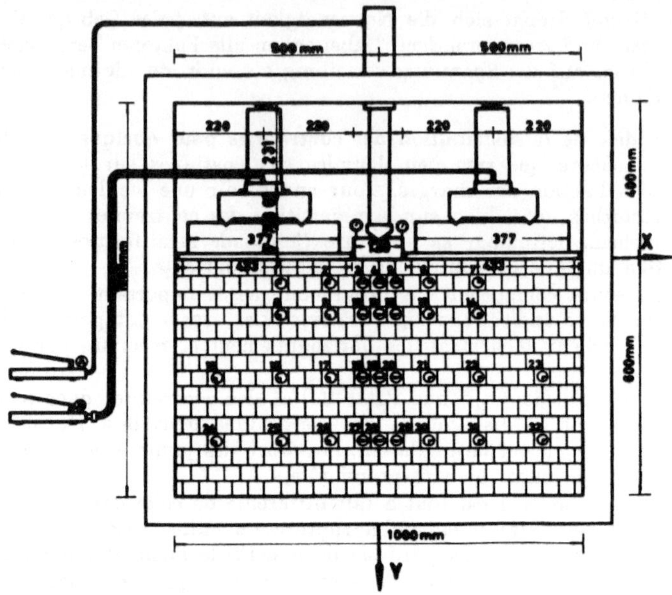

Fig. 1. Schema of model M 3
Schema des Modelles M 3
Schéma du modèle M 3

an angle of 30^0 to the free surface of the semi-plane. The load acted in a direction
perpendicular to the slab and the layers were parallel to it.

A number of tests were also performed on joints, filled and unfilled, for the
purpose of determining the influence of filling material and fissures on the distribution of pressures.

Furthermore, the semi-plane of some models was subjected to pre-loading at
a uniform rate of pressure for the purpose of consolidation, in order to compare
the effects of pre-loading.

Review of Results

Model 1. The lay-out of the model is shown in fig. 2 with comparators for
measuring displacement.

The characteristics of the material used in the construction of the model were
as follows:

modulus of elasticity . . . . $E = 40 \times 10^3$ kg/cm²
angle of internal friction of intact rock . . . . 56^0
angle of friction at joints of elements . . . . 36^0
cohesion of intact hard rock . . . . 3,0 kg/cm²
specific weight . . . . 1,2 t/m²
compression strength . . . . 65 kg/cm²

The load was applied by a concentrated central force exerting pressure through
a slab 16 cm wide, with a low value of absolute rigidity, i. e. the foundation slab
was flexible.

The joints were unfilled and the model had not been subjected to preloading. The results of stress measurements are to be seen on the left of fig. 3.

The load was applied and measurements taken in stages. The pressures exerted through the slab were $p_y = 3.0$, 12.0, 18.0, and 30.0 kg/cm².

In this case, the diagram of stress distribution shows a considerable concentration of stresses along the vertical central axis of the slab, with the result that

Fig. 2. Photograph of model M 1
Photographie des Modelles M 1
Photographie du modèle M 1

pressures at a depth of 0.677 b are 2.6 times as great as those on the contact surface, assuming that they are distributed equally.

The distribution of stresses along the vertical axis of the slab is shown on the left of fig. 3, the dashed lines representing depth stress changes in homogenous semi-plane.

The stress diagram shows a rapid decrease in stress in the same joint, directly to the left and right of the axis of the foundation slab.

A rupture in the material occurred at a pressure of 32,0 kg/cm², whereas the first visible fissures in the blocks appeared at a pressure of 12,0 kg/cm². Visible sinking of slab into blocks occurred at a pressure of 18,0 kg/cm². Fissures and sinking began to appear at loads of approx. 38 per cent, or 56 per cent of the load at rupture.

First signs of rupture were observed at a depth of about 2 b, the crushing of elements in this zone being accompanied by the appearance of oblique surfaces along which the breaking proceeded in the shape of a wedge.

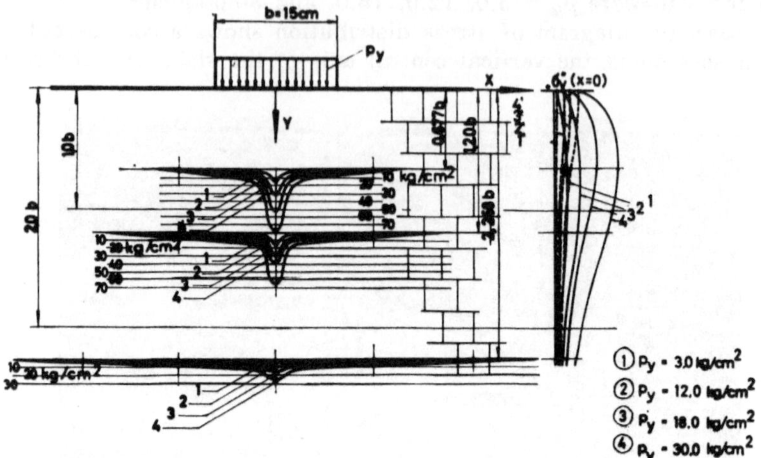

Fig. 3. Distribution of normal stress, σ_y for different loading intensity; model M 1. On the right, stress diagram of cross section, $x = 0$, continuous line. Stresses σ_y for homogeneous and isotropic semi-plane, dashed lines

Spannungsverteilung σ_y für verschiedene Belastung des Modelles M 1. Im Bild rechts: a Spannungsverlauf im Querschnitt $x = 0$ (volle Linie); b σ_y für die homogene und isotrope Halbebene (unterbrochene Linie)

Distribution des contraintes σ_y pour diverses charge sur le modèle M 1 — à droite graphique des contraintes pour $x = 0$ (en traits pleins) contraintes σ_y pour un demi-espace homogène et isotrope (en pointillé)

The factors responsible for the great concentration of the stresses σy in this case were (a) the flexibility of the foundation slab with its consequent inability to transfer the pressures to the contac surface, and (b) the state of joints which,

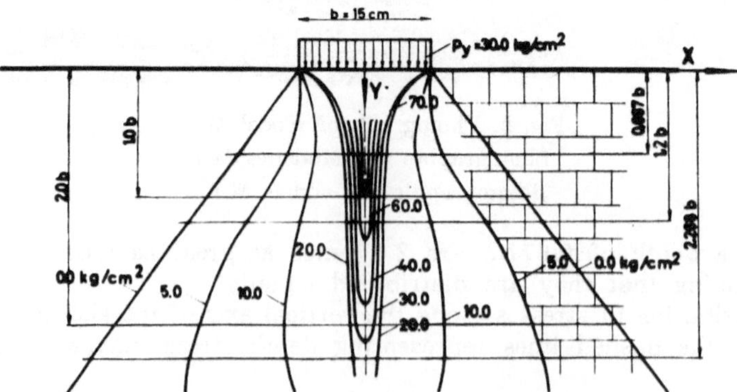

Fig. 4. Lines of equal stresses σ_y for model M 1
Linien gleicher Spannungen des Modelles M 1
Lignes d'égales contraintes σ_y sur le modèle M 1

being unfilled, impeded a more intensive distribution of pressures in the semi-plane. Another contributory cause lay in the circumstance that blocks of ideally identical dimensions could not be constructed, hence the possibility of pressure concentration on tightly pressed blocks and the likelihood of stress decrease in blocks due to the lack of ideal contact. Undoubtedly, the stress concentration was also due in no small measure to the fact that the semi-plane had not been subjected to pre-loading calculated to reduce possible effects of unevenness of some elements and irregular fitting together of blocks.

Fig. 4 shows the lines of equal stresses for the load $p\,y = 30\ \mathrm{kg/cm^2}$ which illustrate clearly the progress and distribution of stress concentration in the model.

With regard to deformations, the displacement diagram, given in fig. 5, clearly shows that settlements or vertical displacements of points are very great,

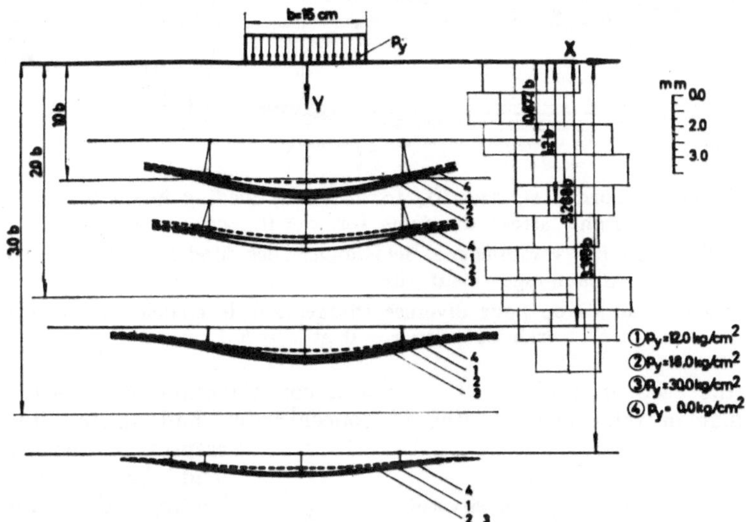

Fig. 5. Deformations in direction of loading for different loading intensity; model M 1
Deformationen in Richtung der Kraftwirkung für verschiedene Belastungen des Modelles M 1
Déformations parallèles à la charge pour diverses charges sur le modèle M 1

the horizontal displacements being insignificant. The maximum vertical displacement of a point on the slab axis, at a depth of $0{,}677\ b$, amounted to 3,62 mm under a stress of 30.00 kg/cm², which in reality would correspond to 4,46 cm.

It might be worth recording that permanent deformations amounted to 40 per cent of the total number of deformations. Dashed lines show the former in fig. 5.

Model 2. Made as the preceding one, this model had the same mechanical characteristics and unfilled joints.

Before the beginning of tests, the free surface of the model was subjected to uniformly distributed loads of 3.0 kg/cm², applied gradually with a view to consolidation.

The load was applied through a plate of high rigidity, which may be regarded as a perfectly rigid slab.

Parts of the free surface of the model to the right and left of the slab were subjected to a pressure equal to 0.05 kg/cm² to off set the influence of the weight of the material itself for a foundation depth twice the width of the foundation slab.

Fig. 6 shows a diagram of the applied normal stresses σy in the joints at the depths of $0.5\ b$, $1.17\ b$, $2.5\ b$ and $3.83\ b$ for three cases of load, namely 10, 20, 30 kg/cm².

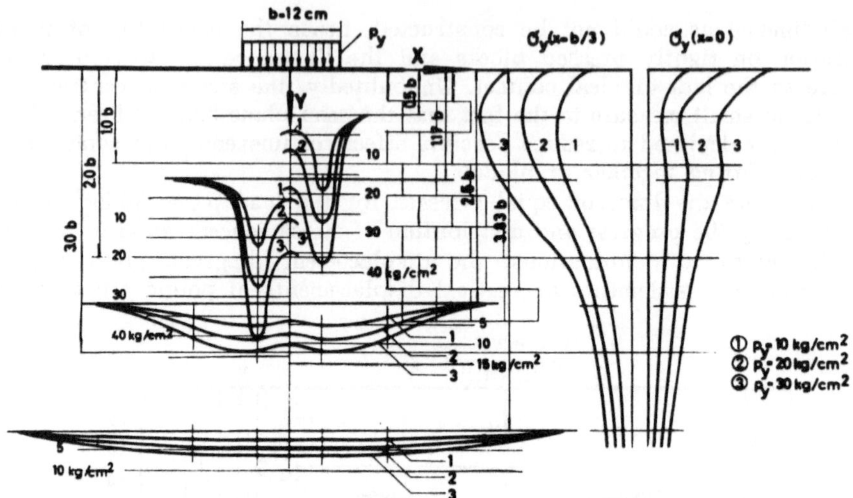

Fig. 6. Distribution of normal stresses, σ_y for different loading intensity; model M 2. On the right, stress diagrams for $x = 0$, and $x = b/3$

Spannungsverteilung σ_y für verschiedene Belastungen des Modelles M 2. Im Bild rechts: Spannungsverlauf für $x = 0$ und $x = b/3$

Distribution des contraintes σ_y pour diverses charges sur le modèle M 2 — à droite contraintes pour $x = 0$ et $x = b/3$

As in the case of model 1, there was a concentration of stresses, with the difference that in this case the line of concentration had shifted by $b/3$ with respect to the axis of the load. As the result of pre-loading the concentration was considerably lower than the one observed in model 1.

The lines of equal stresses for the load $p_y = 30.0$ kg/cm^2 (see fig. 7) show the progress of stress concentration in depth, about the ends of the foundation slab, and the beginning and direction of progressive rupture which concides with the zones of maximum stress concentration.

Vertical displacement of points is shown, for the various stages of loading, in fig. 8. The maximum displacement of a point at a depth of 0.5 b was found to

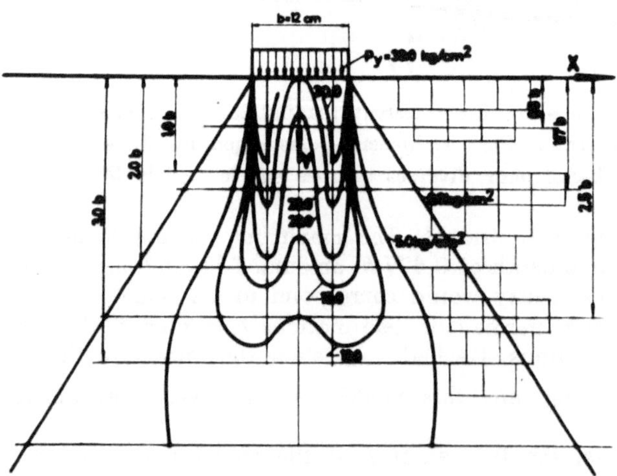

Fig. 7. Lines of equal stresses, σ_y for model M 2

Linien gleicher Spannungen σ_y des Modelles M 2

Lignes d'égales contraintes σ_y sur le modèle M 2

amount to 4.0 mm. Permanent deformations amounted to approx. 40 per cent of all the deformations noted. The greatest measured horizontal displacement did not exceed the value of one tenth of the maximum of vertical displacement, which leads to the conclusion that horizontal shifts and displacement of blocks along the horizontal are negligible.

The rupture occurred at a pressure of 40 kg/cm². The slab was first observed to sink in at a pressure of 30.0 kg/cm². The rupture started with splits and cracks around the edges of the slab and continued to spread towards the axis along the

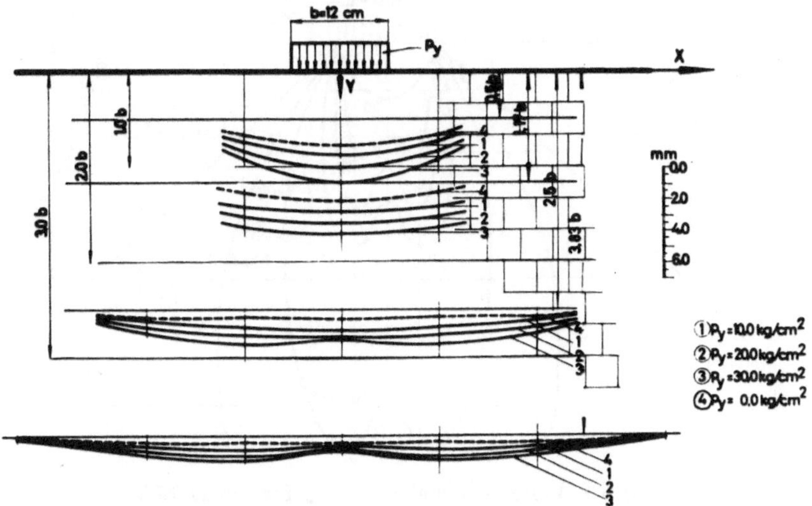

Fig. 8. Deformations in direction of loading for different loading intensity; model M 2
Deformationen in der Richtung der Kraftwirkung für verschiedene Belastungen des Modelles M 2
Déformations parallèles à la charge pour diverses charges sur le modèle M 2

horizontal and also vertically, manifesting itself in the shape of crushed elements in the area under the slab.

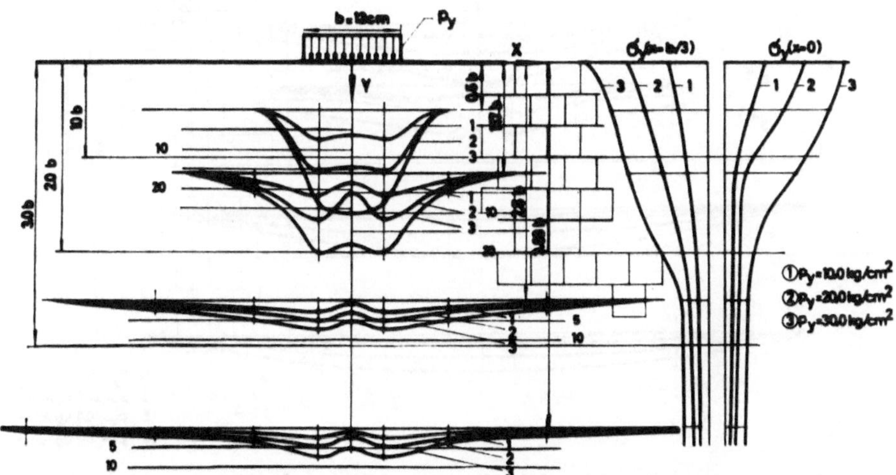

Fig. 9. Distribution of normal stresses, σ_y for different loading intensity; model M 3. On the right, stress diagram for σ_y, and for cross-section $x = 0$ and $x = b/3$
Spannungsverteilung σ_y für verschiedene Belastungen des Modelles M 3. Im Bild rechts: σ_y für die Querschnitte $x = 0$ und $x = b/3$
Distribution des contraintes σ_y pour diverses charges sur le modèle M 3 — à droite graphique des contraintes σ_y pour $x = 0$ et $x = b/3$

Model 3. The model was constructed in the same way as the preceding one. The slab was one of 12.0 cm, as before. The only difference between the two models was in the way the joints were filled. The model was built by placing the

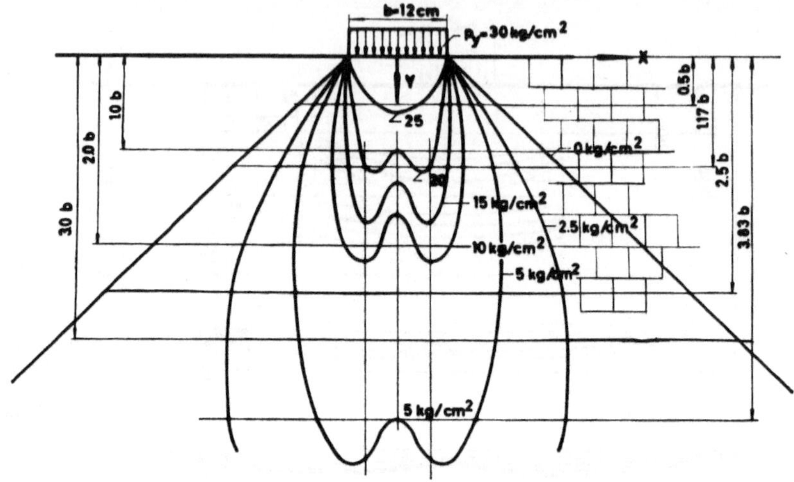

Fig. 10. Lines of equal stress σ_y for model M 3
Linien gleicher Spannungen σ_y des Modelles M 3
Lignes d'égales contraintes σ_y sur le modèle M 3

blocks in mortar of the same composition as the blocks, the joints being 1 to 1.5 mm long. The blocks were first given a coat of lacquer to prevent their joints absorbing

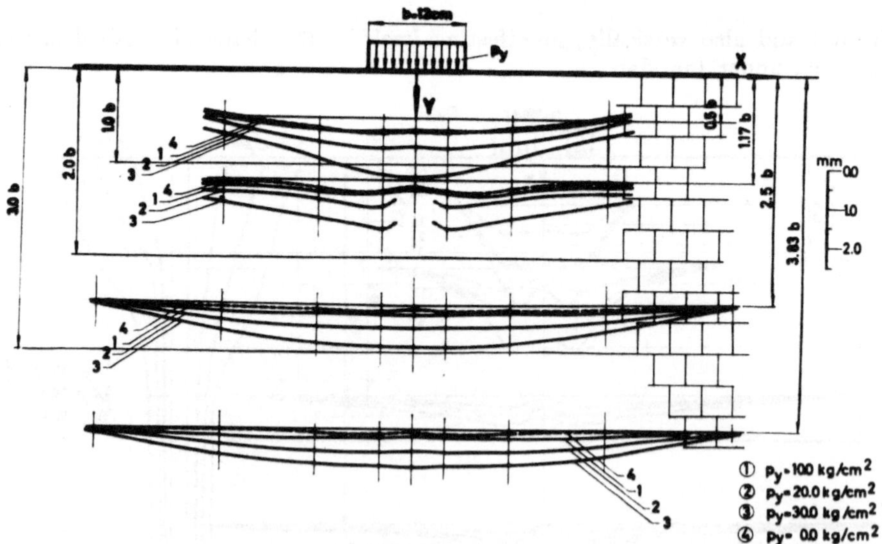

Fig. 11. Deformations in direction of loading for different loading intensity; model M 3
Deformationen in der Richtung der Kraftwirkung für verschiedene Belastungen des Modelles M 3
Déformations parallèles à la charge pour diverses charges sur le modèle M 3

water from the mortar. The tests began 30 days after the completion of the models to allow sufficient time for the mortar within the joints to solidify.

Mechanical characteristics of the material used for this model were as follows:

modulus of elasticity	23 500 kg/cm²
angle of internal friction of intact rock	52⁰
angle of friction at joints of elements	37⁰
cohesion of intact hard rock	3.0 kg/cm²
specific weight	1.2 t/m³
compression strength	37.5 kg/cm²

Before the beginning of tests the entire free surface of the model was subjected to equal pressure of 3.0 kg/cm², applied in 3 successive stages of 1.0 kg/cm² each.

Fig. 12. Photograph of model M 4 after rupture; fissures can be seen between deformeters
Photographie des Modelles M 4 nach dem Bruch. Die Bruchlinien sind zwischen den Dehnungsmessern zu sehen
Photographie du modèle M 4 après cisaillement, avec fissures

The pressure was then removed and the model was ready for slab loads and further testing.

The diagram given in fig. 9 shows stresses in horizontal cross-sections for three successive stages of loading, i. e. 10, 20 and 30 kg/cm². These diagrams show slight stress concentration as the result of the favourable influence of the filling in joints between the elements as well as of the consolidation brought about by preliminary test loads.

In this case there is no stress increase vertically, with the result that the stresses everywhere are lower than the uniform load p_y on the contact surface.

Stress changes in depth (i. e. vertically) are shown in diagrams of fig. 9 for $x = 0$ and $x = b/3$.

The rupture occurred under a load of 45 kg/cm² due to direct crushing and crumbling of elements under the slab.

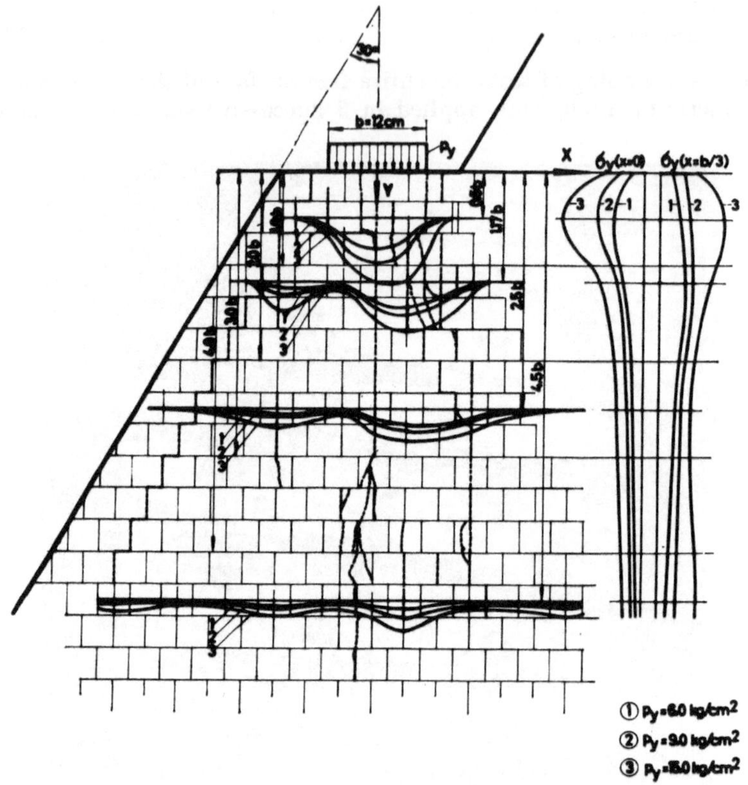

Fig. 13. Distribution of normal stresses, σ_y for different loading cases on model M 4. Dashed lines show the fissures after rupture

Spannungsverteilung σ_y für verschiedene Belastungen des Modelles M 4. Die unterbrochenen Linien zeigen das Rißbild nach dem Bruch

Distribution des contraintes σ_y pour diverses charges sur le modèle M 4. On a dessiné les fissures après la rupture

Lines of equal stresses for a load of $p_y = 30$ kg/cm² are given in fig. 10. The lines show a considerably greater uniformity of stress distribution than in the first two cases.

Vertical displacement of points — is shown in the diagram of fig. 11. The dashed line shows the remaining permanent deformations caused by preliminary test loads. The maximum vertical displacement observed amounted to 3,8 mm.

Of the total number of deformations 15 per cent proved permanent.

Model 4. Made from identical elements and having the same dimensions as the preceding models. The joints were filled with fine sand to ensure uniform fitting, without cohesion between the elements in the joints.

Mechanical characteristics of the material used were the following.

modulus of elasticity . . . . 15 670 kg/cm²
angle of internal friction of intact rock 50⁰
angle of friction at joints of elements 38⁰
cohesion of intact hard rock . . . . 1.0 kg/cm²
unit weight . . . . 1.2 t/m³
compression strength . . . . 28.2 kg/cm²

The model was built so that the load line, which was perpendicular to the slab, made an angle of 30⁰ with the line of free surface of the model. Schematic representation of the model and arrangement of measuring instruments is shown in fig. 12.

The line of layers runs in a direction perpendicular to that of the load. The perpendicular joints overlap by half the width of the element.

The stress diagrams for pressures 6.9, 9.0 and 15.0 kg/cm², given in fig. 13 and referring to the first cycle, have no symmetry as regards the resultant of the

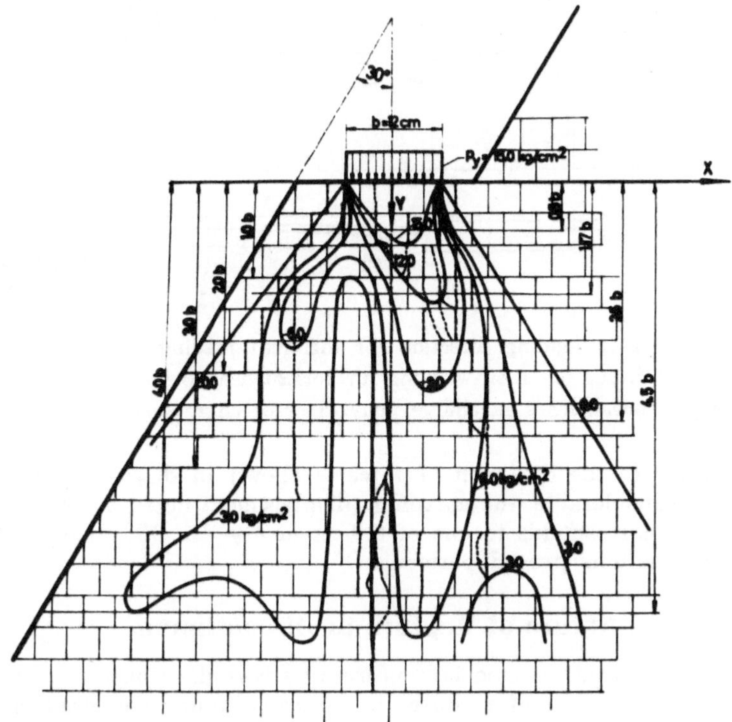

Fig. 14. Lines of equal normal stresses, σ_y; model M 4. Dashed lines show the fissures after rupture

Linien gleicher Spannung σ_y des Modelles M 4. Die unterbrochenen Linien zeigen das Rißbild nach dem Bruch

Lignes d'égales contraintes σ_y sur le modèle M 4. On a dessiné les fissures après la rupture

load and show stresses of considerably greater intensity at the points that lie on the line that passes through the inside edge of the slab and runs parallel to the direction of appered force. The maximum stress noted amounted to 12.8 kg/cm² i. e. 1.42 of the external load p_y.

Fig. 14 represents lines of equal stresses for the load $p_y = 15.0$ kg/cm². The lines show marked divergence and asymmetry, with the result that the line of maximum stresses comes near to the sliding surface of grained material.

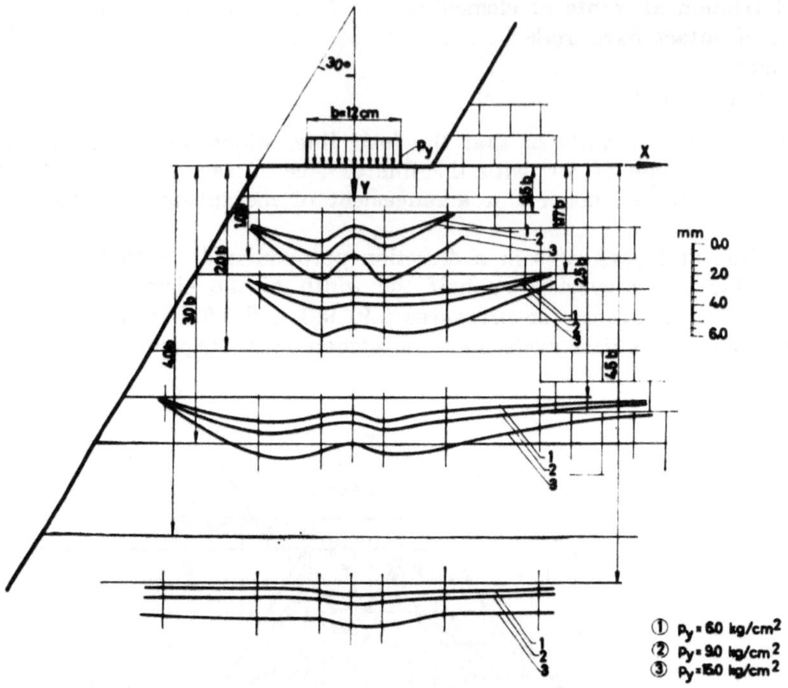

Fig. 15. Deformations in loading direction for different loading intensity on model M 4
Deformationen in Richtung der Kraftwirkung für verschiedene Belastungen des Modells M 4
Déformations parallèles à la charge pour diverses charges sur le modèle M 4

Displacement in the direction of force is shown in diagrams of fig. 15. The displacement is considerable, the maximum displacement noted amounting to 4 mm, which practically corresponds to one of 6.0 cm at $\lambda = 15$. 30 per cent of the total number of deformations proved permanent.

Nature of Rupture in Discontinuum

The results of tests on models concerning stress distribution provide certain data relevant to the nature of ruptures in a discontinuum composed of hard rock elements.

Depending as it does on a number of factors, the nature of a rupture in a discontinuum composed of jointed rock is much more complex than the one that occurs in a granular half space. Among the many factors affecting the nature of rupture and governing the critical bearing capacity, account should be taken of the following: stress state of mountain mass, system and nature of discontinuous surfaces, properties of intact rock, and relation between rigidity of structure and discontinuum.

It is because of all these factors that the rupture in a discontinuum is so very different from the one that occurs in granular material, as regards the manner of its origin, formation of surfaces, types of shear and deformation. The former cannot be compared with the latter — unless the discontinuum has been completely crushed and so much disturbed that no trace remains of a regular arrangement of its blocks (e. g. crushed tectonic zones).

It is common knowledge that the rupture in a discontinuum is a progressive one. The following should be a good instance of the nature of its progressiveness.

Let us assume a half-plane of a discontinuum with the surfaces shown in fig. 16. Assuming, again, the stress state to be as follows: σ_x mainly due to weight, otherwise low; $\sigma_x = \sigma_z$ reduce to zero. The half-plane is subject to a uniformly distributed inclined load over the surface $A\,E$.

The progress of the rupture can be divided into three successive stages.

Stage 1. The load, q, is still small compared with the one at rupture. Load transmission takes place within the limits of a surface marked by the broken lines $A\,F$ and $E\,G$, along which fissures begin to appear. The half-space to the

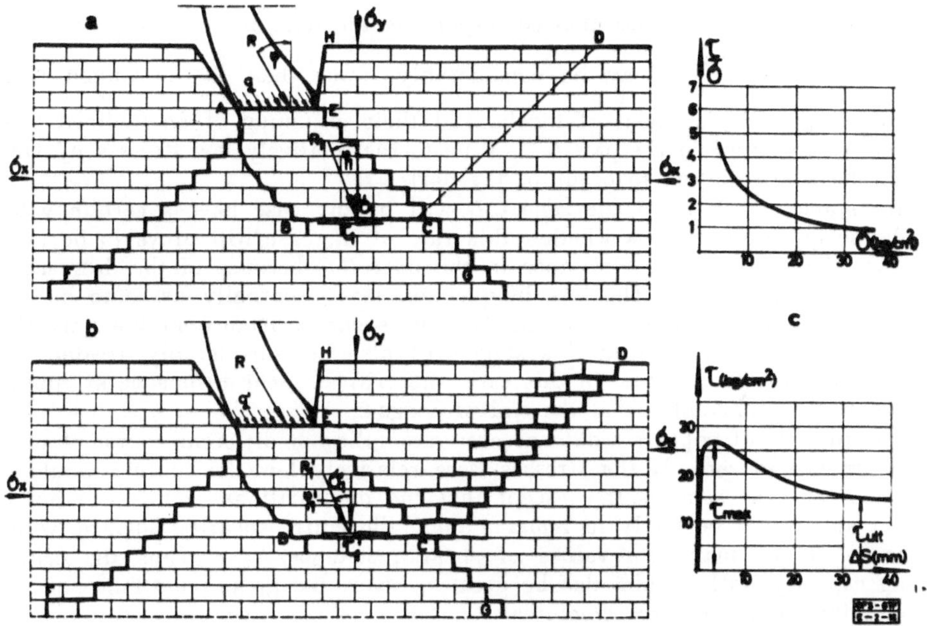

Fig. 16. Schematic representation of rupture for discontinuum investigated; *a* phases I and II; *b* phase III; *c* diagrams for relations $\tau - \Delta s$ and $\dfrac{\tau\,\mathrm{max}}{\sigma} - \sigma$

Schematische Darstellung des Bruches für das untersuche Diskontinuum. *a* Phase I und II; *b* Phase III; *c* Diagramme $\tau - \Delta s$ und $\dfrac{\tau\,\mathrm{max}}{\sigma} - \sigma$

Représentation schématique de la rupture pour le milieu discontinu étudié

right and left of these lines is in no way affected during this stage. So long as the normal stresses on the horizontal surfaces of stratification remain low, the relation $\dfrac{\tau\,\mathrm{max}}{\sigma}$ will remain high (Ref. 1), which ensures a great angle of shear resistance on the horizontal surfaces $A\,E$ and $B\,C$. During this stage there is a total absence of any marked shear, the deformations being mainly due to the compressibility of the rock mass. Shear resistance has not been fully utilized.

Stage 2. An increase in load is accompanied first with a partial break down of the cohesion of the intact rock and the shear resistance along fissures of the surface AB, with the result that shear becomes apparent along the surface. Next, a further increase in load is followed by a corresponding increase in the normal stresses σ along the surface BC and by a reduction in the relation $\dfrac{\tau\,\max}{\sigma}$ (fig. 16 c) until, towards the end of the stage, the limiting equilibrium state on the surface BC has been reached and signs of shear begin to show on the surface.

Stage 3. Shearing is in progress along the surface BC, and it is only now that activation of the surface CD and its participation in the rupture is brought about. Considerable deformations due to shear will have to take place before the zone CD is brought into play. Dependent on the jointing system of the rock mass and the number and width of open fissures, activation of the resistance in the part CD will be provoked either by deformations along the surface BC which correspond to the relation $\dfrac{\tau\,\max}{\sigma}$ or those that correspond to the relation $\dfrac{\tau\,\text{ult}}{\sigma}$ (fig. 16). However, the relation $\dfrac{\tau\,\text{ult}}{\sigma}$ may be considerably lower than the maximum shear resistance $\tau_{\max}$. Displacement of blocks and their assuming of an inclined position may easily occur in the part CD owing to the relatively low stresses p_y. Only then is shearing expected to occur in the part CD.

Resistance to shear in the zone CD will in any case have a certain value, though the question remains whether this rate of increase will prove to be lower or higher than the loss in resistance in the part BC because of the decrease in resistance of $\tau_{\max}$ to τ_{ult}. The loss in question may eventually prove to be considerably greater than the rate of increase in shear resistance in the part CD. Moreover, the deformations at this stage will also be considerably greater and, consequently, less favourable. In this case, the critical shear resistance at rupture may be the one that obtains at the end of Stage 2.

The process of rupture would at any rate follow a different course if τ_x, τ_z did not reduce to zero. If the stresses were relatively low the rupture would on the whole take the same course except that the deformations would then not be so great and would be likely to remain, along the entire shearing surface, within the limits of the maximum values of shear resistance, $\tau_{\max}$. Thus the critical bearing capacity itself would also be greater, hence the great importance of rock consolidation under similar structures.

Conclusion

The tests continue. On the basis of present results regarding the behaviour of hard-jointed rock discontinuum, the following conclusions may be drawn.

Assuming constant properties of hard rock, structure of jointing system, peculiarities and type of load; — it can be stated that the distribution of stresses in a discontinuum depends to a considerable extent on the following factors: the actual stress state of the discontinuum, the preliminary loads, the relation between the rigidity of the structure and the rock mass, and the nature of fillings of fissures and stratification planes.

A particularly great concentration of normal stresses has been found in a discontinuum not pre-loaded but subject to a flexible foundation slab. Due to a great concentration of normal stresses, the gradient of the rate of decrease in shearing stresses is very high, which brings about a diminution of the critical bearing capacity at rupture (fig. 4).

Application of preliminary loads to a discontinuum results in a reduced concentration of normal stresses — the effect of fissures.

The nature of rupture is completely different where granular material is concerned. On this point, there is no reasoning by analogy and comparisons cannot be drawn. The rupture is a progressive one and its nature and the critical bearing capacity depend to a great extent on the strength of intact rock, the actual stress state of the rock mass, the distribution of disconinuous surfaces and their nature, and several other factors. The maximum bearing capacity can at times be mobilized even when only one part of the surface long which shearing is in progress is activated. This might sound paradoxical at first, but the paradox is nevertheless true.

Measurement sometimes fails to give an exact picture of stress distribution and correlation of stresses and strains. This should be attributed to local irregularities due to the different arrangement of blocks in a discontinuum. Such instances, incidentally also occur in nature, for some blocks happen to be in a state of stress which differs from that which prevails in the rock mass as whole.

The many factors affecting the stress distribution and the nature of rupture indicate that each case and each rock mass, should be regarded and studied as a separate problem that calls for careful consideration of all the elements governing the behaviour of a rock mass in the period when it to about to be subjected to load and also during the transference of load from a structure.

Probleme der physikalischen Tektonik

Von

R. Hoeppener, Bonn

Der Inhalt dieses Vortrages wurde in Heft II/1 (1964) der Zeitschrift „Felsmechanik und Ingenieurgeologie" unter dem Titel „Zur physikalischen Tektonik" veröffentlicht.

The Prediction of Rock Movements by Elastic Theory Compared with In-Situ Measurements

By

Peter Hackett. B. Sc., Ph. D., A. I. M. E., F. G. S.*

With 13 Figures

Summary — Zusammenfassung — Résumé

The Prediction of Rock Movement by Elastic Theory Compared with In-situ Measurements. This paper is designed to present the elastic theory of ground movements caused by longwall workings evolved by the author and his colleagues at Nottingham University, Mining Department. It is intended to emphasise the physical aspects rather than the mathematical analysis which has been adequately covered in previous articles. Local departures from elastic behaviour have been ignored but it is not intended to infer that such departures are either unimportant in the mine, or beyond mathematical treatment.

In a two-dimensional analysis, such as the one discussed, it is possible to calculate stress changes and the associated strains and displacements due to mineral extraction. The greater part of the work is concerned with isotropic ground which is probably applicable to some hard rock mining areas (e. g. the gold mines of South Africa). The isotropic case has been shown to be incompatible with the coal fields of Western Europe, but a form of anisotropy has been introduced which goes far towards meeting the conditions in our coal mines.

Comparison of analysis with in-situ measurements shows that under certain conditions it should be possible to predict the changes in stress, strain, and displacement in the ground caused by longwall working.

Über die Voraussage von Gebirgsbewegungen mit Hilfe der Elastizitätstheorie, verglichen mit in-situ-Messungen. In diesem Referat werden die durch den Strebbau mit langen Fronten entstehenden Gebirgsbewegungen elastizitätstheoretisch behandelt. Vom Verfasser und seinen Mitarbeitern an der Universität Nottingham, Abteilung Bergbau, wurde eine Theorie entwickelt, die mehr auf die physikalischen Zusammenhänge eingeht als auf mathematische Betrachtungen, da letztere bereits in vorangegangenen Artikeln angemessen dargelegt wurden. Örtliche Abweichungen vom elastischen Verhalten sind nicht behandelt worden. Daraus soll allerdings nicht abgeleitet werden, daß solche Abweichungen für die Zeche unwesentlich sind oder mit mathematischen Berechnungen nicht erfaßt werden können.

Mittels einer zweidimensionalen analytischen Behandlung, wie sie hier dargelegt wird, ist es möglich, Spannungswechsel und die damit verbundenen Dehnungen und Verschiebungen in Abhängigkeit vom Gestein zu berechnen.

Der größte Teil der Arbeit befaßt sich mit der Theorie des isotropen Bodens, die wahrscheinlich beim Bergbau im Hartgestein, wie zum Beispiel in den Goldbergwerken in Süd-afrika, Anwendung finden kann. Es hat sich jedoch erwiesen, daß beim Kohlebergbau in Westeuropa die Voraussetzung der Isotropie nicht gilt. Allerdings ist eine Art von Anisotropie eingeführt worden, die die Bedingungen in unseren Kohlebergwerken weitgehend erfüllt.

---

* Dr. Peter H a c k e t t, Lecturer in the Department of Mining Engineering, Nottingham University.

Ein Vergleich von Analyse und örtlichen Messungen zeigt, daß es unter bestimmten Bedingungen möglich sein sollte, die Änderung von Spannungen, Dehnungen und Verschiebungen, die im Gebirge infolge Strebbaues mit langen Fronten entstehen, vorauszusagen.

Prédiction des déformations du terrain par la théorie de l'élasticité et comparaison avec les mesures in situ. On présente dans cet article la théorie élastique des déformations du terrain provoquées par l'exploitation des tailles, théorie élaborée par l'auteur et ses collègues du Département des Mines de l'Université de Nottingham. On désire insister sur les aspects physiques plutôt que sur l'analyse mathématique qui a été traitée dans des articles précédents. On a négligé les écarts localisés par rapport au comportement élastique, mais on ne prétend pas que ces écarts soient sans importance dans la mine, ni qu'ils soient inaccessibles au calcul.

Dans un problème à deux dimensions tel que celui qui est traité, on peut calculer les variations de contraintes dues à l'exploitation du minerai et les déformations ou déplacements qui en résultent. La plus grande partie de ce travail concerne un milieu isotrope et doit pouvoir s'appliquer à quelques mines en roches dures (par exemple les mines d'or d'Afrique du Sud). On a montré que l'isotropie ne s'appliquait pas aux bassins houillers d'Europe Occidentale mais on a introduit une forme d'anisotropie qui se rapproche beaucoup des conditions de nos mines de charbon.

La comparaison de cette analyse avec des mesures in situ montre que dans certaines conditions il devrait être possible de prédire les variations de contraintes, de déformation, et de déplacement du terrain, provoquées par l'exploitation des tailles.

Introduction

In the Mining industry there is a need for a greater understanding of the mechanics of ground movements, so that improvements may be made in strata control and a generally more scientific approach to the Art of mining may be made. Coal getting methods are changing rapidly at the present time and the current indications are that this trend will increase in impetus. As such changes occur new problems arise which cannot be solved by the application of experience gained under widely different conditions. Thus, a theoretical understanding of the mechanics of ground movement must prove a useful adjunct to experience. The theory described in this paper is not a panacea for all mining problems and it has distinct limitations which are discussed later. However, the theory does appear to have a wider field of application than many previous writers have imagined. Benefits to be derived from the use of the laws of elasticity are numerous. One is that it is a concept that is widely understood and used by engineers of all disciplines. Also it forms a good basis for investigating a wider range of materials, including those with time-dependent characteristics and those with non-elastic as well as elastic phases. Furthermore, the strains and displacements occuring in the deformed medium are often easily attainable.

There has, of course, been a great number of ground movement theories advanced over the last hundred years, and they fall broadly into two classes. (a) descriptive theories having little or no basis in theoretical mechanics and (b) theories derived from a branch of the mechanics of solids, usually elasticity. An appreciation of earlier theories can be found in articles by Hackett[1] and Berry[2]. However, as far as the author is aware, the only treatment soundly based upon the theory of elasticity (without the aid of models) is due to the theoretical group at the University of Nottingham originating in 1957 with the work of Hackett[3,4]. Several mathematical accounts of the work have been published (see references in the text) and in a recent paper Berry[2] has emphasised some of the physical aspects of the theory. Although much remains to be done to establish a wholly satisfactory theory to account for all phases of ground movement it is hoped that sufficient data now exists to make a worthwhile contribution to the understanding of Rock Movements caused by mining.

Basic Assumptions

An elastic analysis can only be completely satisfactory if the ground behaves in an elastic manner over most of the affected region. Whether this is so, or not, will be largely controlled by the presence or otherwise of faults, intrusions, a large tract of running sand or comminuted material. In the majority of cases where the excavations are fairly deep and there is no great thickness of unconsolidated surface rocks then the behaviour of the coherent rocks may be considered as essentially elastic. Of course, there will be local departures from this behaviour particularly at the excavation and the surface. Near to the excavation there will always be a region of fractured ground, but provided it is small in extent when compared with the whole affected region then by Saint Venant's Principle its effect may be neglected. In the mine, these local distortions are occurring on such an apparently large scale that it is easy to forget their relationship to the overall picture. At the surface the existence of some soil covering will modify the surface strains and displacements but where the soil is shallow the effect will be very slight and may be neglected.

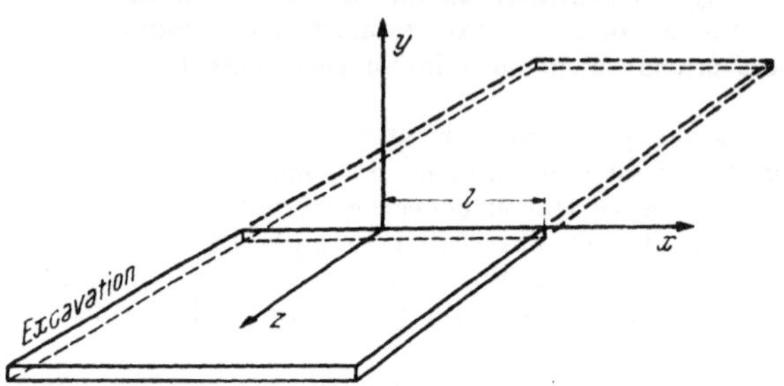

Fig. 1. Axes used in the analysis
Koordinatensystem, das der Spannungsanalyse zugrunde gelegt wurde

It is, perhaps, easier to vizualize ground movement as an elastic phenomenon if the scale is reduced by (for example) 1 in 1000. Imagine for instance a seam 1.5 metres thick at a depth of 500 metres. This would result in a "model" of 500 mms thick with a seam thickness of 1.5 mm where the maximum vertical displacement possible is 1.5 mm — considerably less if packing or stowing were used. Also at this scale it would be difficult to appreciate the effects of local discontinuities in the material and the local variations of stresses produced by them. Recent work with gelatine models[5] (a highly elastic substance) has helped considerably to alleviate misgivings over the validity of elasticity in this context.

Thus, although the theory of elasticity cannot account for much of the local behaviour near to the excavation it can be assumed to be generally applicable and to include the surface, except where great thicknesses of unconsolidated material exist; for example the Limburg Coalfield (Holland).

Only a two dimensional model is considered and the excavation is limited to the longwall type being worked in a horizontal seam (or vein) of small thickness (t).

The Problem

The problem to be solved involves two stress fields. The first is the Primitive stress existing in the ground before mining operations begin. The other is the Induced stress which is added to the Primitive stress through the effect of the mine workings. These Induced stresses are distinct from, but not independent of, the Primitive stress. Thus the stress system in the ground after mining is completed is the sum of the primitive stresses and the induced stresses. In this problem it is the induced stresses which are the most important because it is these stresses which are linked to the actual strains and displacements by the equations of Elasticity. Also it is only these strains which are recorded by measuring devices employed in the mines. Thus all stresses mentioned in this paper will be confined to the Induced stresses unless otherwise stated.

Because we are only concerned with the longwall type of excavation there is no need to argue for a specific magnitude of the horizontal component of the Primitive stress since it does not enter into the solution provided the working is in a horizontal deposit (see Berry[2], Hackett[6]). Should it be necessary to specify the residual stress field then a value for the horizontal component of the primitive stress would be required. In the analysis the vertical component of the Primitive stress $(-P)$ is assumed to be equivalent to one pound per square inch per foot of depth (0.233 kg/cm² per metre of depth).

A full mathematical analysis of all the material presented in this paper may be found by referring to the following papers (a) two-dimensional isotropic treatment[3, 4, 7, 8], (b) general two-dimensional transversely isotropic treatment[9, 10], and (c) three-dimensional treatment[9, 11].

Three states of excavation have been considered. These may be distinguished as: (a) Non-closure, where the roof and floor are not in contact, (b) Partial closure, where some part is in contact, and (c) Complete closure, with contact maintained over the full cross-section of the working. Of these, case (b) is the most likely and (c) is physically impossible. However, the latter state is of importance because it represents a limiting case to which partial closure would tend. These three states lead to different solutions, from which the stress and displacement fields may be

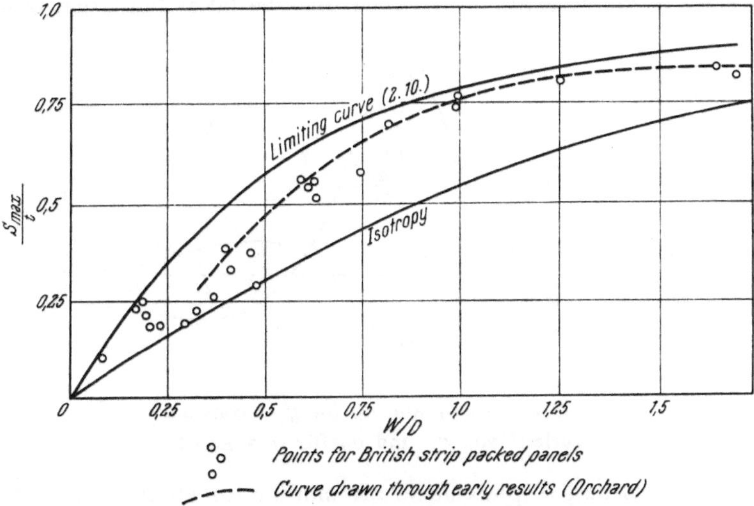

Fig. 2. Maximum subsidence versus width/depth ratio, isotropy and transverse isotropy (2,10) with measured values

Maximale Absenkung über dem Verhältnis von Hohlraumweite zu Teufe, Isotropie und transversale Isotropie (2,10) mit Meßwerten

obtained, but only in the case of complete closure can an exact solution be found. In the isotropic treatment the case of complete closure yields a unique solution in that the displacements at the surface (subsidence) are independent of the elastic constants and dependent only upon the geometry of the problem. This is a useful result and it provides the first proof that the isotropic solution is not compatible with the in-situ measurements from the coalfields of Western Europe. Fig. 2 will illustrate the point. In this diagram the calculated amount of maximum vertical displacement, expressed as a fraction of the seam thickness (and occurring at the centre of the subsidence "trough") is plotted against the width/depth ratio of the panel concerned. On the same graph are plotted a series of points taken from actual surveys of subsidence movements (most of them from a paper by Orchard[12]).

The practical points nearly all lie above the line which represents the maximum possible subsidence (on the isotropic theory). The discrepancy is great enough to make the theory untenable in the field from which the in-situ measurements were taken. However, it has been felt necessary to include some of the detailed results from the isotropic case because not all mining areas have the same well defined stratigraphy as the coalfields of Western Europe. Besides which, in some instances it has not yet been possible to calculate the equivalent information on the transversely isotropic model and the isotropic results serve to give some idea of the work.

It is convenient to discuss the work in two phases. In the first the stresses and displacements underground will be considered and in the second the special case of strain and displacements at the surface will be considered. Finally comparison will be made with certain in-situ measurements and the accuracy of the theory will be discussed.

Stresses and displacements in the Rock Mass
Isotropic case

It must be remembered that three states of excavation exist dependent upon the degree of closure of the roof and floor. Also that the solution is not valid for regions very close to the excavation. Provided that the latter limitation is observed

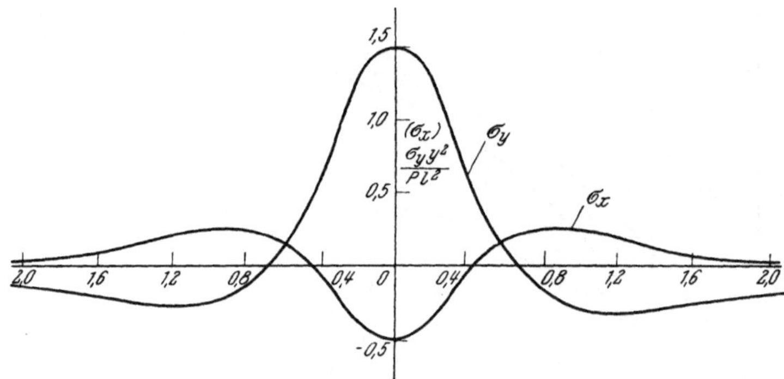

Fig. 3. Graph of σ_x and σ_y for $y =$ constant

Der Verlauf von σ_x und σ_y für $y =$ konst.

the stress distributions and displacements are similar for all three states*, and so consideration will be confined to the non-closure state.

Approximate expressions for the stresses within an infinite ground mass (i. e. no surface) are obtained as follows:

$$\frac{\sigma_x}{P}\frac{y^2}{l^2} = -\frac{K^2-1}{2(K^2+1)^2} + \frac{3K^2-1}{(K^2+1)^3}.$$

$$\frac{\sigma_y}{P}\frac{y^2}{l^2} = -\frac{K^2-1}{2(K^2+1)^2} + \frac{3K^2-1}{(K^2+1)^3},$$

where h = vertical distance from the (horizontal) seam,

l = half width of the excavation,

$K = x/y$ (fig. 1 shows the cartesian system of axes used),

σ_x = horizontal stress component,

σ_y = vertical stress component.

---

* Dependent upon the area of closure within the excavation being equal for all three states, and upon a suitable choice of Elastic constants.

Curves showing the variation of σ_x and σ_y with increase in distance from the excavation are shown in figs. 3 a and 3 b. Over the centre of the excavation the vertical stress component is tensile while the horizontal component is compressive. Both these stresses change in sign over the solid rib sides, both of them also reduce to zero at infinity in all directions. It is interesting to consider the variation of the vertical strain component along a vertical line passing through the unworked portion of the seam at some distance from the excavation, see fig. 4. This strain is easily calculated from the classical elasticity equations, viz:

$$E\,\varepsilon_y = \sigma_y - \nu\,(\sigma_x + \sigma_z),$$

where E = Young's Modulus,
ε_y = strain in y direction,
ν = Poisson's Ratio.

Fig. 4 shows that the region close to the seam is subjected to relatively large compressive strains, while the upper (and also much lower) regions are subjected to a much less severe tensile strain gradually decreasing and becoming zero at infinity. This pattern of strain agrees closely with the findings of workers who have measured strains in vertical shafts due to the action of nearby workings (for example Mohr[13]). The actual magnitude of the strains does, however, depend upon assigning values to Young's Modulus and Poisson's Ratio, a most difficult task which, for reasons stated later, has not yet been accomplished.

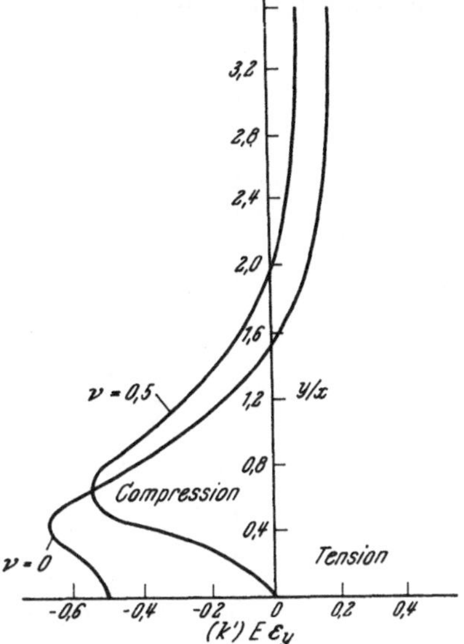

Fig. 4. Graph of vertical strain (ε_y) over x = constant

Der Verlauf der Vertikalverformung (ε_y) über x = konst.

Fig. 5 shows the variation of the horizontal strain component along some horizon at a relatively large distance above the working. This shows that a relatively severe compressive strain occurs over the panel while a lower tensile strain occurs over the unworked ground. Such strains would occur in roadways or levels

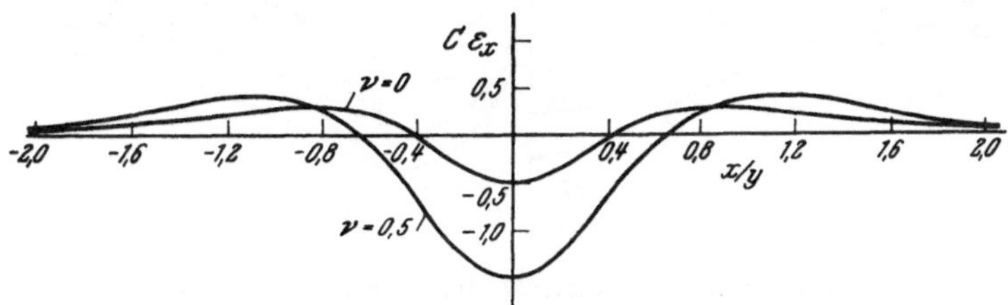

Fig. 5. Graph of horizontal strain (ε_x) over y = constant.
Der Verlauf der Horizontalverformung (ε_x) über y = konst.

above a panel, but again the question of actual magnitudes is difficult to resolve. It is satisfying to observe in such levels the evidence of similar strains, but insufficient records of longitudinal strain exist to make the comparison a quantitative one.

The shear stresses caused by the working are only of large magnitude in the immediate vicinity of the rib side, see fig. 6. (In fact they become infinite at the edge of the excavation.) Since this is partly a region of non-elastic behaviour no broad conclusions can be reached but it is worth while pointing out that the direction and magnitude of these stresses just over the rib side could afford an explanation for the appearance of induced cleavage. Such an explanation seems quite

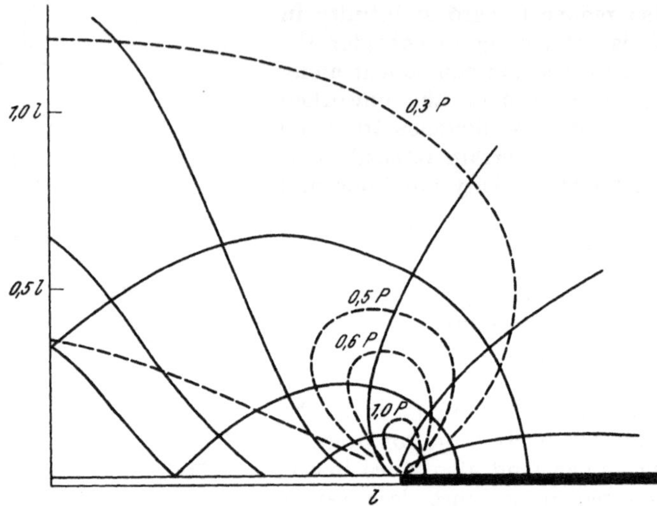

Fig. 6. Shear stresses near the excavation
———— Directions; — — — — Magnitudes
Scherspannungen in der Nähe des Hohlraums

tenable when viewed in the light of the work of Phillips and Faulkner[14], in which they stated that induced cleavage appeared to be an incipient shear fracture.

The displacements occuring within the rock mass are shown in figs. 7, and 8. Fig. 7, shows the variation of vertical displacement and horizontal displacement along a line $x = $ const (greater than l). The curves in fig. 7, demand little comment

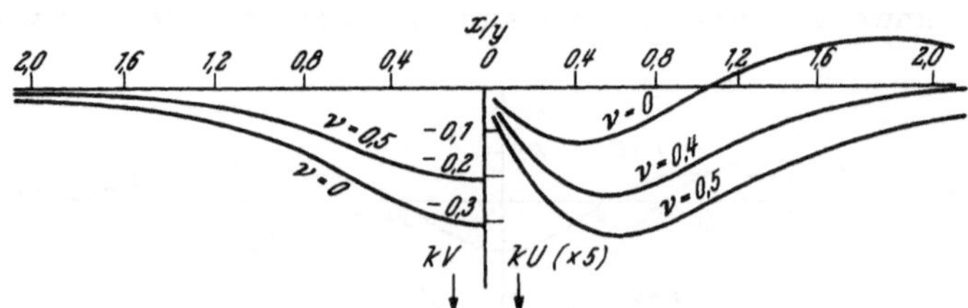

Fig. 7. Horizontal and vertical displacement $y = $ constant
Horizontale und vertikale Verschiebungen $y = $ konst.

except for pointing out the effect of a change in the value of Poisson's Ratio upon the horizontal displacement. It is perhaps pertinent to point out that no In-situ records are known to the author in which lateral movements underground have been reported as being in a direction *away* from the excavation (except at the seam level).

The results in fig. 8 (a) show the possible "leaning" of a shaft affected by mine workings. Also, with regard to the vertical displacement curve (b), it is theoretically possible to have the higher (A) of two points situated on such a vertical line undergoing more subsidence than a lower point (B). This phenomenon was reported by G r o n d from measurements taken in the Dutch Statemines[14].

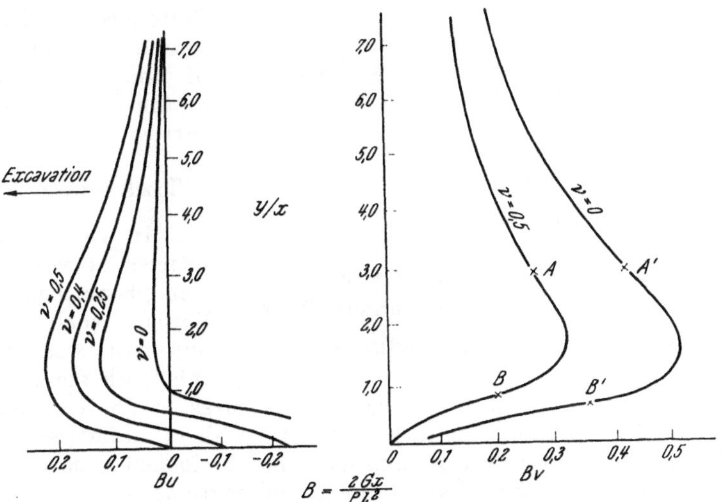

$$B = \frac{2\,G\,x}{P\,\ell^2}$$

Fig. 8. Horizontal and vertical displacement $x = $ constant

Horizontale und vertikale Verschiebungen $x = $ konst.

Stresses and Displacements at the Surface
Isotropic Case

Superposition of the effect of a surface upon the initial solution for an infinite medium does not have a uniform influence. The final solution is dependent upon the state of closure of the excavation. For unclosed, and partially closed excavations a limiting condition is that the surface must be at a large distance from the working — and is thus only valid for small width/depth ratios; whereas the complete closure solution is not subjected to any such restriction. Within the bounds of this limitation, however, the increase in magnitude of the vertical displacement is similar in all three cases and may be approximately written as:

$$\frac{V_s}{V_i} = \frac{8\,(1-\nu)}{(3-2\nu)+K^2\,(1-2\nu)}.$$

where $V_s = $ surface subsidence,

$V_i = $ subsidence within infinite medium at same horizon.

When $\nu = 0.5$ the ratio is independent of K and is equal to 2. There is thus a one hundred per cent increase in the value of the vertical displacement. When $\nu = 0$ the increase varies from $^8/_3$ to zero as $[K]$ varies from zero to infinity. The surface subsidence curve so obtained is incompatible with British survey measurements (and most Western European measurements). The main reason for this is that the curve spreads too far over each side of the undisturbed ground and the trough is still much shallower than actual subsidence profiles.

A similar increase is shown in the horizontal displacement curve with the exception that all movement is now directed towards the excavation irrespective of the value of ν.

The horizontal component of the stress at the surface is increased very significantly. This stress is displayed on fig. 9, where it may be compared with the same component for the infinite medium. The same objection applies to the validity of this curve as applies to the surface displacement curves.

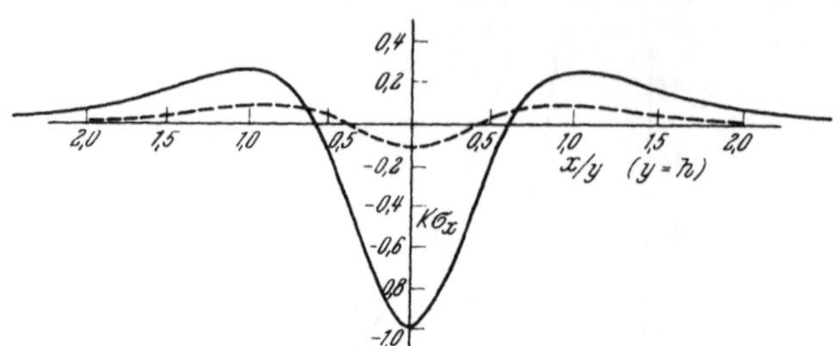

Fig. 9. Horizontal stress at the surface

– – – – Infinite Region; ——— Semi-infinite Region

Horizontalspannungen an der Oberfläche

Transverse Isotropy

In view of the discrepancies between the calculated phenomena recorded in British coalfields a new approach was devised. In this work the ground is considered as a transversely isotropic homogeneous mass. Such a system is valid if the scale of mining operations is large compared with the thickness of the strata and the lamination is regular. The ground may then be looked upon as a laminated material, homogeneous in the mass, although inhomogeneous in small portions.

The behaviour of a transversely isotropic material is dependent upon the value of five elastic constants. These constants may be chosen in a similar manner to that in which the two constants for an isotropic material can be selected from Young's Modulus, Poisson's Ratio, Bulk Modulus and Rigidity Modulus. As for the isotropic case the same three states of closure have been studied. It was found that a crucial part is played in this problem by two constants K_1 and K_2, which may be written as follows:

$$K_1 = (1 - \nu_x^2)^{1/2} (E_x/E_y - \nu_y^2)^{-1/2},$$

$$K_2 = [E_x/2 M - \nu_y (1 + \nu_x)] (E_x/E_y - \nu_y^2)^{-1}.$$

These constants are subject to two theoretical restrictions namely that both K_1 and $K_1 + K_2$ must be positive (K_2 *may* be negative). The expressions for the elastic field can take one of three forms, depending upon whether K_1 is less than, equal to or greater than K_2. From a physical appreciation of the material and the problem, the inequality $K_1 < K_2$ seems the most feasible[15] and the following remarks refer to this case only.

It thus becomes obvious that for a proper analysis of the problem the values of K_1 and K_2 must be known *for the locality under review*. The testing of small samples of the constituent rocks of a stratigraphical series in the laboratory is unlikely to lead to a practical solution. The testing of large tracts of ground is promising[16] but is beset by difficulties of experimental design and interpretation. It may be possible, however, to determine the magnitudes of K_1 and K_2 from an actual subsidence curve, provided that it has been recorded under suitable conditions. To achieve this a certain approximation to the exact expression for the surface displacement is made (which involves a negligible loss of accuracy). If the width to depth ratio is fairly small the vertical subsidence at the surface due to a horizontal excavation is given by

$$-V_0 = \frac{A\,h^3 \sqrt{(K_1 + K_2)\,2}}{\pi\,(K_1^2\,x^4 + 2\,K_2\,h^2\,x^2 + h^4)} \tag{1}$$

where A is the cross-sectional area of closure. If closure were complete then A would be equal to $W . t$ where $W =$ width of the excavation and t the seam thickness. However, closure is likely to be much less than this because of packing of the waste and the incomplete compression of fractured material — particularly close to the rib-sides. It is the relative convergence of the intact "Main roof" and the floor which really determines A.

If equation (1) is written as

$$\frac{\pi h V_0}{A \sqrt{2(K_1 + K_2)}} = V = \left[K_1^2 \left(\frac{x}{h}\right)^4 + 2 K_2 \left(\frac{x}{h}\right)^2 + 1 \right]^{-1}, \tag{2}$$

then V is a dimensionless quantity which is unity when $x = 0$. From this relationship it may be shown that

$$K_1 = \frac{h^2}{x_1^2} \left[2 x \frac{dV}{dx} - 1 \right]^{1/2} x = x_1,$$

$$K_2 = \frac{h^2}{x_1^2} \left[1 - n \frac{dV}{dx} \right] x = x_1, \tag{3}$$

where $[x] = x_1$ are the points on the curve; where $\nu = 0.5$ (i. e. point of half subsidence).

These expressions may be used for determining K_1 and K_2 for a particular subsidence profile. There are, of course, certain limitations imposed upon the selection of such an actual curve. The panel over which the subsidence is measured must lie in a fairly deep, almost horizontal seam with a fairly small W/D ratio. The ends of the panel must be far enough away from the subsidence line to have no further effect upon it and, of course, the initial levellings must have been taken before the face had a chance to affect the levelling stations. The levelling must also extend well beyond the rib-sides of the excavation so that the measurable limits of surface subsidence may be defined. Finally there should be no possibility of previous workings affecting the stability of the ground over which levellings are taken. If these conditions are fulfilled then a smooth symmetrical curve should be obtained. This curve may then be considered to correspond approximately to the curve of V given by equation (2). The points $\pm x_1$ may now be found by establishing the position on the curve when $V = 0.5$. At these points the slope of the curve may be measured* and by substitution into equations (3) values for K_1 and K_2 may be found. Using these values a curve for V against x can be constructed to be compared with the actual subsidence curve.

It will be realized that very few panels are in such a position that all the foregoing requirements may be fulfilled. Thus so far it has only been possible to isolate one subsidence profile which fulfils all the requirements. An earlier report by Berry and Sales[10] giving full details of the application of the above method was unfortunately based upon a curve which was found to be an incorrect *forecast* of the final profile made at an early stage of the panel development. The comments below are based upon the *true* subsidence profile and surface strain measurements.

The subsidence was caused by a longwall panel working the Top Hard seam at Manton Colliery. The seam is nearly horizontal and lies at a depth of 1973 feet and was 550 feet wide with an extraction thickness of 52 inches. The coal measures overlying the seam consist of thin shales and sandstones with one 75 ft. thick sandstone bed some 1140 ft. above the seam. A Permotriassic capping consisting of 150 ft. of the lower limestone series, 210 ft. of interbedded sandstones and marls and finally 240 ft. of massive jointed sandstones exists unconformably upon the coal measures.

---

* The curve should be symmetrical, but if it is assymmetrical the average of the two slopes may be taken.

The double-unit panel (see fig. 10) advanced at a rate of 100 ft. a month.
Gate side packs of 6 yards each were built, the rest of the waste being supported
on 3 yard strip packs with 10 yard wastes. The position of the levelling stations
(nos. 60—100) is clearly shown in the figure to ocuppy a line roughly at right
angles to the direction of face advance. The vertical subsidence curve shown in

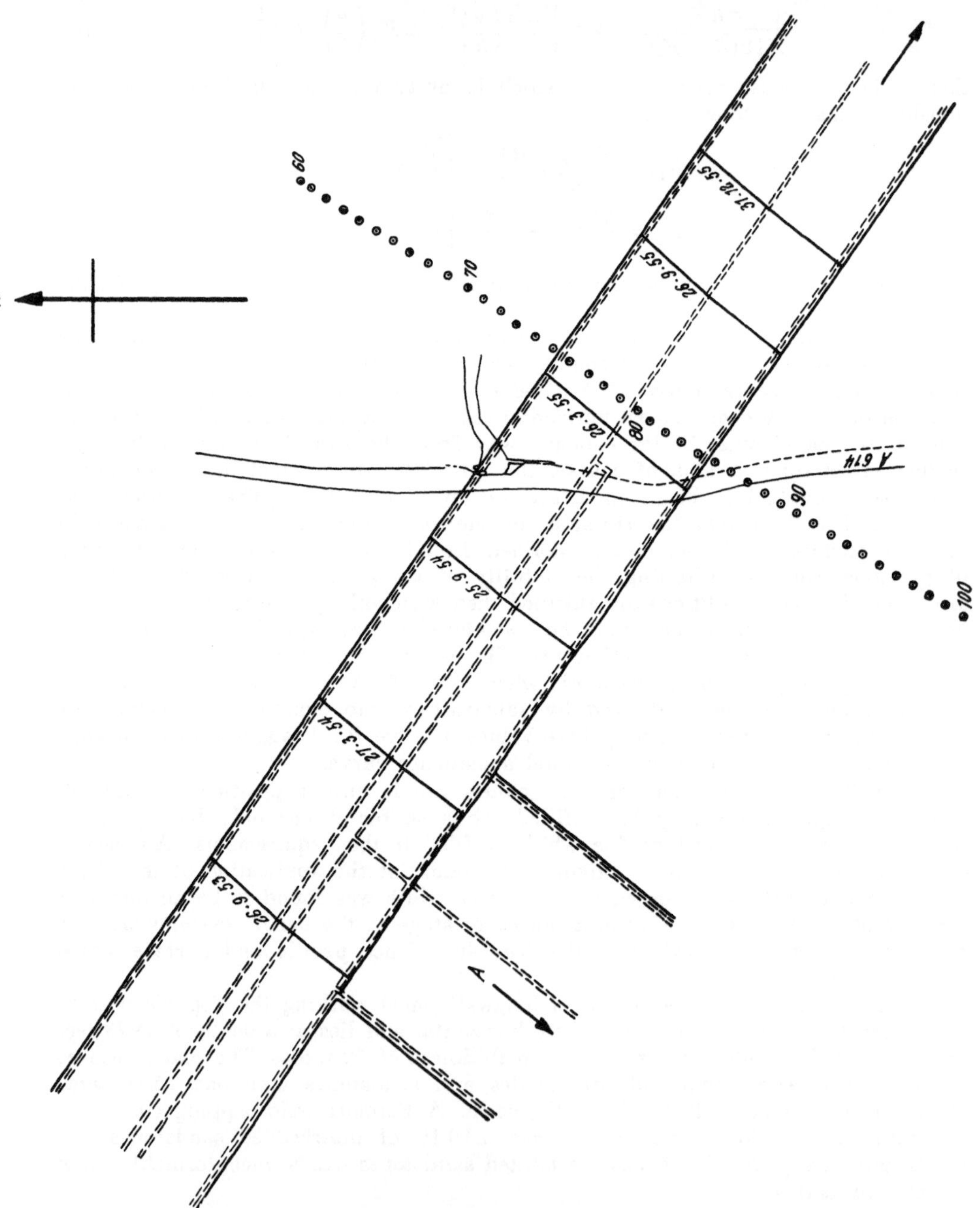

Fig. 10. Plan of mine workings (Manton) showing subsidence measuring points
Grubenbild (Manton) mit Senkungsmeßpunkten

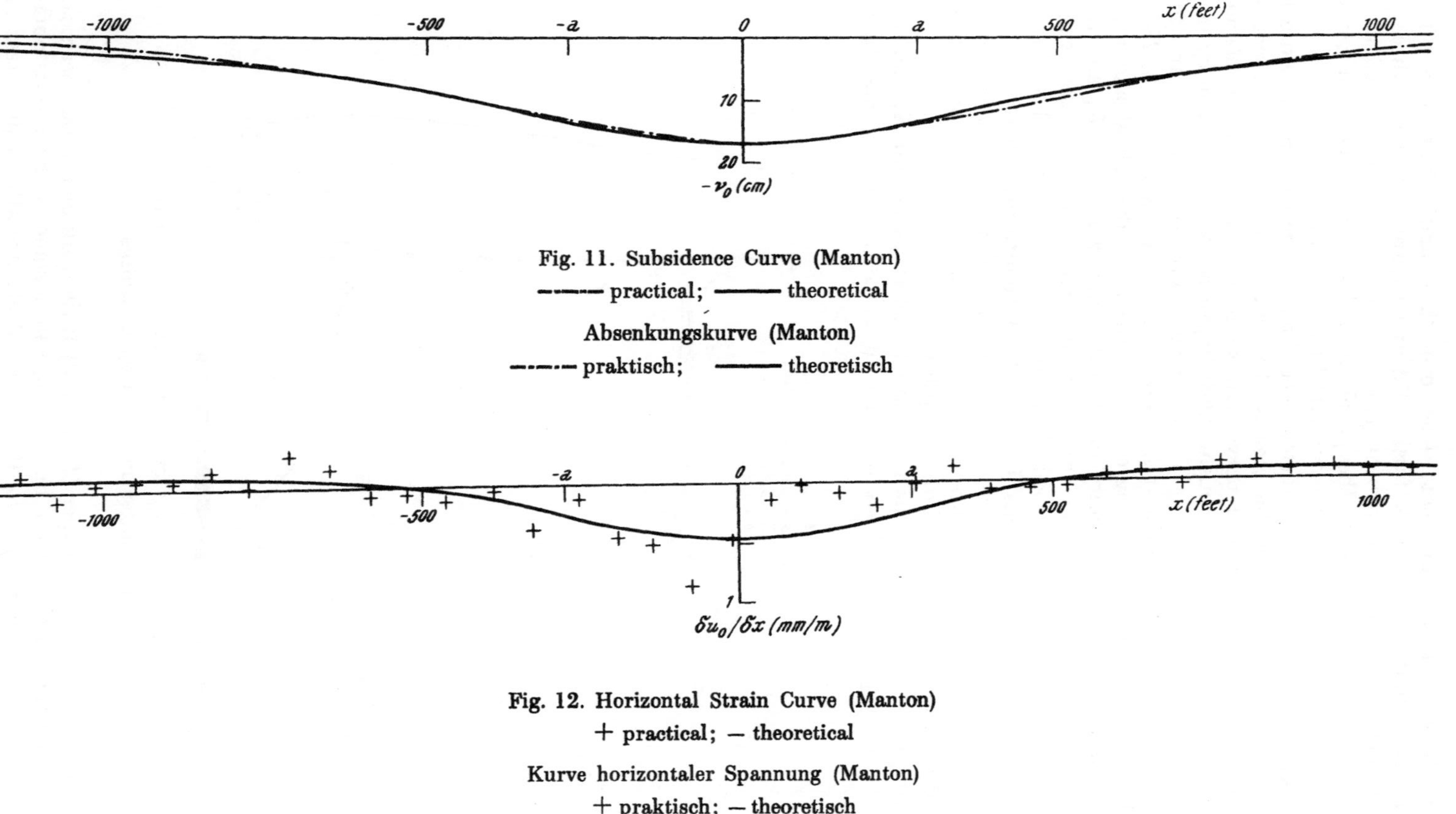

Fig. 11. Subsidence Curve (Manton)
—·—·— practical; ——— theoretical
Absenkungskurve (Manton)
—·—·— praktisch; ——— theoretisch

Fig. 12. Horizontal Strain Curve (Manton)
+ practical; — theoretical
Kurve horizontaler Spannung (Manton)
+ praktisch; — theoretisch

fig. 11, is drawn from measurements taken over three and a half years after the face had passed under the line. A further set of measurements taken three years later show that little further movement had taken place and that it had all been confined to the Southern side of the panel. It is likely that the later workings (A in the diagram) were responsible for this further movement.

The values of K_1 and K_2 were determined from the actual profile and a theoretical curve was drawn using the results. The curve so obtained did not agree closely enough with the practical curve and so improvements in the method of determination of K_1 and K_2 were sought. Such improvements were made by making the supposition that the ratio of the Young's moduli involved was not far removed from unity. Thus values for K_1 and K_2 were found. They were $K_1 = 2$, $K_2 = 10$. Using these values there is good agreement between the theoretical and practical curves. Further proof of the validity of the calculated values for K_1 and K_2 was to hand when it was found that the surface *strains* calculated using these values of K_1 and K_2 were a surprisingly good fit to the practical curves. The theoretical curves mentioned are shown in figs. 11, and 12. A significant fact to emerge from further investigation was that the subsidence curve at the surface is again largely independent of the *degree* of closure at the excavation. This would only appear to hold for deep panels with a low width to depth ratio.

Having satisfied the conditions necessary to select values for K_1 and K_2 it is possible to make predictions about the state of stress within the rock mass. Up

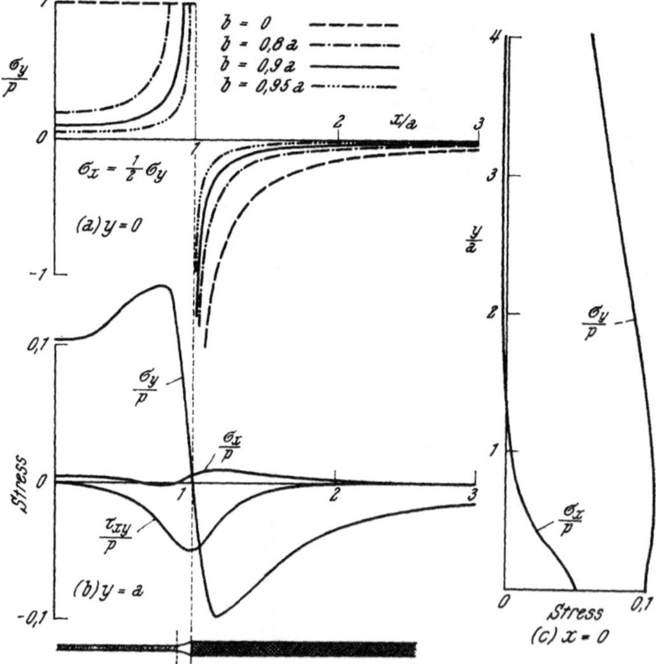

Fig. 13. Stresses in an anisotropic rock mass

Spannungen im anisotropen Gebirge

to the time of submitting this paper (June 1963) little work had in fact been carried out on defining the state of stress but fig. 13, shows some of the results obtained. Apart from the values of K_1 and K_2 the degree of closure affects the stress distribution, particularly near to the excavation. The degree of closure is denoted by the ratio b/l where $b =$ half the closed width and $l =$ half the width of the excavation $(= W/2)$. Some curves for σ_x and σ_y are shown for a selected value of $b/l = 0.90$ (a reasonable fraction). Comparison of these curves with those for the Isotropic case shows that the greatest difference is to be found close to the excavation. However, it is necessary to emphasise that the strains associated with these stresses may differ significantly from those in the isotropic case, since the equations relating stresses and strains now becomes

$$E_x \varepsilon_x = (1 + \nu_x) \left[(1 - \nu_x)\, \sigma_x - \nu_y\, \sigma_y\right],$$
$$E_x \varepsilon_y = (1 + \nu_x) \left[-\nu_y\, \sigma_x + (1 - \nu_x)\, \sigma_y / K_1{}^2\right].$$

Conclusions

This paper presents a possible method of predicting ground movements based upon an assumption that the ground behaves in a predominantly elastic manner. There are, of course, some instances where such an approach cannot be justified but there are others where even if there is some non-elastic behaviour the method will suffice. It is now necessary to obtain more practicall displacement and strain curves in order to furnish further proof of the validity of the theory.

Also to improve upon the theory it will be necessary to know more about the nature of the ground movements and associated stress fields around the excavation. Also to delineate the fracture zones that exist around the excavations and the effect of surface soils.

Besides developing and refining the elastic analysis work is proceeding at Nottingham on several other aspects. One is a visco-elastic generalisation of the elastic treatment. Another is developing a technique to take into account the effect of yielding materials so that soil effects and fracture zones may be considered. On the laboratory side, investigations are being carried on into the behaviour of rocks under stress, particularly at rupture.

As has been pointed out earlier, it is becoming increasingly difficult to see whether much use can be made in such a theory of the values of elastic constants found by performing small scale experiments in the laboratory; particularly if the constants for a transversely isotropic medium are required.

Finally it is confidently thought that the theory described can lay the foundation for a scientific analysis and prediction of rock movements caused by mining.

Acknowledgements

The author wishes to acknowledge the help given by his colleagues in the preparation of the paper. In particular to Dr. D. S. Berry who initiated and developed the transversely isotropic treatment. Also to Dr. T. W. Sales who carried out much of the analysis and to Professor F. B. Hinsley for his continued interest and guidance.

Also to the National Coal Board who have provided funds for much of the work.

Bibliography

[1] Hackett, P.: Rock Mechanics and the Mining Engineer. Mine and Quarry Engineering, May 1962, Vol. 28, No. 5, p. 215.

[2] Berry, D. S.: Ground Movement Considered as an Elastic Phenomenon. Min. Engr. (To be published.)

[3] Hackett, P.: Ph. D. Thesis 1957, University of Nottingham.

[4] Hackett, P.: Trans Instn. Min. Engrs. 1959, *118*, 421.

[5] Whetton, J. T. and H. J. King: Proceedings of the European Congress on Ground Movement. 1957, University of Leeds, p. 27.

[6] Hackett, P.: Elasticity and Rock Mechanics Mine and Quarry Engineering. (To be published.)

[7] Berry, D. S.: J. Mech. Phys. Solids, 1960, 8, 280.

[8] Sales, T. W.: M. Sc. Thesis 1959, University of Nottingham.

[9] Sales, T. W.: Ph. D. Thesis 1961, University of Nottingham.

[10] Berry, D. S. and T. W. Sales: J. Mech. Phys. Solids. 1961, 9, 52.

[11] Berry, D. S. and T. W. Sales: Ibid 1962. 10, 73.

[12] Orchard, R. J.: Trans. Instn. Min. Engrs. 1957, 116, 942.

[13] Mohr, F.: The effect of extraction on the roof with special reference to the action on shafts. Bergb. Arch., 1951, 14 (1), 1.

[14] Grond, G. J. A.: The precise topographical measurements in underground works. Paper A 3. Int. Conf. Liege, 1951.

[15] Berry, D. S.: J. Mech. Phys. Solids, 1963. II. (To be published.)

[16] Habib, P.: Determination of the modulus of elasticity of rocks in situ. Sols et Foundations No. 3, 1950.

[17] Phillips, D. W. and Faulkner: Cleavage induced by mining T. I. M. E. Vol. 89, 1934—35, p. 265.

Die Anwendung des Modells eines herumirrenden Teilchens auf die Probleme der Mechanik rolliger Medien

Von

J. Litwiniszyn, Krakow, Polen

Wegen verspäteten Eintreffens der Abbildungsunterlagen wird dieser Aufsatz voraussichtlich in einem der nächsten Hefte der Zeitschrift „Felsmechanik und Ingenieurgeologie" erscheinen.

Ein rechnerischer Weg
zur Ermittlung der Standsicherheit von Böschungen in Fels
mit durchgehenden, ebenen Absonderungsflächen

Von

W. Wittke[*]

Mit 14 Textabbildungen

Zusammenfassung — Summary — Résumé

Ein rechnerischer Weg zur Ermittlung der Standsicherheit von Böschungen in Fels mit durchgehenden, ebenen Absonderungsflächen. In der Arbeit wird mit Hilfe der Vektoranalysis ein rechnerisches Verfahren zur Ermittlung der Standsicherheit von Felsböschungen unter folgenden Annahmen entwickelt:

Im Gestein sind mehrere Systeme ebener, paralleler Absonderungsflächen (wie Schicht-, Kluft- und Schieferungsflächen) ausgebildet, die das Gestein vollkommen in einzelne Blöcke zerlegen. Der gegenseitige Abstand der Absonderungsflächen eines Systems und die Lage der Systeme im Raum sind beliebig. In den Flächen können einer Relativbewegung der Gesteinsblöcke nur Reibungskräfte entgegenwirken. Ist in den Absonderungsflächen außer der Reibung noch Schubfestigkeit vorhanden, so wird ihr Einfluß als zusätzliche, durch die Rechnung nicht erfaßte Sicherheit angesehen.

Das Verfahren beschränkt sich auf Gesteine, deren Festigkeit im Verhältnis zur Schubfestigkeit in den Absonderungsflächen groß ist, so daß man die Gesteinsblöcke als starr ansehen kann. Das bedeutet, daß bei einer bestimmten Gesteinsfestigkeit die Böschungshöhe ein bestimmtes Maß nicht überschreiten darf.

Es können ebene Böschungen untersucht werden, die bis zu einer Höhe H unter dem Winkel $0 \leq \alpha < \pi$ und oberhalb von H unter einem Winkel von $0 \leq \delta < \pi/2$ gegen die Horizontale geneigt sind.

Außer der Schwerkraft wird eine waagrechte Erdbebenbeschleunigung als Belastung der Felsböschung berücksichtigt.

Es wird zunächst untersucht, ob eine Bewegung der Felsblöcke in Richtung der Böschung kinematisch möglich ist. Diese Bewegung kann in einer Drehung oder in einem Gleiten in Richtung der Böschung bestehen. Weiterhin wird geprüft, ob eine Drehung unter der jeweiligen Belastung der Böschung auftreten kann. Die Sicherheit gegen Gleiten wird schließlich als Quotient aus treibenden und haltenden Kräften definiert. Im Falle einer Rutschung kann deren Reichweite berechnet werden. Restspannungen im Gebirge wurden entsprechend den eingangs genannten Annahmen nicht berücksichtigt.

Es ist beabsichtigt, das Rechenverfahren zu erweitern, um weitere Einflüsse, wie zum Beispiel den des Kluftwassers, zu berücksichtigen.

A Numerical Method of Calculating the Stability of Slopes in Rocks with Systems of Plane Joints. A numerical method of calculating the stability of rock slopes is developed by means of vector analysis with the following assumptions concerning the properties of the rock and the shape of the slope:

The rock is intersected by several joint systems e. g. strata, planes of schistosity and fissures etc. These joints traverse the entire rock mass and divide it into separate blocks.

---

[*] Dr.-Ing. Walter W i t t k e, Institut für Bodenmechanik und Grundbau der Technischen Hochschule Fridericiana in Karlsruhe, Deutschland.

No assumption is necessary concerning either the mutual distance between joints of one system or their strike and dip. Relative movements of the blocks of rock mentioned can be resisted by frictional forces only. Additional cohesion strength within the joints represents an additional amount in the factor of safety and is not considered in the calculations. The investigations are limited to rocks whose strength is high compared to the shear strength within the joints, so that in the calculations the rock can be considered rigid. For a given compressive strength of the rock this assumption is valid up to a certain limiting height of slope.

The computations can be applied to plane slopes with an angle of inclination of $0 \leq \alpha < \pi$ up to a certain height H. Above this height the inclination can be $0 \leq \delta < \pi/2$.

Horizontal acceleration caused by an earthquake can be taken in account, in addition to gravity. Based on the above assumption, investigations are made to determine whether movement of the rock blocks is possible kinematically. In the event of movement the rocks may slide and rotate.

Considering the manner of loading of the blocks the possibility of rotation can be established for a special case. If sliding only is possible the factor of safety can be determined by dividing the restraining frictional forces by the components of gravity and earthquake forces in the direction of sliding. The extent of the sliding area can also be determined. Residual stresses which exist in the rock are not taken into account.

It is proposed to extend the investigations and to consider other variables as, for instance, the static pressure of water in the joints.

Un procédé de calcul pour l'étude de la stabilité des versants rocheux divisés par des surfaces de séparation plans. Ce travail utilise l'analyse vectorielle comme procédé de calcul de la stabilité des versants rocheux dans les hypothèses suivantes:

La roche est parcourue par plusieurs systèmes de surfaces plans parallèles (plans de stratification, de diaclases et de schistosité) qui la divisent en blocs complètement séparés. L'écartement des surfaces d'un même système, et la situation du système dans l'espace sont quelconques. Sur les surfaces, seules les forces de frottement peuvent s'opposer au déplacement relatif des blocs rocheux. S'il existe une autre sorte de résistance en plus du frottement, on considérera son influence comme une sécurité supplémentaire non prise en compte dans le calcul.

Le procédé est limité à une roche dont la résistance est grande par rapport à celle des surfaces de séparation de sorte qu'on puisse considérer les blocs rocheux comme rigides. Ceci signifie que pour une résistance donnée de la roche, la hauteur du versant ne doit pas dépasser une valeur déterminée.

On étudie des versants plans, inclinés sur l'horizontale d'un angle α compris entre 0 et π jusqu'à une hauteur H et d'un angle δ compris entre 0 et $\pi/2$ au-dessus de cette hauteur.

Les forces agissant sur le versant sont la pesanteur et l'accélération horizontale due aux tremblements de terre.

On étudie d'abord si le déplacement du bloc rocheux dans le sens de la pente est cinématiquement possible. Ce déplacement peut être une rotation ou un glissement suivant la pente.

On vérifie ensuite si une rotation peut se produire pour chaque système de charge du versant. La sécurité au glissement est finalement définie comme le rapport des forces agissantes aux forces résistantes. Dans le cas d'un éboulement, on peut calculer son ampleur. On n'a pas tenu compte des contraintes initiales dans le massif conformément aux hypothèses énoncées au début.

On peut envisager d'élargir ce procédé de calcul pour tenir compte d'autres influences, par exemple des souspressions.

Hauptstichworte (Descriptors)[1]. Felsmechanik (rock mechanics) — Böschungen (slopes) — Standsicherheitsnachweis (stability analysis) — Vektoranalysis (vector analysis) — Böschungsrutschungen (landslides) — Schwerkraft (gravity) — Erdbeben (earthquakes).

Stichworte (Identifiers). Zerklüfteter Fels (jointed rock) — Rechenbeispiele (examples of calculation).

I. Einleitung

Das mechanische Verhalten von Felsgestein wird in der Mehrzahl aller Fälle im wesentlichen durch das Festigkeitsverhalten in den Absonderungsflächen, wie z. B. Kluft-, Schieferungs- oder Schichtflächen bestimmt. Müller[2,3], Pacher[3], Terzaghi[4], de Quervain[5] und andere haben z. B. mehrfach darauf hingewiesen, daß die Standfestigkeit von Felsböschungen neben einigen anderen Einflußfaktoren sehr wesentlich von der Lage und der Festigkeit der Kluft-, Schieferungs- und Schichtflächen beeinflußt wird.

In der folgenden Arbeit soll versucht werden, mit Hilfe der Vektoranalysis einen Rechenweg zur Ermittlung der Standsicherheit von Böschungen zu entwickeln. In diesem Verfahren, das für eine Reihe praktischer Fälle anwendbar sein dürfte, wird das Festigkeitsverhalten in den Absonderungsflächen berücksichtigt.

II. Aufgabenstellung und Annahmen

Um die Möglichkeiten der Anwendung des folgenden Verfahrens klar abgrenzen zu können, sollen die Aufgabenstellung und die Annahmen, die den Überlegungen zugrunde liegen, zunächst umrissen werden.

Im Gestein sollen mehrere Systeme paralleler Absonderungsflächen ausgebildet sein.

In der Abb. 1 ist eine Böschung mit zwei solchen Systemen dargestellt.

Das Verfahren soll zunächst für zwei Systeme abgeleitet werden, die Erweiterung auf mehrere Systeme wird dann später angedeutet. Die dargestellten Absonderungsflächen in der Abb. 1 sind eben. Gewisse Unebenheiten, wie sie in der Natur immer vorkommen, schließen die Anwendung des Verfahrens jedoch nicht aus. Die Lage der Flächen im Raum ist beliebig. Die Flächen sind durchgehend und trennen das Gestein daher vollkommen in einzelne Blöcke, deren Relativbewegung wie bei einem Trockenmauerwerk nur durch Reibungskräfte verhindert werden kann. Dabei ist es jedoch möglich, auch Fälle zu untersuchen, bei denen die Flächen nicht ganz durchgehen, bei denen es sich also um Mehrkörpersysteme nach der von Müller[2] gegebenen Definition handelt. In solchen Fällen ist in den Ebenen außer der Reibung noch Schubfestigkeit wirksam, und die im folgenden ermittelten Standsicherheiten liegen dann lediglich auf der sicheren Seite. Es scheint unter Umständen möglich, bei derartigen Systemen zunächst als

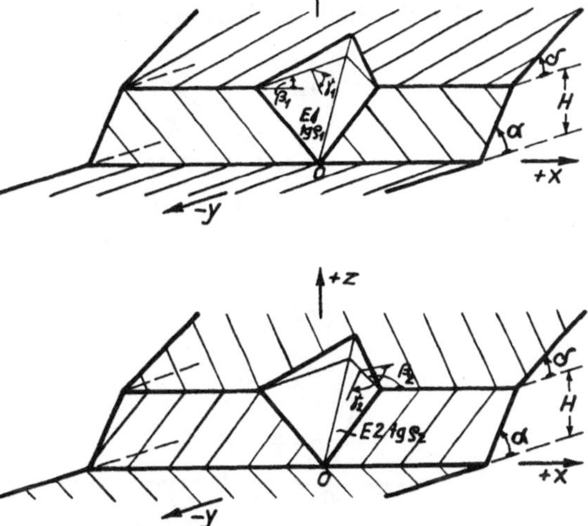

Abb. 1. Felsböschung mit 2 Systemen durchgehender, paralleler, ebener Absonderungsflächen

Rock slope with 2 systems of parallel plane joints, traversing the entire rock mass

Versant rocheux traversé par 2 systèmes de surfaces de séparation planes et parallèles

grobe Annäherung mit Hilfe von Korrekturfaktoren oder der Einführung von fiktiven Reibungsbeiwerten die höhere Standfestigkeit infolge der Schubfestigkeit rechnerisch abzuschätzen. Bei derarigen Versuchen könnten Arbeiten, wie sie von

Krsmanović und Langof[6] durchgeführt wurden, sehr wertvoll sein. Das in dieser Arbeit untersuchte System ist jedoch ein Vielkörpersystem.

Der gegenseitige Abstand der Ebenen eines Systems ist beliebig. Nach unten ist er lediglich dadurch begrenzt, daß angenommen wird, daß sich das Gestein nicht wie ein Lockersediment verhält, und für die obere Grenze wird die Annahme getroffen, daß die Schnittlinien der Ebenen beider Systeme mindestens teilweise aus der Böschung austreten. Weiterhin wird angenommen, daß die Reibungsfestigkeit in den Absonderungsflächen das Bruchverhalten des Gesteins bestimmt und die Gesteinsblöcke, ohne verformt zu werden, Biege- und Druckspannungen aufnehmen können. Diese Annahme, die auch ein Fließen des Gesteins, z. B. im Bereich des Böschungsfußes, ausschließt, kann im allgemeinen bei festen Gesteinen bis zu Böschungshöhen von mehreren hundert Metern als erfüllt angesehen werden. Sie schließt außerdem ein Ausknicken von unter hohen Spannungen stehenden, seitlich nicht gehaltenen Schichtpaketen aus. Das Raumgewicht des Gesteins ist auf die nachfolgenden Betrachtungen ohne Einfluß. Weiterhin werden Restspannungen im Gebirge und der Einfluß von Kluftwasserschub nicht berücksichtigt. Es ist zwar grundsätzlich möglich, das Verfahren zu erweitern. Eine Berücksichtigung der vernachlässigten Einflüsse führt jedoch teilweise zu einem erheblich größeren Rechenaufwand.

III. Geometrie

In der Abb. 1 sind aus den Systemen der Absonderungsflächen zwei Ebenen, $E\,1$ und $E\,2$ herausgegriffen, die aus der Böschung einen pyramidenförmigen Felskörper herausschneiden. Die Schnittlinie beider Ebenen tritt in Punkt 0 in Höhe des Böschungsfußes aus der Böschung aus. Der Punkt 0 stellt gleichzeitig den Nullpunkt des rechtshändigen, kartesischen Koordinatensystems x, y und z dar. In den Ebenen sollen die Reibungsbeiwerte $\mathrm{tg}\,\varrho_1$ und $\mathrm{tg}\,\varrho_2$ wirksam sein. Die Lage der Ebe-

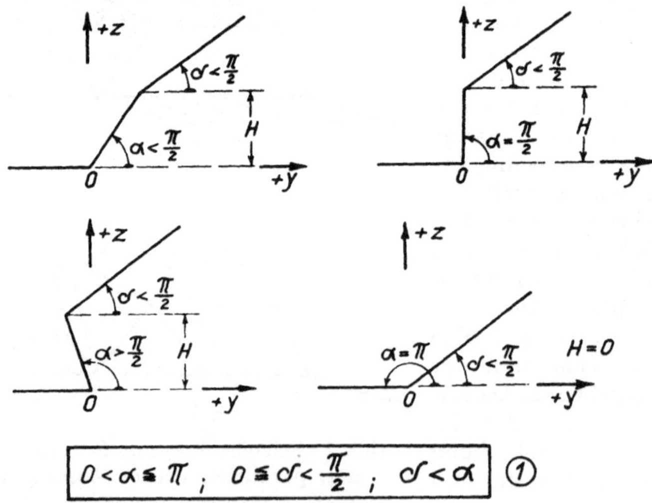

Abb. 2. Böschungsformen und Neigungswinkel, für die das Verfahren anwendbar ist
Shape and inclination of slopes considered by the stability analysis
Formes de versants et angles pour lesquels le procédé est applicable

nen im Raum ist festgelegt durch die in einer waagerechten Ebene gemessenen Winkel β_1 und β_2 und durch die in senkrecht auf $E\,1$ und $E\,2$ stehenden Ebenen gemessenen Winkel γ_1 und γ_2. Diese Winkel können aus den von Geologen im Feld gemessenen Winkeln des Streichens und Einfallens ermittelt werden. Die dargestellte Böschung ist bis zu einer Höhe H unter einem Winkel α und oberhalb unter δ gegen die Horizontale geneigt. Sie verläuft, wie dargestellt, parallel zur x-Achse und ist gerade.

In der Abb. 2 sind die Bereiche dargestellt, in denen α und δ liegen können. Nach Abb. 1 kann α alle Werte zwischen 0 und π annehmen, während δ auf Werte zwischen 0 und $\frac{\pi}{2}$ beschränkt ist.

$$0 < \alpha \leqq \pi; \ 0 \leqq \delta < \frac{\pi}{2}; \ \delta < \alpha. \tag{1}$$

Der Wert $\delta = 0$, der eine waagrechte Böschung oberhalb der Höhe H bedeutet, ist selbstverständlich einbegriffen. Weiterhin muß der Winkel δ kleiner als der Winkel α sein. Aus den angegebenen Winkelbereichen geht hervor, daß auch über-

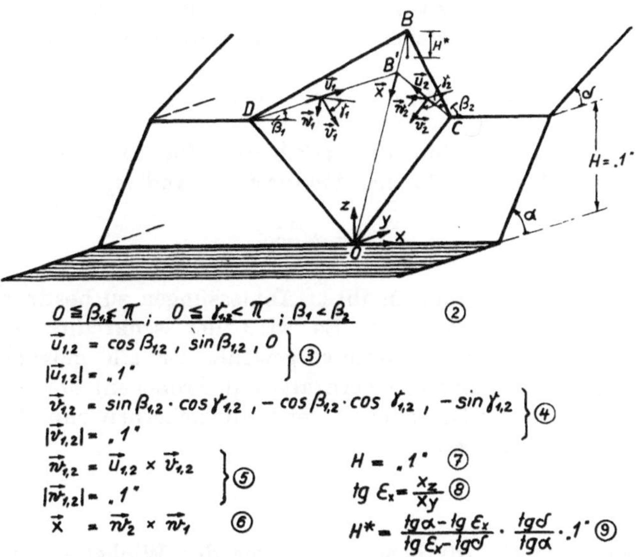

$$0 \leqq \beta_{1,2} \leqq \pi; \ 0 \leqq \gamma_{1,2} < \pi; \ \beta_1 < \beta_2 \qquad ②$$

$$\left. \begin{aligned} \vec{u}_{1,2} &= \cos\beta_{1,2}, \ \sin\beta_{1,2}, \ 0 \\ |\vec{u}_{1,2}| &= .1˚ \end{aligned} \right\} ③$$

$$\left. \begin{aligned} \vec{v}_{1,2} &= \sin\beta_{1,2}\cdot\cos\gamma_{1,2}, \ -\cos\beta_{1,2}\cdot\cos\gamma_{1,2}, \ -\sin\gamma_{1,2} \\ |\vec{v}_{1,2}| &= .1˚ \end{aligned} \right\} ④$$

$$\left. \begin{aligned} \vec{n}_{1,2} &= \vec{u}_{1,2} \times \vec{v}_{1,2} \\ |\vec{n}_{1,2}| &= .1˚ \end{aligned} \right\} ⑤$$

$$\vec{x} = \vec{n}_2 \times \vec{n}_1 \qquad ⑥$$

$$H = .1˚ \qquad ⑦$$

$$\operatorname{tg}\varepsilon_x = \frac{x_z}{x_y} \qquad ⑧$$

$$H^* = \frac{\operatorname{tg}\alpha - \operatorname{tg}\varepsilon_x}{\operatorname{tg}\varepsilon_x - \operatorname{tg}\delta} \cdot \frac{\operatorname{tg}\delta}{\operatorname{tg}\alpha} \cdot .1˚ \qquad ⑨$$

Abb. 3. Beschreibung der Geometrie durch Vektoren
Vector description of geometry
Représentation vectorielle

hängende Böschungen, die ja im Fels, im Gegensatz zum Lockersediment, standsicher sein können, untersucht werden können.

Im folgenden soll jetzt die Ableitung des Rechenweges für zwei Ebenen, $E\,1$ und $E\,2$, näher beschrieben werden. Dazu benötigen wir zunächst eine Beschreibung der geometrischen Verhältnisse mit Hilfe von Vektoren. Eine gute Einführung in die Vektorrechnung, die für den vorliegenden Fall vollkommen ausreichend ist, gibt Hay[7].

In Abb. 3 sind die Eckpunkte des pyramidenförmigen Felskörpers mit O, B, C und D bezeichnet. OD und OC stellen die Schnittlinien der Ebenen $E\,1$ und $E\,2$ mit dem unter α geneigten Böschungsteil dar. CB und DB sind die Schnittlinien im oberen Böschungsbereich. DB' und CB' sind durch D und C gehende Höhenlinien der Ebenen $E\,1$ und $E\,2$.

$$0 \leqq \beta_{1,2} < \pi; \ 0 \leqq \gamma_{1,2} < \pi; \ \beta_1 < \beta_2. \tag{2}$$

Nach der Gl. (2) können die Winkel β und γ Werte zwischen 0 und π und den Wert 0 selbst annehmen. Außerdem müssen die Ebenen so bezeichnet werden, daß $\beta_1 < \beta_2$ ist. Nach den Gln. (3) und (4) erhält man die Vektoren der Höhen- und der Fallinien der Ebenen $E\,1$ und $E\,2$ wie folgt:

$$\vec{u}_{1,2} = \cos\beta_{1,2'} \sin\beta_{1,2'}\, 0, \tag{3}$$

$$\vec{v}_{1,2} = \cos\beta_{1,2} \cdot \cos\gamma_{1,2'} - \cos\beta_{1,2} \cdot \cos\gamma_{1,2'} - \sin\gamma_{1,2}. \tag{4}$$

Man erkennt, daß sich die drei Komponenten der Vektoren in x-, y- und z-Richtung durch Winkelfunktionen von β und γ ausdrücken lassen.

Die Beträge der Vektoren sind:

$$|\vec{u}_{1,2}| = |\vec{v}_{1,2}| = \text{„}1\text{“}.$$

Aus dem Vektorprodukt der Gl. (5) erhält man die Einheitsvektoren, die senkrecht auf den Ebenen 1 und 2 stehen, und deren Betrag wiederum gleich „1" ist.

$$\vec{w}_{1,2} = \vec{u}_{1,2} \times \vec{v}_{1,2}. \tag{5}$$

Der Richtungsvektor der Schnittlinie OB beider Ebenen ergibt sich aus Gl. (6), wiederum als äußeres Produkt aus den Vektoren w_2 und w_1.

$$\vec{x} = \vec{w}_2 \times \vec{w}_1. \tag{6}$$

Um die dargestellte Pyramide in ihren Abmessungen zu bestimmen, benötigen wir weiterhin eine Längenabmessung. Es wird der senkrechte Abstand H der Böschungskrone vom Koordinatennullpunkt gewählt. Da alle möglichen derartigen Pyramiden sowie ihre Belastung (Schwerkraft und Erdbeben) ähnlich sind, ist die absolute Größe von H für die Berechnung der Standsicherheit ohne Belang. Es wird daher nach Gl. (7) $H = \text{„}1\text{“}$ gewählt.

$$H = \text{„}1\text{“}. \quad - \tag{7}$$

Weiterhin wird die Gl. (8) benötigt, in der der Winkel ε_x die Neigung der Parallelprojektion der Schnittlinie OB in Richtung der x-Achse gegen die Horizontale bedeutet, darin sind x_z und x_y die Komponenten des soeben dargestellten Schnittlinienvektors in Richtung der z- und y-Achse.

$$\operatorname{tg}\varepsilon_x = \frac{x_z}{x_y}. \tag{8}$$

Der senkrechte Abstand der Punkte B und B' ergibt sich durch geometrische Überlegungen aus der Gl. (9) zu:

$$H^+ = \frac{\operatorname{tg}\alpha - \operatorname{tg}\varepsilon_x}{\operatorname{tg}\varepsilon_x - \operatorname{tg}\delta} \cdot \frac{\operatorname{tg}\delta}{\operatorname{tg}\alpha} \cdot \text{„}1\text{“}. \tag{9}$$

Darin bedeutet „1" die Höhe der Böschung.

Für die nachfolgend erläuterten Kippsicherheitsuntersuchungen wird weiterhin die Bestimmung des Schwerpunktes der Pyramide erforderlich. Durch geometrische Überlegungen erhält man den entsprechenden Vektor vom Nullpunkt zum Schwerpunkt nach der Abb. 4 auf folgendem Wege:

Es werden zunächst die Vektoren der Geraden OD und OC nach den Gln. (10) und (11) erhalten.

$$\overrightarrow{OD} = -\frac{1}{\operatorname{tg}\gamma_1 \cdot \sin\beta_1} + \frac{1}{\operatorname{tg}\alpha \cdot \operatorname{tg}\beta_1}, \frac{1}{\operatorname{tg}\alpha}, \text{„}1\text{“}, \tag{10}$$

$$\overrightarrow{OC} = -\frac{1}{\operatorname{tg}\gamma_2 \cdot \sin\beta_2} + \frac{1}{\operatorname{tg}\alpha \cdot \operatorname{tg}\beta_2}, \frac{1}{\operatorname{tg}\alpha}, \text{„}1\text{“}. \tag{11}$$

Mit Hilfe von (10) und (11) ergibt sich der Vektor $\overrightarrow{OA}$ aus der Gleichung:

$$\overrightarrow{OA} = \tfrac{1}{2}\,(\overrightarrow{OD} + \overrightarrow{OC}). \tag{12}$$

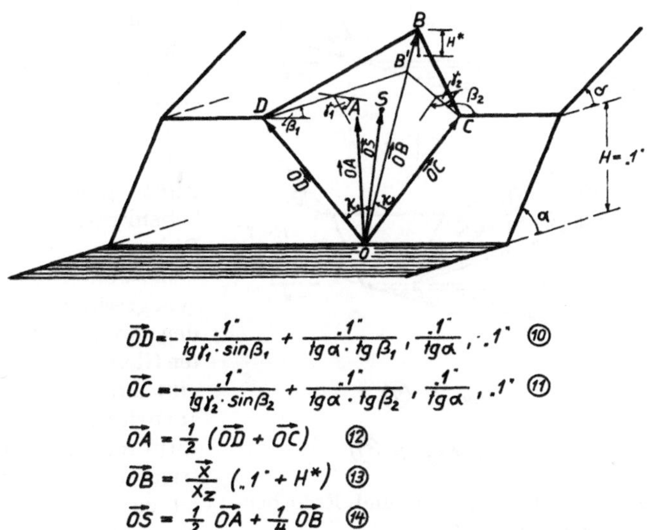

$$\overrightarrow{OD} = -\frac{.1\,'}{tg\gamma_1 \cdot \sin\beta_1} + \frac{.1\,'}{tg\alpha \cdot tg\beta_1}, \frac{.1\,'}{tg\alpha}, .1\,' \quad ⑩$$

$$\overrightarrow{OC} = -\frac{.1\,'}{tg\gamma_2 \cdot \sin\beta_2} + \frac{.1\,'}{tg\alpha \cdot tg\beta_2}, \frac{.1\,'}{tg\alpha}, .1\,' \quad ⑪$$

$$\overrightarrow{OA} = \tfrac{1}{2}\,(\overrightarrow{OD} + \overrightarrow{OC}) \quad ⑫$$

$$\overrightarrow{OB} = \frac{\vec{x}}{x_z}\,(.1\,' + H^*) \quad ⑬$$

$$\overrightarrow{OS} = \tfrac{1}{2}\,\overrightarrow{OA} + \tfrac{1}{4}\,\overrightarrow{OB} \quad ⑭$$

Abb. 4. Bestimmung des Schwerpunktes der Pyramide $(\overrightarrow{OS})$
Determination of the centre of gravity of the pyramid $(\overrightarrow{OS})$
Détermination du centre de gravité de la pyramide $(\overrightarrow{OS})$

Der Punkt A liegt auf der Mitte der Geraden zwischen D und C. Den dargestellten Vektor $\overrightarrow{OB}$ erhält man aus Gl. (13) zu:

$$\overrightarrow{OB} = \frac{\vec{x}}{x_z} \cdot (1 + H^+) \tag{13}$$

und schließlich den gesuchten Vektor von O zum Schwerpunkt (S)

$$\overrightarrow{OS} = \tfrac{1}{2}\,\overrightarrow{OA} + \tfrac{1}{4}\,\overrightarrow{OB}. \tag{14}$$

IV. Belastung der Böschung

Mit Hilfe der Abb. 5 wollen wir jetzt die Belastung des pyramidenförmigen Felskörpers darstellen und in Vektorform ausdrücken.

Die an der Pyramide angreifende Kraft ist das Eigengewicht. Weil die absolute Größe des Felskeiles und das Raumgewicht auf die nachfolgenden Betrachtungen ohne Einfluß sind, wird der Gewichtsvektor mit Gl. 15 durch die im Schwerpunkt wirkende Gravitationsbeschleunigung g ausreichend beschrieben.

$$\vec{G} = 0, 0, -g. \tag{15}$$

In Erdbebengebieten liegende Felsböschungen werden durch Erdbebenstöße zusätzlich belastet. Diese Einwirkungen können als waagrechte und senkrechte Beschleunigungen wechselnden Vorzeichens berücksichtigt werden. Der Rechenaufwand, den eine solche Annahme nach sich zieht, ist jedoch erheblich. Da die Angaben über die absolute Größe der Erbebenbeschleunigung recht unsicher sind und der Einfluß der horizontalen Komponente im allgemeinen ausschlaggebend ist, erscheint es daher gerechtfertigt, die vertikale Beschleunigung zu vernachlässigen. Die Richtung der horizontalen Erdbebenkraft kann je nach der Richtung der auftretenden Erdbebenwellen um 360^0 in der Waagrechten schwanken. Für den im Abschnitt 7 erläuterten Gleitsicherheitsnachweis ergibt sich die ungünstigste Beanspruchung, wenn man ihre Richtung je nach den geometrischen Verhältnissen parallel zu den Vektoren $\vec{x}'$ bzw. $\vec{v}'_{1,2}$ annimmt. Diese Vektoren erhält man aus den bekannten Vektoren $\vec{x}$ und $\vec{v}_{1,2}$, wenn man deren z-Komponente zu Null macht. Es ist üblich, die absolute Größe der horizontalen Beschleunigung in Bruchteilen der Gravitationsbeschleunigung $(a' . g)$ anzugeben. — Für a' werden in dem Deutschen Normblatt DIN 4149[8] z. B. Werte in der Größenordnung von 0 bis 0,2 empfohlen. —

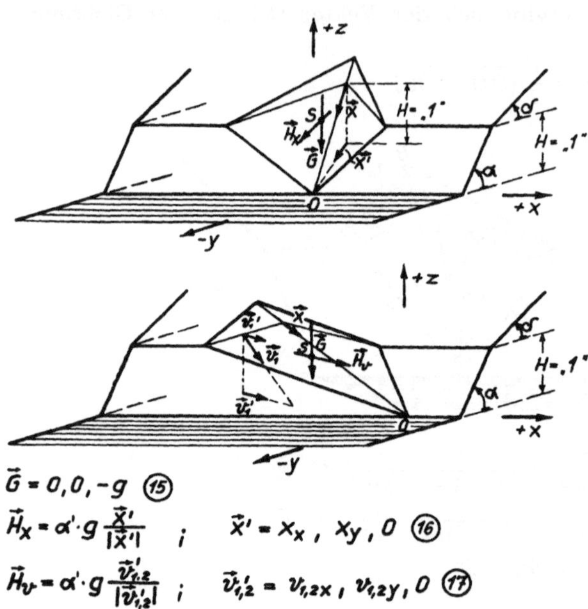

$$\vec{G} = 0, 0, -g \;\; ⑮$$

$$\vec{H}_x = a'\cdot g\,\frac{\vec{x}'}{|\vec{x}'|} \;\; ; \qquad \vec{x}' = x_x,\, x_y,\, 0 \;\; ⑯$$

$$\vec{H}_v = a'\cdot g\,\frac{\vec{v}'_{1,2}}{|\vec{v}'_{1,2}|} \;\; ; \qquad \vec{v}'_{1,2} = v_{1,2x},\, v_{1,2y},\, 0 \;\; ⑰$$

Abb. 5. Belastung infolge Eigengewicht und Erdbebenbeschleunigung

Load due to gravity and earthquake acceleration

Charges dues au poids propre et à l'accélération des tremblements de terre

Die Belastungen aus Erdbeben $\vec{H}_x$ und $\vec{H}_v$ erhält man dann aus den Gln. (16) und (17)

$$\vec{H}_x = a' g\,\frac{\vec{x}'}{|\vec{x}'|}; \;\; \vec{x}' = x_x, x_y, 0, \tag{16}$$

$$\vec{H}_v = a' g\,\frac{\vec{v}'_{1,2}}{|\vec{v}_{1',2}|}; \;\; \vec{v}'_{1,2} = v_{1,2x}, v_{1,2y}, 0. \tag{17}$$

Für den Nachweis der Kippsicherheit (Abschnitt 9) ergibt sich die ungünstigste Richtung der Erdbebenkraft $(\vec{H}_k)$ aus der Forderung, daß das Kippmoment (M_k) um die mögliche Drehachse $(\vec{d})$ maximal wird.

Für das Kippmoment gilt die Beziehung:

$$M_k = \vec{d} . (\overrightarrow{OS} \times \vec{H}_k).$$

In dieser Gleichung sind für ein Beispiel die Vektoren $\vec{d}$ und $\overrightarrow{OS}$ bekannt und für H_k gilt:

$$\vec{H}_k = H_{kx}, H_{ky}, 0,$$

$$\left| \sqrt{H_{kx}{}^2 + H_{ky}{}^2} \right| = \alpha'^2 g^2.$$

Das Skalar M_k läßt sich dann mit Hilfe dieser Gleichung als Funktion von H_{kx} ausdrücken.

$$M_k = f(H_{kx})$$

und man kann mit einer Extremwertbetrachtung finden, für welches H_{kx} also $\vec{H}_k$ das Kippmoment maximal wird.

Wegen des höheren Rechenaufwands, den der angedeutete Weg mit sich bringt, und wegen der erwähnten Unsicherheiten, die die Annahme einer Erdbebenkraft ohnehin beinhaltet, wird im folgenden auch für die Kippuntersuchungen die Erdbebenkraft aus den Gln. (16) bzw. (17) ermittelt.

V. Nachweis der Sicherheit gegen Gleiten

A. Gleitbedingungen

Aus der Abb. 6 geht hervor, daß die Standsicherheit der Böschung der beschriebenen Art z. B. durch ein Abgleiten des Felskörpers O, B, C, D gefährdet sein kann. Ob ein solches Abgleiten kinematisch überhaupt möglich ist, hängt von

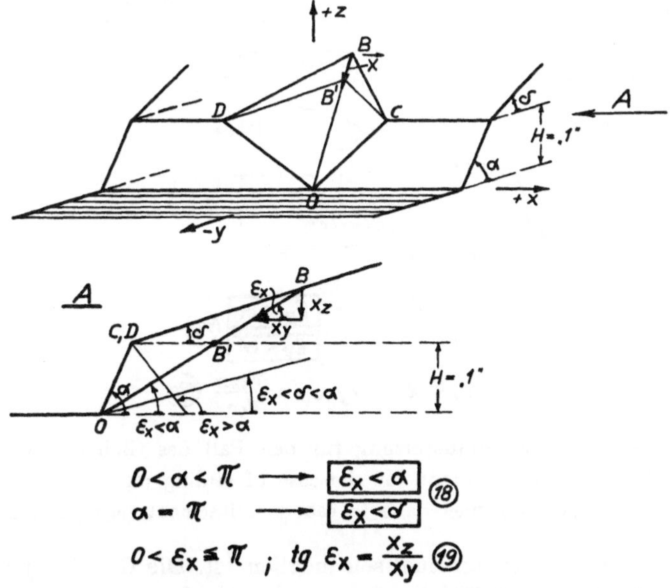

Abb. 6. Geometrische Voraussetzung für den Fall des Gleitens
Geometrical condition for the case of sliding
Hypothèses géométriques dans le cas de glissement

der Lage der Ebenen $E\,1$ und $E\,2$ im Raume ab, oder genauer ausgedrückt, von der Lage der Schnittlinie $\bar{x}$ beider Ebenen im Verhältnis zur Böschung.

Aus der Ansicht A, die die Richtung der negativen x-Achse hat, läßt sich das genauer erkennen. Danach muß für Werte 0 kleiner α und kleiner π der in x-

Richtung projizierte Neigungswinkel der Schnittlinie (ε_x) kleiner als α sein. Für den Wert $\alpha = \pi$ muß ε_x kleiner als δ sein. Für alle anderen Werte von ε_x ist ein Abgleiten von Felskeilen kinematisch nicht möglich, weil sich kein Körper aus der Böschung herauslösen kann. Das ist aus den Darstellungen anderer Winkelverhältnisse in dieser Abb. anschaulich zu ersehen. Diese Gleitbedingungen sind in der Gl. (18) formuliert.

$$0 < \alpha < \pi \rightarrow \varepsilon_x < \alpha,$$
$$\alpha = \pi \rightarrow \varepsilon_x < \delta. \tag{18}$$

Den Winkel ε_x, der Werte zwischen 0 und π annehmen kann, erhält man aus der Gleichung

$$\operatorname{tg} \varepsilon_x = \frac{x_z}{x_y}. \tag{19}$$

In der Abb. 7 ist noch ein Sonderfall dargestellt, in dem entweder die Ebene E_1 oder E_2 parallel zur Böschung streicht, was sich durch die Gleichung $u_{1,2z} = u_{1,2y} = 0$ ausdrücken läßt. In diesem Fall bestimmt die Neigung der Fallinie ε_v

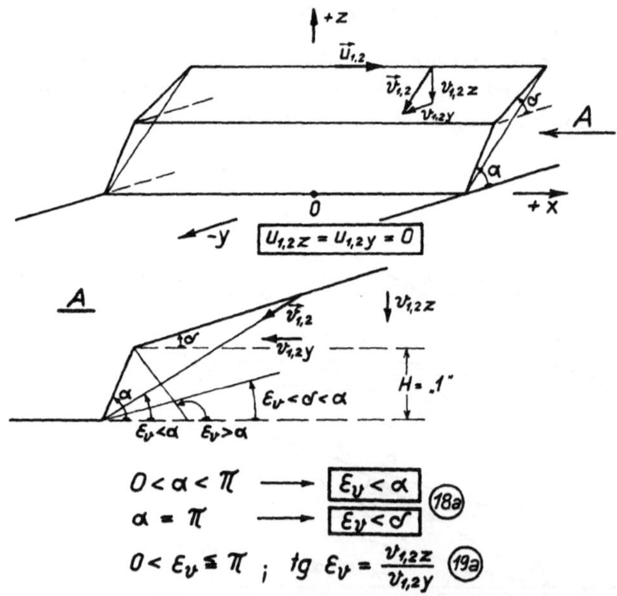

Abb. 7. Geometrische Voraussetzung für den Fall des Gleitens (Sonderfall)
Geometrical conditions for the case of sliding (special case)
Hypothèses géométriques dans le cas de glissement (cas particulier)

dieser Ebene, ob ein Gleiten kinematisch möglich ist. Die Gl. (18 a) gibt die Gleitbedingungen für diesen Sonderfall. Sie hat die gleiche Form wie die bereits erläuterte Gl. (18).

$$0 < \alpha < \pi \rightarrow \varepsilon_v < \alpha,$$
$$\alpha = \pi \rightarrow \varepsilon_v < \delta. \tag{18 a}$$

Den Winkel ε_v erhält man aus der Gl. (19 a).

$$0 < \varepsilon_v \leqq \pi; \quad \operatorname{tg} \varepsilon_v = \frac{v_{1,2z}}{v_{1,2y}}. \tag{19 a}$$

B. Mögliche Gleitrichtungen

Ergibt sich nach den Gln. (18) oder (18 a), daß sich ein Felskeil O, B, C, D aus der Böschung herauslösen kann, so ist es für den Nachweis der Gleitsicherheit erforderlich, die mögliche Gleitrichtung des Felskörpers zu kennen.

Wenn es kinematisch möglich ist, wird ein auf einer schiefen Ebene gelagerter, durch sein Eigengewicht belasteter Körper, das Bestreben haben, dem größten Gefälle folgend parallel zu den Fallinien abzugleiten.

Die Fallinien sind in der Abb. 8 wiederum mit den Vektoren v_1 und v_2 dargestellt. Schneidet die positive Richtung beider Fallinien jedoch die Schnittlinie $\vec{x}$

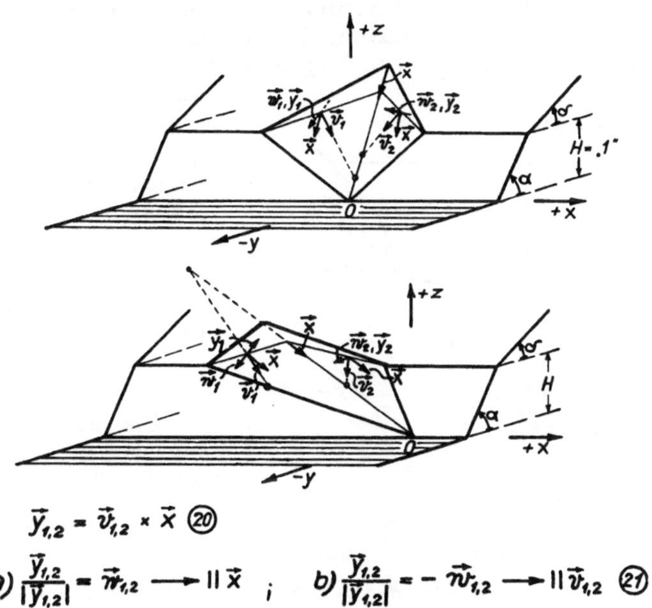

Abb. 8. Ermittlung der möglichen Gleitrichtung
Determination of the possible direction of sliding
Détermination de la direction possible du glissement

beider Ebenen — dieser Fall ist im oberen Bild der Abb. 8 dargestellt —, so ist ein Gleiten parallel zu den Fallinien kinematisch nicht möglich. Der Felskörper wird dann zwischen beiden Ebenen eingekeilt und kann nur parallel zur Richtung $\vec{x}$ gleiten. Schneidet, wie im unteren Bild der Abb. 8 dargestellt, eine der beiden Fallinien, in diesem Falle ist es v_1, die Schnittlinie x in positiver Richtung nicht, so kann der Körper in Richtung des größten Gefälles $\vec{v}_1$ abgleiten. Mathematisch lassen sich diese Aussagen durch die Vektoren $\vec{y}_1$ und $\vec{y}_2$ beschreiben, die man aus der Gl. (20) erhält:

$$\vec{y}_{1,2} = \vec{v}_{1,2} \times \vec{x}. \tag{20}$$

Gl. (21) gibt die mögliche Gleitrichtung an:

$$\frac{\vec{y}_{1,2}}{|\vec{y}_{1,2}|} = \vec{w}_{1,2} \rightarrow \parallel \vec{x}, \tag{21 a}$$

$$\frac{\vec{y}_{1,2}}{|\vec{y}_{1,2}|} = -\vec{w}_{1,2} \rightarrow \parallel \vec{v}_{1,2}. \tag{21 b}$$

Zeigen die Vektoren $\vec{y}_{1,2}$ in die Richtung der Vektoren $\vec{w}_{1,2}$, so ist Gleiten parallel zur Schnittlinie ($\vec{x}$) möglich (21 a). Zeigt einer der Vektoren $\vec{y}_{1,2}$ in die entgegengesetzte Richtung von $\vec{w}_{1,2}$, so kann der Körper parallel zu einer der Falllinien $\vec{v}_{1,2}$ abgleiten.

Im Sonderfall der Abb. 7, in dem eine Ebene parallel zur Böschung streicht, kann das Gleiten selbstverständlich nur parallel zur Fallinie dieser Ebene erfolgen.

C. Gleitsicherheitsnachweis für den Fall des Gleitens parallel zur Fallinie $\vec{v}$

Der Gleitsicherheitsnachweis wird zunächst für die Fälle erbracht, bei denen das Gleiten parallel zu einer der beiden Falllinien $\vec{v}_{1,2}$ erfolgen kann. Eine solche Möglichkeit ist in der Abb. 9 in den Bildern *a* und *b* dargestellt.

Die ungünstigste Erdbebenbelastung ist in diesem Fall der Vektor $\vec{H}_v$ nach Gl. (17), der parallel zur senkrechten Projektion der Gleitrichtung wirkt. Man erhält den Betrag der parallel zu $\vec{v}$ wirkenden Komponente der angreifenden Kräfte $\vec{H}_v$ und $\vec{G}$ aus der Gl. (22) (s. Ansicht *A*, Abb. 9).

$$|\vec{T}| = |(\vec{G} + \vec{H}_v) \cdot \vec{v}|. \quad (22)$$

Der Vektor dieser Komponente ergibt sich aus Gl. (22 a).

$$\vec{T} = |\vec{T}| \cdot \vec{v}. \quad (22\ a)$$

Die normal zur Gleitrichtung wirkende Komponente ist nach Gl. (23):

$$N = (\vec{G} + \vec{H}_v) \cdot \vec{w}. \quad (23)$$

Die dem Gleiten widerstehende Kraft ergibt sich durch Multiplikation von N mit dem Reibungsbeiwert tg ϱ aus der Gl. (24).

$$W = N \cdot \mathrm{tg}\,\varrho. \quad (24)$$

Ergibt die Gl. (21) als Gleitrichtung den Vektor $\vec{v}_2$, so muß man die Normalkomponente (N) aus der Gl. (23 a) ermitteln.

$$N = (\vec{G} + \vec{H}_v)\,(-w_2). \quad (23\ a)$$

In diese Gleichung ist der Vektor $\vec{w}_2$ mit einem negativen Vorzeichen eingeführt, weil er aus der Ebene $E\,1$ herauszeigt und sich nach (23) ein negativer Wert für eine Druckkraft N ergeben würde. — Das gilt nicht für den Sonderfall (Abb. 9, Bild a), weil hier der Vektor w nach Voraussetzung immer in die Ebene hineinzeigt (s. Gl. [2] für $\beta_{1,2}$). —

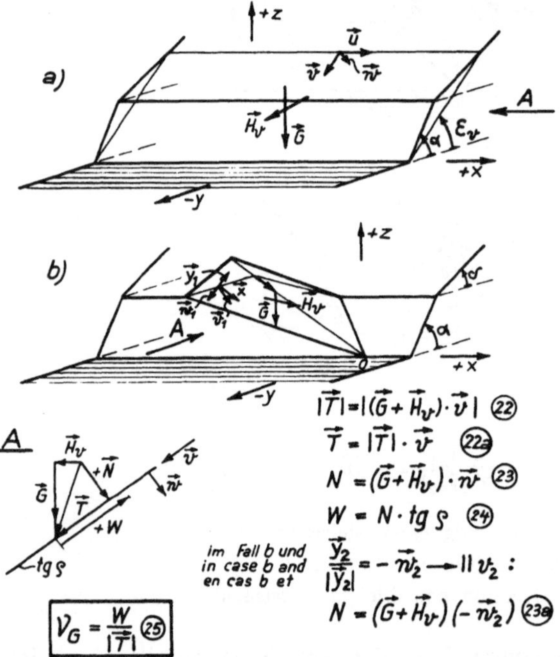

Abb. 9. Nachweis der Gleitsicherheit für den Fall des Gleitens parallel zu $\vec{v}$

Calculation of the factor of safety in the case of sliding parallel to $\vec{v}$

Vérification de la sécurité au glissement dans le cas du glissement parallèle à $\vec{v}$

Vorzeichen eingeführt, weil er aus der Ebene $E\,1$ herauszeigt und sich nach (23) ein negativer Wert für eine Druckkraft N ergeben würde. — Das gilt nicht für den Sonderfall (Abb. 9, Bild a), weil hier der Vektor w nach Voraussetzung immer in die Ebene hineinzeigt (s. Gl. [2] für $\beta_{1,2}$). —

Die Sicherheit gegen Gleiten wird dann als Quotient aus den widerstehenden und treibenden Kräften ermittelt und nach Gl. (25) erhalten:

$$\nu_G = \frac{W}{|T|}. \quad (25)$$

Ergibt sich für W, also auch für ν_G ein negativer Wert, so bedeutet das, daß die Normalkräfte N Zugkräfte sind, der Körper sich also von seiner Unterlage abheben muß, oder anders ausgedrückt, aus der Böschung herausfällt. Das kann bei überhängenden Böschungen der Fall sein, und es ergibt sich dann folgerichtig eine Gleitsicherheit von $\nu_G < 0$.

D. Gleitsicherheitsnachweis für den Fall des Gleitens parallel zur Schnittlinie $\vec{x}$

Ist die mögliche Gleitrichtung durch den Vektor $\vec{x}$ definiert, so wird der Gleitsicherheitsnachweis nach Abb. 10 durchgeführt.

Man erhält den Betrag des Vektors der treibenden Kraft aus der Gl. (26), wobei als Erdbebenkraft in diesem Falle der Vektor $\vec{H}_x$ nach Gl. (16) am ungünstigsten ist (Abb. 10, Ansicht A).

$$|\vec{T}| = (\vec{G} + \vec{H}_x) \cdot \frac{\vec{x}}{|\vec{x}|}. \quad (26)$$

Es ist dann der Vektor $\vec{T}$ nach der Gl. (27) gleich:

$$\vec{T} = \frac{1}{|\vec{x}^2|} \cdot |(\vec{G} + \vec{H}_x) . \vec{x}| . \vec{x}. \quad (27)$$

Die normal dazu gerichtete Komponente, den Vektor $\vec{N}$, erhält man als Differenz aus:

$$\vec{N} = (\vec{G} + \vec{H}_x) - \vec{T}. \quad (28)$$

Diese Kraft muß in diesem Fall, um die widerstehenden Kräfte zu erhalten, in die beiden Richtungen senkrecht zu den Ebenen E_1 und E_2 zerlegt werden (Abb. 10, Ansicht B, parallel zu $\vec{x}$):

$$\vec{N} = \vec{V}_1 + \vec{V}_2 = c_1 . \vec{w}_1 + c_2 . (-\vec{w}_2). \quad (29)$$

Aus dieser Gleichung für $\vec{N}$ kann man die Unbekannten c_1 und c_2 bestimmen, wenn man sie für die drei Komponenten in x-, y- und z-Richtung anschreibt. Der Betrag von c_1 ist gleich dem Betrag des Vektors $\vec{V}_1$ und der Betrag von c_2 ist gleich dem Betrag des Vektors $\vec{V}_2$,

$$|c_1| = |\vec{V}_1|, \quad |c_2| = |\vec{V}_2|. \quad (30)$$

Die widerstehende Kraft W ergibt sich dann zu:

$$W = c_1 . \operatorname{tg} \varrho_1 + c_2 . \operatorname{tg} \varrho_2. \quad (31)$$

Und die Gleitsicherheit zu:

$$\nu_G = \frac{W}{|\vec{T}|}. \quad (25)$$

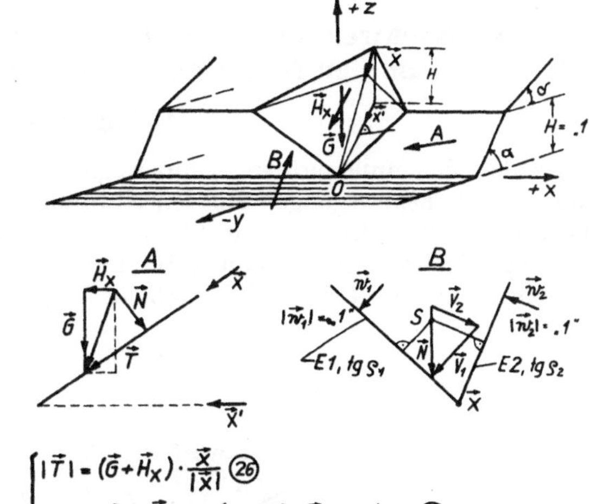

Abb. 10. Nachweis der Gleitsicherheit für den Fall des Gleitens parallel zu $\vec{x}$

Calculation of the factor of safety in the case of sliding parallel to $\vec{x}$

Vérification de la sécurité au glissement dans le cas du glissement parallèle à $\vec{x}$

Auch in diesem Falle bedeutet ein negativer Wert für die Gleitsicherheit ν_G, daß der Felskeil aus der Böschung herausfällt. Ist keine Erdbebenwirkung vorhanden, so vereinfacht sich die Rechnung. Es kann dann entweder $\vec{H}_v$ oder der Vektor $\vec{H}_x$ gleich 0 gesetzt werden.

VI. Nachweis der Sicherheit gegen Rotationsbewegungen um den Nullpunkt (0-)Kippsicherheitsnachweis

A. Allgemeines

Außer den untersuchten Gleitbewegungen der Felskeile können in bestimmten Fällen Rotationsbewegungen um den Nullpunkt (0) auftreten. Die Möglichkeiten für eine Rotation sind in der Abb. 11 an Hand von zwei Beispielen dargestellt.

Eine Drehbewegung kann z. B. nach dem linken Bild der Abb. 11 in einem Abdrehen des Felskeiles auf der Ebene $E\,1$, das ist die Ebene ODB, erfolgen. Die

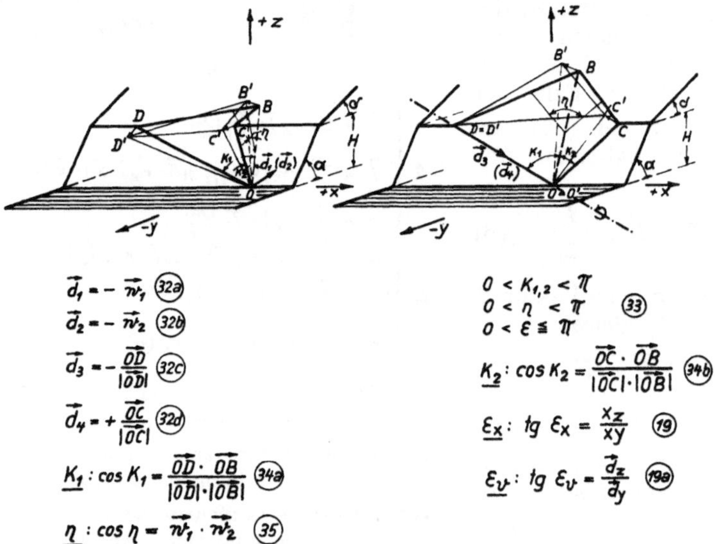

$$\vec{d}_1 = -\,\vec{n}_1 \qquad (32a)$$
$$\vec{d}_2 = -\,\vec{n}_2 \qquad (32b)$$
$$\vec{d}_3 = -\frac{\vec{OD}}{|\vec{OD}|} \qquad (32c)$$
$$\vec{d}_4 = +\frac{\vec{OC}}{|\vec{OC}|} \qquad (32d)$$
$$K_1: \cos K_1 = \frac{\vec{OD}\cdot\vec{OB}}{|\vec{OD}|\cdot|\vec{OB}|} \qquad (34a)$$
$$\eta: \cos\eta = \vec{n}_1\cdot\vec{n}_2 \qquad (35)$$

$$0 < K_{1,2} < \pi$$
$$0 < \eta < \pi \qquad (33)$$
$$0 < \varepsilon \lessgtr \pi$$
$$K_2: \cos K_2 = \frac{\vec{OC}\cdot\vec{OB}}{|\vec{OC}|\cdot|\vec{OB}|} \qquad (34b)$$
$$\varepsilon_x: tg\,\varepsilon_x = \frac{x_z}{x_y} \qquad (19)$$
$$\varepsilon_v: tg\,\varepsilon_v = \frac{d_z}{d_y} \qquad (19a)$$

Abb. 11. Mögliche Drehbewegungen des Felskeiles
Possibilities of rotation of the rock pyramid
Rotation possible de la pyramide

Drehachse läßt sich durch den Vektor $\vec{d}_1$ beschreiben, der durch den Nullpunkt geht und senkrecht auf $E\,1$ steht. Der Drehvektor $\vec{d}_1$, dessen Richtung gleichzeitig den Drehsinn festlegt, ist nach Gl. (32 a) gleich dem negativen Vektor $\vec{w}_1$:

$$\vec{d}_1 = -\,\vec{w}_1. \qquad (32\text{ a})$$

Der Drehung entgegen wirken in diesem Falle außer der Komponente der Belastung, die senkrecht zur Drehachse gerichtet ist, Reibungskräfte in der Fläche ODB. Die gleiche Möglichkeit des Abdrehens über eine Fläche besteht natürlich auch für die Ebene E_2. Den zugehörigen Drehvektor $(\vec{d}_2)$ erhält man aus Gl. (32 b):

$$\vec{d}_2 = -\,\vec{w}_2. \qquad (32\text{ b})$$

Außer diesen angedeuteten Möglichkeiten kann der Körper in manchen Fällen auch über eine der Schnittlinien zwischen der Böschung und den Ebenen E_1 und

E_2, das sind die Linien OD und OC, kippen. In diesem Fall würde sich der Felskeil von beiden Ebenen abheben, wie es im rechten Bild der Abb. 11 für die Kante OD dargestellt ist. Die Drehachsen können durch die Gln. (32 c und d) beschrieben werden:

$$\vec{d}_3 = -\frac{\overrightarrow{OD}}{|\overrightarrow{OD}|},$$

(32 c)

$$\vec{d}_4 = +\frac{\overrightarrow{OC}}{|\overrightarrow{OC}|}.$$

(32 d)

B. Rotationsbedingungen

Die Geometrie des Felskeiles $OBCD$ bestimmt, welche der skizzierten vier Drehbewegungen mit den Zwangsbedingungen des Systems verträglich sind. Eine Drehung um die Achse $\vec{d}_1$ bzw. $\vec{d}_2$ ist z. B. nur dann kinematisch möglich, wenn sich der Körper von den Ebenen E_2 bzw. E_1 abheben kann und sich dabei nicht verklemmt. Eine Rotation um die Achsen $\vec{d}_3$ und $\vec{d}_4$ setzt voraus, daß sich der Felskeil dabei von beiden Ebenen E_1 und E_2 abheben kann. Die Kenntnis der in der Abb. 11 dargestellten Winkel K_1, K_2, η und $\varepsilon_{x,v}$ erlaubt es, die im jeweiligen Fall kinematisch möglichen Drehachsen zu bestimmen.

Die möglichen Schwankungsbereiche dieser Winkel sind in der Gl. (33) angegeben:

$$0 < K_{1,2} < \pi,$$
$$0 < \eta < \pi,$$
$$0 < \varepsilon_{x,v} \lessgtr \pi.$$

(33)

Man erhält die Winkel K_1 und K_2 aus den Gln. (34 a) und (34 b).

$$\cos K_1 = \frac{\overrightarrow{OD} \cdot \overrightarrow{OB}}{|\overrightarrow{OD}| \cdot |\overrightarrow{OB}|},$$

(34 a)

$$\cos K_2 = \frac{\overrightarrow{OC} \cdot \overrightarrow{OB}}{|\overrightarrow{OC}| \cdot |\overrightarrow{OB}|}.$$

(34 b)

Die Vektoren $\overrightarrow{OD}$, $\overrightarrow{OC}$ und $\overrightarrow{OB}$ können mit Hilfe der bekannten Beziehungen (10), (11) und (13) ermittelt werden. Den Winkel η bestimmt die Gl. (35)

$$\cos \eta = \vec{w}_1 \cdot \vec{w}_2$$

(35)

und die Winkel ε_x und ε_v wurden bereits vorher durch die Gln. (19) und (19 a) festgelegt.

Außer den Gleitbedingungen des Abschnitts V, A muß für eine Rotation zunächst die folgende Beziehung erfüllt sein:

$$\varepsilon_{x,v} > \delta.$$

(36)

Ist der Winkel ε_x (Gleiten parallel $\vec{x}$) bzw. der Winkel ε_v (Gleiten parallel v) kleiner als der Neigungswinkel δ, so liegt der Schwerpunkt des Gleitkörpers sehr weit innerhalb der Böschung und ein Kippen ist ausgeschlossen. In der Tabelle der Abb. 12 sind weiterhin in der ersten bis dritten Spalte verschiedene Bereiche für die Winkel K_1, K_2 und η angegeben und in der letzten Spalte die zu diesen Winkelbereichen gehörenden kinematisch möglichen Drehachsen. Die vierte enthält einige zusätzliche Bedingungen, die erfüllt sein müssen, wenn die Drehachsen der letzten Spalte möglich sein sollen.

Bei der Durchrechnung eines Beispiels ermittelt man nun die Winkel $K_{1,2}$ und η, stellt fest, in welchem Winkelbereich die berechneten Werte liegen und findet in der letzten Spalte der Tabelle der Abb. 12, welche Drehachsen kinematisch möglich

$$\boxed{\varepsilon_{x,v} > \sigma} \quad \text{\textcircled{36}}$$

K_1	K_2	η	weitere Bedingungen further conditions conditions supplémentaires	$\vec{d}$
$< \dfrac{\pi}{2}$	$< \dfrac{\pi}{2}$	$< \dfrac{\pi}{2}$	—	$\vec{d_1}, \vec{d_2}$
$< \dfrac{\pi}{2}$	$< \dfrac{\pi}{2}$	$> \dfrac{\pi}{2}$	—	$\vec{d_1}, \vec{d_2}, \vec{d_3}, \vec{d_4}$
$> \dfrac{\pi}{2}$	$> \dfrac{\pi}{2}$	$0 < \eta < \pi$	—	—
$> \dfrac{\pi}{2}\left(< \dfrac{\pi}{2}\right)$	$< \dfrac{\pi}{2}\left(> \dfrac{\pi}{2}\right)$	$< \dfrac{\pi}{2}$	—	$\vec{d}_{1(2)}$
$> \dfrac{\pi}{2}\left(< \dfrac{\pi}{2}\right)$	$< \dfrac{\pi}{2}\left(> \dfrac{\pi}{2}\right)$	$> \dfrac{\pi}{2}$	$tg\,K_{2(1)} \lessgtr tg(\pi - K_{1(2)})\cos(\pi - \eta)$	$\vec{d}_{1(2)}, \vec{d}_{4(3)}, \vec{d}_{3(4)}$
$> \dfrac{\pi}{2}\left(< \dfrac{\pi}{2}\right)$	$< \dfrac{\pi}{2}\left(> \dfrac{\pi}{2}\right)$	$> \dfrac{\pi}{2}$	$tg\,K_{2(1)} \lessgtr \dfrac{tg(\pi - K_{1(2)})}{\cos(\pi - \eta)}$	$\vec{d}_{1(2)}, \vec{d}_{4(3)}$

$$a)\ K_1 = K_2 \lesseqgtr \frac{\pi}{2},\ \left.\begin{array}{l} \beta_1 = \pi - \beta_2 \\ \gamma_1 = \pi - \gamma_2 \end{array}\right\} \longrightarrow \vec{d}_5 = 1,0,0$$

$$b)\ u_{1,2z} = u_{1,2y} = 0 \longrightarrow \vec{d}_5 = 1,0,0$$

Abb. 12. Mögliche Drehbewegungen $(\vec{d_n})$ in Abhängigkeit von den Winkeln $\varepsilon_{x,v}$, $K_{1,2}$ and η

Possible rotation $(\vec{d_n})$ as a function of the angles $\varepsilon_{x,v}$, $K_{1,2}$ and η

Rotation possible en fonction des angles $\varepsilon_{x,v}$, $K_{1,2}$ et η

sind. Zwei Sonderfälle, die unten auf der Abb. 12 (Gln. a und b) behandelt werden, sind auf Abb. 13 nochmals ausführlich dargestellt.

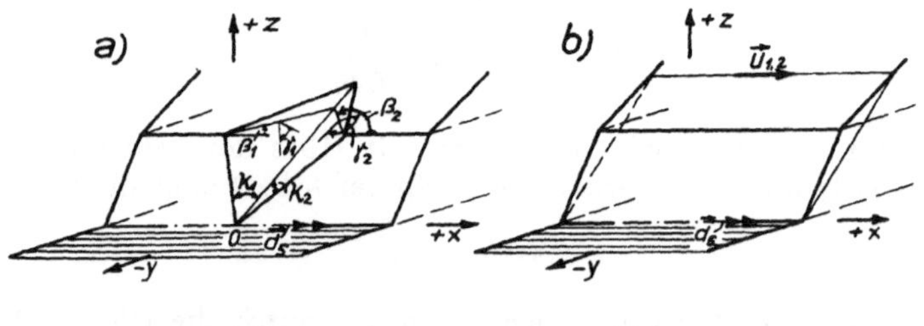

$$\left.\begin{array}{l} K_1 = K_2 \lesseqgtr \dfrac{\pi}{2} \\ \beta_1 = \pi - \beta_2 \\ \gamma_1 = \pi - \gamma_2 \end{array}\right\} \vec{d}_5 = 1,0,0 \qquad u_{1,2z} = u_{1,2y} = 0 \longrightarrow \vec{d}_5 = 1,0,0$$

Abb. 13. Mögliche Drehbewegungen $(\vec{d_n})$ für zwei Sonderfälle

Possible rotation $(\vec{d_n})$ for two special cases

Rotation possible pour deux cas particuliers

Die Abb. 13a soll einen symmetrischen Gleitkörper zeigen, bei dem $K_1 = K_2 \leqq \dfrac{\pi}{2}$ sind und außerdem $\beta_1 = \pi - \beta_2$ und $\gamma_1 = \pi - \gamma_2$ sind. In diesem Fall würde bei Belastung durch Eigengewicht ein Kippen aus Symmetriegründen über die angedeutete waagerechte Drehachse ($\vec{d}_5 = 1, 0, 0$) durch den Nullpunkt erfolgen, und der Körper sich von beiden Ebenen E_1 und E_2 gleichmäßig abheben.

In der Abb. 13 b liegen ähnliche Verhältnisse vor. Der hier dargestellte Felskörper wurde auch schon bei Gleitbetrachtungen als Sonderfall $u_{1,2z} = u_{1,2y} = 0$ abgehandelt. Hier streicht eine der beiden Ebenen parallel zur Böschung. Die mögliche Drehachse ist dann ebenfalls Vektor $\vec{d}_5 = 1, 0, 0$.

Auf eine Herleitung der in der Abb. 12 angegebenen Winkelbereiche und der zugehörigen kinematisch möglichen Drehachsen wird verzichtet, weil sie über den Rahmen eines Vortrags hinausgeht. In einer in Kürze erscheinenden weitergehenden Veröffentlichung des Verfassers[9] werden die Bereiche kinematisch möglicher Drehachsen für die verschiedenen Fälle mathematisch abgegrenzt und begründet. Man erhält, daß auch Lagen zwischen den erwähnten vier Drehachsen ($\vec{d}_1 - \vec{d}_4$) mit den Zwangsbedingungen des Systems verträglich sind. Diese sind jedoch bei der Belastung des Felskeiles aus Eigengewicht weniger stabil und scheiden daher aus.

C. Einschränkung der kinematisch möglichen Drehachsen

Durch eine zusätzliche statische Überlegung kann man die Zahl der möglichen Drehachsen noch weiter einschränken. In Abb. 14 a ist die Ansicht eines Felskörpers $OBCD$ in Richtung des Vektors $-\vec{x}$ dargestellt. Liegt der Schwerpunkt des Körpers

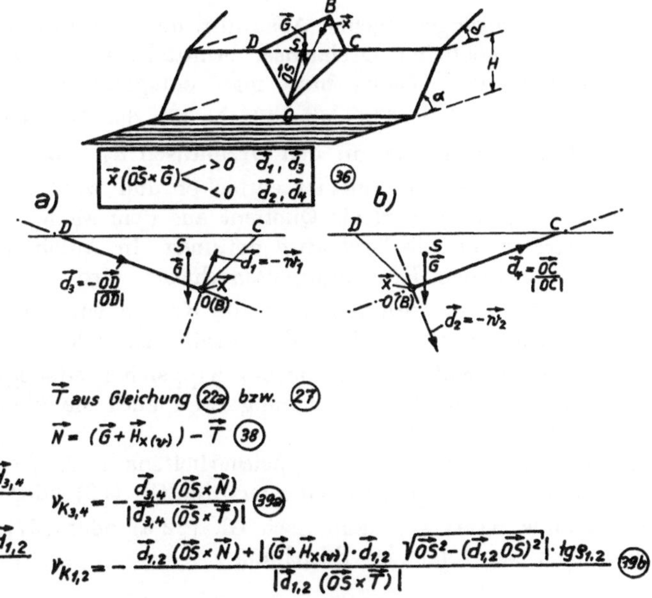

Abb. 14. Einschränkung der kinematischen möglichen Drehbewegungen und Nachweis der Sicherheit gegen Kippen des Felskeiles

Restriction of the possible rotations, and calculations of the factor of safety in the case of rotation

Limitation des rotations possibles au point de vue cinématique et vérification de la sécurité de la pyramide au basculement

nun, wie in der Darstellung auf der gleichen Seite der Schnittlinie ($\vec{x}$), wie der Punkt D, so durchstößt der Vektor des Gewichtes ($\vec{G}$) die Ebene E_1. In diesem Fall brauchen nur die Drehachsen $\vec{d}_1$ und $\vec{d}_3$ näher betrachtet zu werden, soweit sie auf Grund der vorangegangenen Überlegungen kinematisch möglich sind.

Bei einer Rotationsbewegung um die Achsen $\vec{d}_2$ oder $\vec{d}_4$ müßte sich der Körper von der Ebene E_1 abheben. Das Gewicht $\vec{G}$ übt jedoch bei dieser Lage ein Moment um die Schnittlinie aus, das den Körper auf die Ebene E_1 zurückdrückt.

Der umgekehrte, im Bild b der Abb. 14 dargestellte Fall liegt vor, wenn das Gewicht auf derselben Seite der Schnittlinie liegt, wie der Punkt C. In diesem Fall muß die Sicherheit gegen Rotation um die Achsen $\vec{d}_2$ und $\vec{d}_4$ nachgewiesen werden.

Mathematisch können diese beiden Fälle ohne die Anschauung mit Hilfe der Gl. (37) unterschieden werden.

$$\vec{x} \cdot (\overrightarrow{OS} \times \vec{G}) \quad \begin{cases} > 0 \to \vec{d}_1, \vec{d}_3, \\ \\ < 0 \to \vec{d}_2, \vec{d}_4. \end{cases} \tag{37}$$

Ist danach das Moment des im Schwerpunkt angreifenden Gewichts um den Vektor $\vec{x}$ positv, so kann — wenn kinematisch möglich — eine Drehung um $\vec{d}_1$ und $\vec{d}_3$ erfolgen. Ist das Moment negativ, so kommen die Drehachsen $\vec{d}_2$ und $\vec{d}_4$ in Betracht.

D. Nachweis der Sicherheit gegen Rotation

Ergibt sich aus dem vorangegangenen Abschnitt, daß eine Rotation kinematisch möglich ist, und sind die möglichen Drehachsen ermittelt, so muß der Kippsicherheitsnachweis erbracht werden. Dabei muß man entsprechend Abschnitt VI, A zwischen den Fällen des Abdrehens über eine Fläche mit den Drehachsen $\vec{d}_1$ und $\vec{d}_2$ und des Abkippens über eine Kante mit den Drehachsen $\vec{d}_3$ und $\vec{d}_4$ unterscheiden.

Der Nachweis für den letzten Fall ist einfacher und wird daher zuerst behandelt. Die Kippsicherheit (ν_K) wird als Quotient aus dem Moment der haltenden und der treibenden Kräfte um die Drehachse definiert. In Anlehnung an den bei Fundamentgründungen üblichen Kippsicherheitsnachweis wird als haltende Kraft die senkrecht zur möglichen Gleitrichtung wirkende Komponente ($\vec{N}$) der Belastung angenommen. Die treibende Kraft ist dann die parallel zur Gleitrichtung gerichtete Komponente ($\vec{T}$). Eine eingehende Diskussion des Kippsicherheitsbegriffs erfolgt in der bereits erwähnten weiteren Veröffentlichung, weil auch sie über den Rahmen dieses Vortrags hinausgehen würde.

Nach Abschnitt 4 wird auch für den Kippsicherheitsnachweis die Belastung aus Erdbeben je nach der möglichen Gleitrichtung aus der Gl. (16) oder (17) ermittelt. Man erhält die treibende Kraft ($\vec{T}$) dann nach Gl. (22 a) oder (27). Die haltende Komponente ergibt sich aus folgender Beziehung:

$$\vec{N} = (\vec{G} + \vec{H}_{v,x}) - \vec{T}. \tag{38}$$

Die Kippsicherheit ist dann:

$$\nu_{K3,4} = - \frac{\vec{d}_{3,4}\,(\overrightarrow{OS} \times \vec{N})}{|\vec{d}_{3,4}\,(\overrightarrow{OS} \times \vec{T})|}. \tag{39 a}$$

Die Sicherheit gegen Drehung um die Achsen $\vec{d}_1$ und $\vec{d}_2$ erhält man aus der Gl. (39 b):

$$\nu_{K1,2} = - \frac{\vec{d}_{1,2}(\overrightarrow{OS} \times \vec{N}) + \left|(\vec{G} + \vec{H}_{x,v}) \cdot \vec{d}_{1,2}\sqrt{\overrightarrow{OS}^2 - (\vec{d}_{1,2} \cdot \overrightarrow{OS})^2}\right| \cdot \mathrm{tg}\,\varrho_{1,2}}{|\vec{d}_{1,2} \cdot (\overrightarrow{OS} \times \vec{T})|}. \tag{39 b}$$

Die Beziehung unterscheidet sich von der Gl. (39 a) dadurch, daß zum haltenden Moment das Moment aus der in der Ebene E_1 bzw. E_2 wirkende Reibungskraft hinzukommt. Diese Reibungskraft erhält man, indem man die parallel $\vec{d}_{1,2}$ wirkende Komponente der Belastung mit dem wirksamen Reibungsbeiwert multipliziert.

$$(\vec{G} + \vec{H}_{x,v}) \cdot \vec{d}_{1,2} \cdot \operatorname{tg} \varrho_{1,2}.$$

Als Hebelarm dieser Reibungskraft wird der senkrechte Abstand des Schwerpunktes von der Drehachse angenommen:

$$\sqrt{\overrightarrow{OS}^2 - (\vec{d}_{1,2} \cdot \overrightarrow{OS})^2}.$$

Für die in Abb. 13 dargestellten Sonderfälle $(\vec{d}_5)$ kann der Kippsicherheitsnachweis nach Gl. (39 a) erbracht werden.

Ergibt sich bei den bisher erläuterten Untersuchungen keine ausreichende Standsicherheit, so läßt sich die maximale in y-Richtung gemessene Reichweite (R) möglicher Rutschungen oder Kippungen auf Grund geometrischer Betrachtungen aus der Gl. (40) bestimmen:

$$R_{\mathrm{max}} = -\frac{x_y}{|\vec{x}|} \cdot (H + H^+). \tag{40}$$

Ist keine ausreichende Kippsicherheit, wohl aber die Gleitsicherheit gegeben, so braucht sich bei stark zerklüfteten Felskörpern die maximale Reichweite des Kippens nicht einzustellen.

Durch das Abkippen von einzelnen kleineren Felskörpern am Rande der Böschung verändert sich die Form des gesamten Körpers, der dann zum Teil stehen bleiben kann. In diesen Fällen wären Sondernachweise erforderlich, wenn genauere Angaben über die Reichweite erforderlich sind.

VII. Erweiterung auf mehr als zwei Systeme von Absonderungsflächen

In der Praxis muß man sich sehr häufig mit der Standsicherheit von Böschungen im Gebirge mit mehr als zwei Absonderungsflächen auseinandersetzen. Das angedeutete Verfahren läßt sich auf diese Fälle ausdehnen, indem man z. B. bei drei Absonderungsflächen den Nachweis der Gleitsicherheit für Felskörper führt, die durch die Ebenen E_1/E_2, E_2/E_3 und E_1/E_3 begrenzt werden. Der Gleitsicherheitsnachweis muß in solchen Fällen also für $\frac{n!}{2}$ Möglichkeiten geführt werden. Das heißt jeweils immer so viele Möglichkeiten, wie Gleitkörperformen durch die verschiedenen Absonderungsflächen entstehen können.

Entsprechend muß man auch beim Nachweis der Kippsicherheit vorgehen. Es ist hierbei jedoch zu berücksichtigen, daß in Böschungsnähe, besonders bei überhängenden Böschungen, durch drei oder mehr Absonderungsflächen Felskörper abgetrennt werden können, für die der Nachweis der Kippsicherheit im Rahmen des geschilderten Verfahrens nicht erbracht werden kann.

In solchen Fällen muß die Kippsicherheit in böschungsnahen Bereichen speziell nachgewiesen werden. Hinweise für derartige Nachweise finden sich bei Müller[2].

VIII. Rechenbeispiele

Im folgenden soll die Anwendung des skizzierten Rechenweges an einigen Beispielen erläutert werden. Die Berechnungen wurden mit dem Rechenstab durchgeführt. Das Verfahren läßt sich jedoch auch für Elektronenrechner programmieren, so daß es mit verhältnismäßig geringem Aufwand möglich sein dürfte, auch mehrere Varianten eines bestimmten Problems zu behandeln.

In den Beispielen bedeuten die in Klammern angegebenen Zahlen Gleichungsnummern, auf die sich der entsprechende Rechengang bezieht.

Beispiel 1: „Herauslösen von Felskörpern aus der Böschung kinematisch nicht möglich."

Aufgabenstellung:

Böschung:

$$\alpha = 90^0, \ \delta = 10^0 \tag{1}$$

Absonderungsflächen:

$$\beta_1 = 70^0, \ \gamma_1 = 130^0, \ \beta_2 = 120^0, \ \gamma_2 = 80^0, \ \varrho_1 = 20^0, \ \varrho_2 = 30^0 \tag{2}$$

Belastung:

$$\alpha' = 0 \rightarrow \text{keine Belastung aus Erdbeben.}$$

Geometrie:

$$\vec{u}_1 = \quad 0,34, \quad 0,94, \quad 0 \tag{3}$$
$$\vec{u}_2 = -0,50, \quad 0,87, \quad 0$$

$$\vec{v}_1 = -0,60, \quad 0,22, -0,77 \tag{4}$$
$$\vec{v}_2 = \quad 0,15, \quad 0,09, -0,99$$

$$\vec{w}_1 = -0,72, \quad 0,26, \quad 0,64 \tag{5}$$
$$\vec{w}_2 = -0,85, -0,50, -0,17$$

$$\vec{x} \ = -0,27, \quad 0,67, -0,58 \tag{6}$$

Gleitbedingungen:

$$0 < \varepsilon_x \lesseqgtr \pi, \ \operatorname{tg} \varepsilon_x = -0,87, \ \varepsilon_x = 139^0 \tag{19}$$
$$0 < \alpha = \pi, \ \varepsilon_x < \alpha \tag{18}$$

Gleitbedingung nicht erfüllt, es kann sich kein Felskörper aus der Böschung herauslösen. (Sonderfall nach **Abb. 7** liegt nicht vor; $u_{1,2y} \neq 0$.)

Beispiel 2: „Gleiten **parallel** $\vec{x}$, Gleitsicherheit $v_G < 1$."

Aufgabenstellung:

Böschung:

$$\alpha = 80^0, \ \delta = 0 \tag{1}$$

Absonderungsflächen:

$$\beta_1 = 50^0, \ \gamma_1 = 50^0, \ \beta_2 = 110^0, \ \gamma_2 = 140^0, \ \varrho_1 = 20^0, \ \varrho_2 = 25^0 \tag{2}$$

Belastung:

$$\alpha' = 0,15 \rightarrow \text{Belastung aus Erdbeben.}$$

Geometrie:

$$\vec{u}_1 = \quad 0,64, \quad 0,77, \quad 0 \tag{3}$$
$$\vec{u}_2 = -0,34, \quad 0,94, \quad 0$$

$$\vec{v}_1 = \quad 0,49, -0,41, -0,77 \tag{4}$$
$$\vec{v}_2 = -0,72, -0,26, -0,64$$

$$\vec{w_1} = \quad 0{,}59, \quad 0{,}49, -0{,}64 \tag{5}$$

$$\vec{w_2} = -0{,}60, -0{,}22, \quad 0{,}77$$

$$\vec{x} = -0{,}24, -0{,}84, -0{,}42, \quad |\vec{x}| = 0{,}97 \tag{6}$$

$$H = \text{„1"} \tag{7}$$

$$H^+ = 0, \text{ weil } \delta = 0 \tag{9}$$

Belastung:

$$G = 0, 0, -g \tag{15}$$

$$\alpha' = 0{,}15$$

$$\vec{x}' = 0{,}24, -0{,}84, 0 \tag{16}$$

$$|\vec{x}'| = 0{,}86$$

$$\vec{H_x} = -0{,}04\,g, -0{,}15\,g, 0$$

Gleitbedingungen:

$$0 < \varepsilon_x \leqq \pi, \ \text{tg}\,\varepsilon_x = 0{,}50, \ \varepsilon_x = 27^0 \tag{19}$$

$$0 < \alpha < \pi, \ \varepsilon_x < \alpha = 80^0 \tag{18}$$

Gleiten ist möglich. (Sonderfall nach Abb. 7 liegt nicht vor; $u_{1,2y} \neq 0$.)

Gleitrichtungen:

$$\vec{y_1} = -0{,}47, \quad 0{,}39, -0{,}51, \quad |\vec{y_1}| = 0{,}80 \tag{20}$$

$$\vec{y_2} = -0{,}45, -0{,}15, \quad 0{,}54, \quad |\vec{y_2}| = 0{,}72$$

$$\left. \begin{array}{l} \dfrac{\vec{y_1}}{|\vec{y_1}|} \approx w_1 \\[2ex] \dfrac{\vec{y_2}}{|\vec{y_2}|} \approx w_2 \end{array} \right\} \quad \text{Gleichrichtung} \parallel \vec{x} \tag{21}$$

Gleitsicherheitsnachweise:

$$|\vec{T}| = \quad 0{,}57\,g \tag{26}$$

$$\vec{T} = -0{,}14\,g, -0{,}49\,g, -0{,}25\,g \tag{27}$$

$$\vec{N} = \quad 0{,}10\,g, \quad 0{,}34\,g, -0{,}75\,g \tag{28}$$

$$c_1 = \quad 0{,}43\,g, \ c_2 = 0{,}59\,g \tag{29}$$

$$W = \quad 0{,}42\,g \tag{31}$$

$$\nu_G = \quad 0{,}74 \tag{25}$$

(Die Gleitsicherheit ist nicht gewährleistet.)

$$R_{\max} = 0{,}87 \cdot H \ \text{(Reichweite)} \tag{40}$$

Beispiel 3: „Gleiten parallel $\vec{v_1}$ und Kippen um $\vec{d_1}$ möglich ($\nu_G > 1$; $\nu_k > 1$)."

Aufgabenstellung:

Böschung:

$$\alpha = 80^0, \ \delta = 0 \tag{1}$$

Absonderungsflächen:

$$\beta_1 = 50^0, \ \gamma_1 = 32^0, \ \beta_2 = 150^0, \ \gamma_2 = 80^0, \ \varrho_1 = \varrho_2 = 35^0 \tag{2}$$

Belastung:

$$\alpha' = 0 \rightarrow \text{keine Erdbebenkräfte}$$

Geometrie:

$$\vec{u}_1 = 0,64, 0,77, 0 \tag{3}$$
$$\vec{u}_2 = -0,87, 0,50, 0$$

$$\vec{v}_1 = 0,65, -0,55, -0,53 \tag{4}$$
$$\vec{v}_2 = 0,09, 0,15, -0,99$$

$$\vec{w}_1 = -0,41, 0,34, -0,85 \tag{5}$$
$$\vec{w}_2 = -0,49, -0,86, -0,17$$

$$\vec{x} = 0,78, -0,35, -0,52, \; |\vec{x}| = 1 \tag{6}$$

$$H = 1 \tag{7}$$

$$H^+ = 0, \text{ weil } \delta = 0 \tag{9}$$

$$\overrightarrow{OD} = -1,94, 0,18, 1, \; |\overrightarrow{OD}| = 2,25 \tag{10}$$

$$\overrightarrow{OC} = -0,66, 0,18, 1, \; |\overrightarrow{OC}| = 1,22 \tag{11}$$

$$\overrightarrow{OA} = -1,30, 0,18, 1 \tag{12}$$

$$\overrightarrow{OB} = -1,52, 0,68, 1, \; |\overrightarrow{OB}| = 1,93 \tag{13}$$

$$\overrightarrow{OS} = -1,03, 0,26, 0,75 \tag{14}$$

Belastung:

$$G = 0, 0, -g \tag{15}$$
$$\alpha' = 0 \quad \rightarrow \quad \vec{H}_x = 0, \; \vec{H}_v = 0 \tag{16} + \text{(17)}$$

Gleitbedingungen:

$$0 \leqq \varepsilon_x \leqq \pi, \; \text{tg}\,\varepsilon_x = 1,48, \; \varepsilon_x = 56^0 \tag{19}$$
$$0 \leqq \alpha < \pi, \; \varepsilon_x < \alpha = 80^0 \tag{18}$$

Gleiten ist möglich. (Sonderfall nach Abb. 7 liegt nicht vor; $u_{1,2\,y} \neq 0$.)

Gleitrichtungen:

$$\vec{y}_1 = 0,10, -0,08, 0,20, \; |\vec{y}_1| = 0,24 \tag{20}$$
$$\vec{y}_2 = -0,42, -0,73, -0,15, \; |\vec{y}_2| = 0,85$$

$$\frac{\vec{y}_1}{|\vec{y}_1|} = -\vec{w}_1 \rightarrow \text{Gleitrichtung} \parallel \vec{v}_1 \tag{21}$$

$$\frac{\vec{y}_2}{|\vec{y}_2|} = +\vec{w}_2$$

Gleitsicherheitsnachweise:

$$|\vec{T}| = 0,53\,g \tag{22}$$
$$N = 0,85\,g \tag{23}$$
$$W = 0,60\,g \tag{24}$$
$$\nu_G = 1,12 \tag{25}$$

Gleitsicherheit ist gegeben.

Rotationsbedingungen:

$$K_1 = 21^0 < \frac{\pi}{2} \tag{34 a}$$

$$K_2 = 26^0 < \frac{\pi}{2} \tag{34 b}$$

$$\eta = 87^0 < \frac{\pi}{2} \tag{35}$$

Nach Abb. 12 sind folgende Drehachsen möglich:

$$\vec{d_1} = -\vec{w_1} = +0{,}41, -0{,}34, +0{,}85 \tag{32 a}$$

$$\vec{d_2} = -\vec{w_2} = +0{,}49, +0{,}86, +0{,}17 \tag{32 b}$$

Einschränkung der kinematisch möglichen Drehachsen:

$$\vec{x} \cdot (\overrightarrow{OS} \times \vec{G}) = 0{,}16\,g > 0 \rightarrow \vec{d_1} \tag{37}$$

$$\nu_{k1} = \frac{0{,}21\,g + 0{,}77\,g}{0{,}27\,g} = 3{,}6 > 1 \tag{39 b}$$

Beispiel 4: „Gleiten parallel $\vec{x}$ und Abdrehen über Ebene E_1 möglich $(\vec{d_1})$ $(\nu_G > 1;\ \nu_k < 1)$."

Aufgabenstellung:

Böschung:

$$\alpha = 87^0,\ \delta = 0^0 \tag{1}$$

Absonderungsflächen:

$$\beta_1 = 20^0,\ \gamma_1 = 75^0,\ \beta_2 = 30^0,\ \gamma_2 = 85^0,\ \varrho_1 = 20^0,\ \varrho_2 = 30^0 \tag{2}$$

Belastung:

$$\alpha' = 0 \rightarrow \text{keine Belastung aus Erdbeben}$$

Geometrie:

$$\vec{u_1} = 0{,}94,0{,}34,0 \tag{3}$$
$$\vec{u_2} = 0{,}87,0{,}50,0$$
$$\vec{v_1} = 0{,}09, -0{,}24, -0{,}97 \tag{4}$$
$$\vec{v_2} = 0{,}04, -0{,}08, -0{,}99$$
$$\vec{w_1} = -0{,}33,0{,}91, -0{,}26 \tag{5}$$
$$\vec{w_2} = -0{,}50,0{,}86, -0{,}10$$
$$\vec{x} = -0{,}15, -0{,}10, -0{,}17,\ |\vec{x}| = 0{,}24 \tag{6}$$
$$H = 1 \tag{7}$$
$$H^+ = 0,\ \text{weil}\ \delta = 0 \tag{9}$$
$$\overrightarrow{OD} = -0{,}65, 0{,}05, 1, \tag{10}$$
$$\overrightarrow{OC} = -0{,}08, 0{,}05, 1, \tag{11}$$
$$\overrightarrow{OA} = -0{,}36, 0{,}05, 1, \tag{12}$$
$$\overrightarrow{OB} = 0{,}81, 0{,}62, 1 \tag{13}$$
$$\overrightarrow{OS} = 0{,}02, 0{,}18, 0{,}75 \tag{14}$$

Belastung:

$$G = 0, 0, -g \tag{15}$$

$$\alpha' = 0 \rightarrow \text{keine Erdbebenbelastung}$$

$$\vec{H_x} = \vec{H_v} = 0 \tag{16}$$

Gleitbedingungen:

$$0 < \varepsilon_x \leqq \pi, \ \text{tg}\,\varepsilon_x = 1{,}6, \ \varepsilon_x \approx 58^0 \tag{19}$$

$$0 < \alpha < \pi, \ \varepsilon_x < \alpha = 87^0 \tag{18}$$

Gleiten ist möglich. (Sonderfall nach Abb. 7 liegt nicht vor; $u_{1,2y} \neq 0$.)

Gleitrichtungen:

$$\vec{y_1} = -0{,}06, 0{,}16, -0{,}04, \ |\vec{y_1}| = 0{,}17$$

$$\vec{y_2} = -0{,}09, 0{,}15, -0{,}02, \ |\vec{y_2}| = 0{,}18 \tag{20}$$

$$\left. \begin{aligned} \frac{\vec{y_1}}{|\vec{y_1}|} &= \vec{w_1} \\[1em] \frac{\vec{y_2}}{|\vec{y_2}|} &= \vec{w_2} \end{aligned} \right\} \quad \text{Gleitrichtung} \parallel \vec{x} \tag{21}$$

Gleitsicherheitsnachweise:

$$|\vec{T}| = \quad 0{,}73\,g \tag{26}$$

$$\vec{T} = -0{,}43\,g, -0{,}33\,g, -0{,}53\,g, \tag{27}$$

$$\vec{N} = \quad 0{,}43\,g, \quad 0{,}33\,g, -0{,}47\,g \tag{28}$$

$$c_1 = \quad 3{,}40\,g, \ c_2 = 3{,}10\,g \tag{29}$$

$$W = \quad 3{,}10\,g \tag{31}$$

$$\nu_G = \quad 4{,}25 \tag{25}$$

Gleitsicherheit ist **gewährleistet**.

Rotationsbedingungen:

$$K_1 \approx 73^0 < \frac{\pi}{2} \tag{34 a}$$

$$K_2 \approx 48^0 < \frac{\pi}{2} \tag{34 b}$$

$$\eta \ \approx 14^0 < \frac{\pi}{2} \tag{35}$$

Nach Abb. 12 sind folgende Drehachsen möglich:

$$\vec{d_1} = -\vec{w_1} = +0{,}33, -0{,}91, +0{,}26 \tag{32 a}$$

$$\vec{d_2} = -\vec{w_2} = +0{,}50, -0{,}86,' +0{,}10 \tag{32 b}$$

Einschränkung der kinematisch möglichen Drehachsen:

$$\vec{x} \cdot (\overrightarrow{OS} \times \vec{G}) \approx +0{,}03\,g > 0 \rightarrow \vec{d_1} \tag{37}$$

$$\nu_{k1} = 0{,}97 < 1 \tag{39 b}$$

Kippsicherheit ist **nicht gewährleistet**.

Beispiel 5: „Felskeil kippt über die Kante $\overrightarrow{OD}$."

Aufgabenstellung:

Böschung:

$$\alpha = 120^0, \ \delta = 0 \tag{1}$$

Absonderungsflächen:

$$\beta_1 = 60^0, \ \gamma_1 = 30^0, \ \beta_2 = 160^0, \ \gamma_2 = 110^0, \ \varrho_1 = 35^0, \ \varrho_2 = 30^0 \tag{2}$$

Belastung:

$$\alpha' = 0 \to \text{ keine Erdbebenkräfte}$$

Geometrie:

$$\vec{u}_1 = \quad 0{,}50, \quad 0{,}87, \quad 0$$

$$\vec{u}_2 = -0{,}94, \quad 0{,}34, \quad 0 \tag{3}$$

$$\vec{v}_1 = \quad 0{,}75, -0{,}43, -0{,}50$$

$$\vec{v}_2 = -0{,}12, -0{,}32, -0{,}94 \tag{4}$$

$$\vec{w}_1 = -0{,}43, +0{,}25, -0{,}86$$

$$\vec{w}_2 = -0{,}32, -0{,}88, +0{,}34 \tag{5}$$

$$\vec{x} = \quad 0{,}67, -0{,}42, -0{,}46, \ |\vec{x}| = 0{,}92 \tag{6}$$

$$H^+ = 1 \tag{7}$$

$$H^+ = 0, \text{ weil } \delta = 0 \tag{9}$$

$$\overrightarrow{OD} = -2{,}33, -0{,}58, 1 \tag{10}$$

$$\overrightarrow{OC} = +2{,}64, -0{,}58, 1 \tag{11}$$

$$\overrightarrow{OA} = +0{,}16, -0{,}58, 1 \tag{12}$$

$$\overrightarrow{OB} = -1{,}46, \quad 0{,}92, 1 \tag{13}$$

$$\overrightarrow{OS} = -0{,}29, -0{,}06, 0{,}75 \tag{14}$$

Belastung:

$$\vec{G} = 0, 0, -g \tag{15}$$

$$\alpha' = 0 \quad \to \quad \vec{H}_x = 0, \ \vec{H}_v = 0 \tag{16 + 17}$$

Gleitbedingungen:

$$0 < \varepsilon_x \leqq \pi, \ \text{tg}\,\varepsilon_x = 1{,}09, \ \varepsilon_x \approx 48^0 \tag{19}$$

$$0 < \alpha < \pi, \ \varepsilon_x < \alpha = 120^0 \tag{18}$$

Gleiten ist möglich. (Sonderfall nach Abb. 7 liegt nicht vor; $u_{1.2} \neq 0$.)

Gleitrichtung:

$$\vec{y}_1 = -0{,}01, +0{,}01, -0{,}03, \ |\vec{y}_1| = 0{,}03$$

$$\vec{y}_2 = -0{,}25, -0{,}69, +0{,}27, \ |\vec{y}_2| = 0{,}79 \tag{20}$$

$$\left.\begin{array}{l} \dfrac{\vec{y}_1}{|\vec{y}_1|} \approx \vec{w}_1 \\[2ex] \dfrac{\vec{y}_2}{|\vec{y}_2|} \approx \vec{w}_2 \end{array}\right\} \quad \text{Gleitrichtung} \parallel \vec{x} \tag{21}$$

Gleitsicherheitsnachweise:

$$|\vec{T}| = +0{,}50\,g \tag{26}$$

$$\vec{T} = +0{,}37\,g, -0{,}23\,g, -0{,}25\,g \tag{27}$$

$$\vec{N} = -0{,}37\,g, +0{,}23\,g, -0{,}75\,g \tag{28}$$

$$c_1 = \quad 0{,}87\,g, \quad c_2 = -0{,}04\,g \tag{29}$$

$$W = \quad 0{,}59\,g \tag{31}$$

$$\nu_G = \quad 1{,}18 \tag{25}$$

Rotationsbedingungen:

$$K_1 \approx \quad 42^0 < \frac{\pi}{2}, \quad \mathrm{tg}\,K_1 \approx 0{,}90 \tag{34 a}$$

$$K_2 \approx 126^0 > \frac{\pi}{2} \tag{34 b}$$

$$\eta \approx 112^0 > \frac{\pi}{2} \tag{35}$$

Nach Abb. 12 ist:

$$\mathrm{tg}\,K_1 > \mathrm{tg}\,(\pi - K_2)\,.\,\cos\,(\pi - \eta) = 0{,}53 \rightarrow \vec{d_4}$$

$$\mathrm{tg}\,K_1 < \frac{\mathrm{tg}\,(\pi - K_2)}{\cos\,(\pi - \eta)} = 3{,}64 \rightarrow \vec{d_2} \text{ und } \vec{d_3}$$

sind als Drehachsen möglich.

Einschränkung der kinematisch möglichen Drehachsen:

$$\vec{x}\,.\,(\overrightarrow{OS} \times \vec{G}) \approx +0{,}16\,g \rightarrow \vec{d_3} \tag{37}$$

$$\nu_{k3} = 0{,}75 \tag{39 a}$$

Felskeil kippt über die Kante $\overrightarrow{OD}$.

Beispiel 6: „Felskeil fällt aus der Böschung heraus."

Aufgabenstellung:

Böschung:

$$\alpha = 130^0, \quad \delta = 0 \tag{1}$$

Absonderungsflächen:

$$\beta_1 = 20^0, \quad \gamma_1 = 120^0, \quad \beta_2 = 150^0, \quad \gamma_2 = 65^0, \quad \varrho_1 = \varrho_2 = 20^0 \tag{2}$$

Belastung:

$$\alpha' = 0 \rightarrow \text{keine Belastung aus Erdbeben}$$

Geometrie:

$$\vec{u_1} = \quad 0{,}94, \quad 0{,}34, \quad 0$$
$$\vec{u_2} = -0{,}87, \quad 0{,}50, \quad 0 \tag{3}$$

$$\vec{v_1} = -0{,}17, +0{,}47, -0{,}87$$
$$\vec{v_2} = +0{,}21, +0{,}37, -0{,}90 \tag{4}$$

$$\vec{w_1} = -0{,}29, +0{,}82, +0{,}50$$
$$\vec{w_2} = -0{,}45, -0{,}78, -0{,}41 \tag{5}$$

$$\vec{x} = -0,05, +0,35, -0,60, \ |\vec{x}| = 0,69 \tag{6}$$

$$H = 1 \tag{7}$$

$$H^+ = 0, \ \text{weil} \ \delta = 0 \tag{9}$$

Belastung:

$$\vec{G} = 0, 0, -g \tag{15}$$

$$\alpha' = 0 \to \text{keine Erdbebenkräfte}$$

$$\vec{H}_x = \vec{H}_v = 0 \tag{16}$$

Gleitbedingungen:

$$0 < \varepsilon_x \leqq \pi, \ \operatorname{tg} \varepsilon_x = -1,70, \ \varepsilon_x \approx 121^0 \tag{19}$$

$$0 < \alpha < \pi, \ \varepsilon_x < \alpha = 130^0 \tag{18}$$

Gleiten nicht möglich. (Sonderfall nach Abb. 7 liegt nicht vor; $u_{1,2y} \neq 0$.)

Gleitrichtung:

$$\vec{y}_1 = 0,02, -0,06, -0,03, \ |\vec{y}_1| = 0,07$$
$$\vec{y}_2 = 0,10, \quad 0,17, \quad 0,09, \ |\vec{y}_2| = 0,23 \tag{20}$$

$$\left.\begin{array}{l} \dfrac{\vec{y}_1}{|\vec{y}_1|} = -\vec{w}_1 \\[2mm] \dfrac{\vec{y}_2}{|\vec{y}_2|} = -\vec{w}_2 \end{array}\right\} \quad \text{Gleitrichtung entweder} \ \| \ \vec{v}_1 \ \text{oder} \ \| \ \vec{v}_2 \tag{21}$$

Gleitsicherheitsnachweise $\| \vec{v}_1$:

$$|\vec{T}| = \quad 0,87\,g \tag{22}$$

$$N = -0,5 \ \ g \tag{23}$$

$$W = -0,18\,g \tag{24}$$

$$\nu_G = -0,21 \tag{25}$$

Körper fällt aus der Böschung heraus.

Literatur

[1] United States Department of the Interior, Bureau of Reclamation, Thesaurus of Descriptors, Keywords and Cross-References for Indexing and Retrieving the Literatur of Water Resources Development. Denver USA, Eigenverlag, Tentative Edition.

[2] Müller, L.: Der Felsbau. Band 1. Ferdinand-Enke-Verlag, Stuttgart 1963.

[3] Müller, L. und F. Pacher: Die Sicherung des Talsperrenaushubs mit talwärts fallenden Gleitschichten. Geologie und Bauwesen, Jg. 23, H. 2, 1957.

[4] Therzaghi, K.: Stability of steep slopes on hard unweathered rock. Norges Geotekniske Institut, Publikaasjon Nr. 50, Oslo 1963.

[5] Quervain, F.: Der Fels als Gesteins-Großbereich. Schweizerische Bauzeitung, Jg. 81, H. 6, 1963.

[6] Krsmanović, D. und L. Langof: Large Scale Laboratory Tests of the Shear Strength of Rocky Material. XIV. Salzburger Felsmechanik-Kolloquium, 1963 (s. S. 20 dieses Supplementbandes).

[7] Hay, C. E.: Vector and tensor analysis. Dover Publications, Inc. New York 1953.

[8] DIN 4149: Bauten in deutschen Erdbebengebieten. Richtlinien für Bemessung und Ausführung. Beuth-Vertrieb G. m. b. H., Berlin W 15 und Köln 1957.

[9] Wittke, W.: Berechnung der Standsicherheit von Felsböschungen. XV. Salzburger Felsmechanik-Kolloquium, 1964, Felsmechanik und Ingenieurgeologie, Suppl. II, 1965.

Einige felsmechanische Meßergebnisse aus dem Druckschacht des Kaunertalkraftwerkes

Von

G. Seeber[*]

Mit 25 Textabbildungen

Zusammenfassung — Summary — Résumé

Einige felsmechanische Meßergebnisse aus dem Druckschacht des Kaunertalkraftwerkes.
Während der Vorarbeiten für den Bau des Kaunertalkraftwerkes, die auch den Ausbruch
einiger Sondierstollen umfaßten, sowie auch während der Vortriebe wurden mit der Radialpresse der TIWAG zahlreiche Messungen durchgeführt, die ein gutes Bild über die mechanischen Eigenschaften des Gebirges gaben. Im vorliegenden Bericht werden einige Meßergebnisse bekanntgegeben.

**Some Rock Mechanics Results of Measurements in the Pressure Chamber of the
Kaunertal Power Plant.** During preparatory work for the construction of the Kaunertal
power plant, together with the excavation of some sondage tunnels, and during the head
works numerous measurements were carried out with the radial jack of the TIWAG. These
measurements provided a fair idea of the mechanical properties of the rock. This report
contains results of some measurements.

**Quelques mesures de mécanique des roches pour le puits en charge de l'aménagement
du Kaunertal.** Pendant les travaux préparatoires de l'aménagement hydro-électrique du
Kaunertal, comprenant la perforation de galeries de sondage, et pendant les travaux
définitifs, de nombreuses mesures ont été effectuées avec le vérin radial de la TIWAG.
Ce rapport présente quelques uns de ces résultats, qui ont permis d'apprécier convenable-
ment les qualités mécaniques du massif rocheux.

I. Übersicht

Der Druckschacht des Kaunertalkraftwerkes, dessen Innendruck bis rd. 100 atü
ansteigt, liegt in einem Felsrücken, der aus Gesteinen der Bündner Schiefer auf-
gebaut ist. Der untere Bereich umfaßt Kalkschiefer, die in Mitte des Steilschachtes
teilweise stark durchbewegt und weich, in der Flachstrecke aber relativ fest sind.

Im mittleren Bereich, um das Zwischenfenster Kample, treffen wir auf verschiedene Serien der bunten Bündner Schiefer: Dolomit, Serizitschiefer, teilweise
mit gelben, sandigen Rauhwacken, sowie Gips. Dolomit und Gips sind standfest,
die Serizitschiefer werden, sobald sie mit Wasser in Berührung kommen, breiig
weich.

200 m oberhalb des Zwischenfensters schließt ein mehr oder weniger homo-
gener Bereich von Kalkphylliten an, die mäßig fest und in größeren Abständen von
starken Störungen durchzogen sind. Diese Kalkphyllite reichen noch etwa 3,5 km
in den Druckstollen.

---

[*] Dipl.-Ing. Dr. techn. Gerhard S e e b e r, Tiroler Wasserkraftwerke A. G., TIWAG,
Innsbruck, Landhausplatz 2.

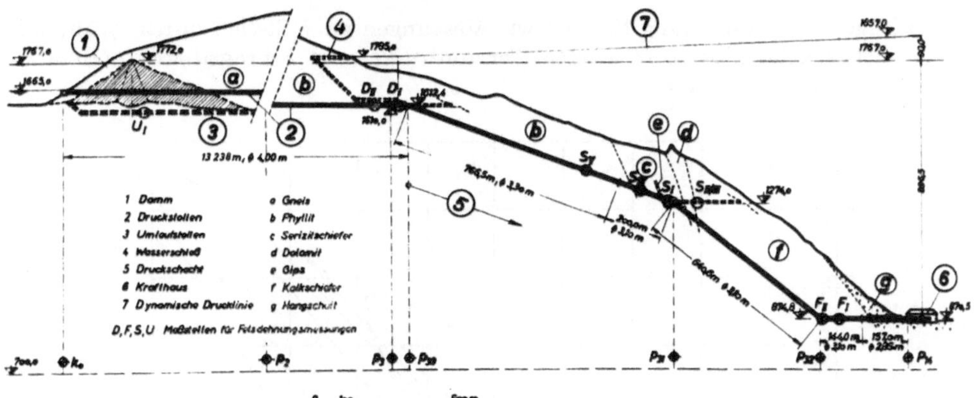

Abb. 1. Kaunertalkraftwerk, Höhenplan Druckschacht
Kaunertal power plant, longitudinal profile of the pressure shaft
Centrale du Kaunertal, profil en long du puits en charge

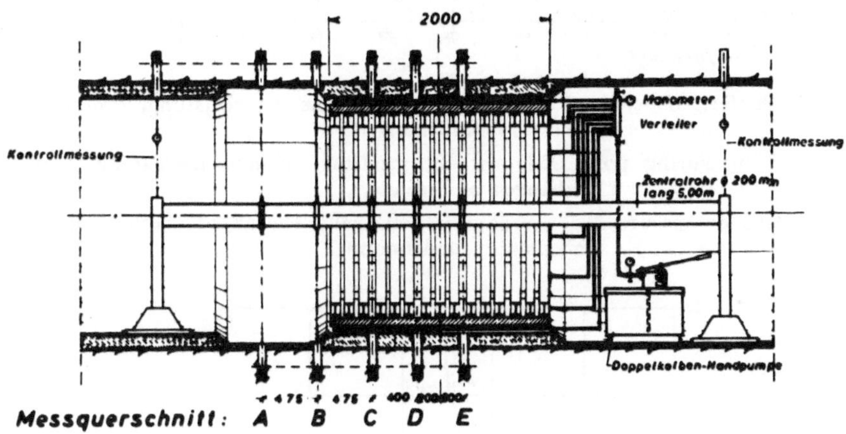

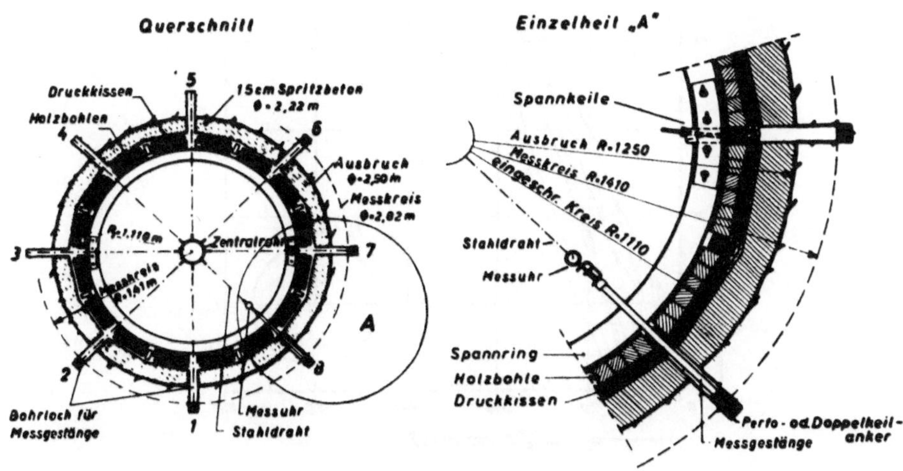

Abb. 2. Radialpresse der TIWAG
TIWAG radial jack
Vérin cylindrique de la TIWAG

Im Zuge der Erschließung führten wir Messungen an 7 Stellen durch. Während des Vortriebes gelang es trotz großen Termindruckes, im Schrägschacht noch drei weitere Messungen unterzubringen.

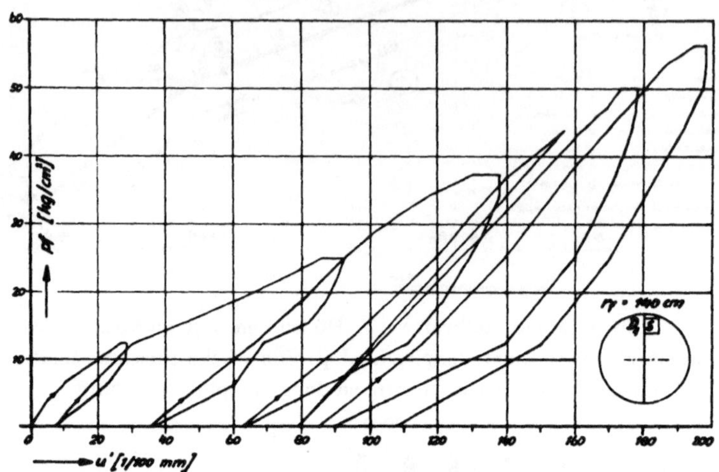

Abb. 3 a. Kalkschiefer, Meßstelle *F* I, nicht injiziert, ohne Vorbelastung, größte Durchmesserverschiebungen

Calcareous schist, measuring point *F* I, without grouting, without preloading, maximum changes of diameter

Calcschiste — point de mesure *F* I, avant injection, sans charge préalable, déformation maximum du diamètre

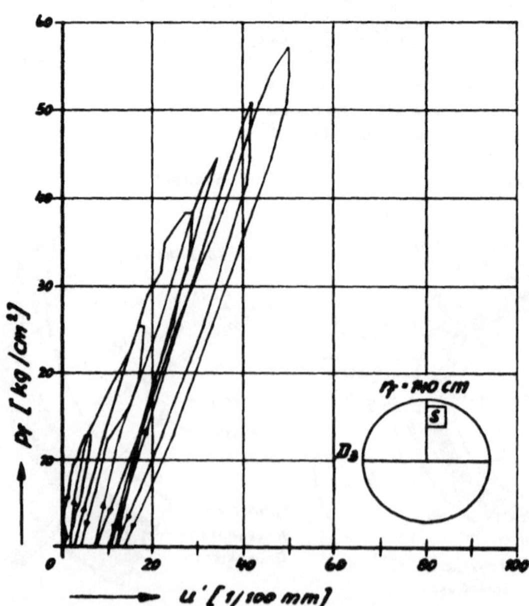

Abb. 3 b. Desgleichen, kleinste Durchmesserverschiebungen

As per 3 a, minimum changes of diameter

Idem — déformation minimum du diamètre

Das Meßgerät und die Durchführung der Messungen sollen hier nicht behandelt werden[1]. Zum besseren Verständnis der Meßergebnisse zeige ich lediglich ein Bild unserer Radialpresse. Die Felsbelastung wird über Druckkissen aufgebracht (max. Kissendruck rd. 100 atü). Gemessen wird die Verschiebung der Ankerköpfe von Felsankern, die zwischen den Druckkissen in den Fels reichen, gegen ein außerhalb des Einflußbereiches der Belastung verankertes Zentralrohr.

II. Meßergebnisse

A. Kalkschiefer der grauen Bündner Schiefer

Die Meßstrecke $F\,\text{I}$ in der Flachstrecke wurde nur als Einfach-Meßstrecke eingerichtet. Nach der ersten Messung wurde mit einem Druck von 20 kg/cm² injiziert

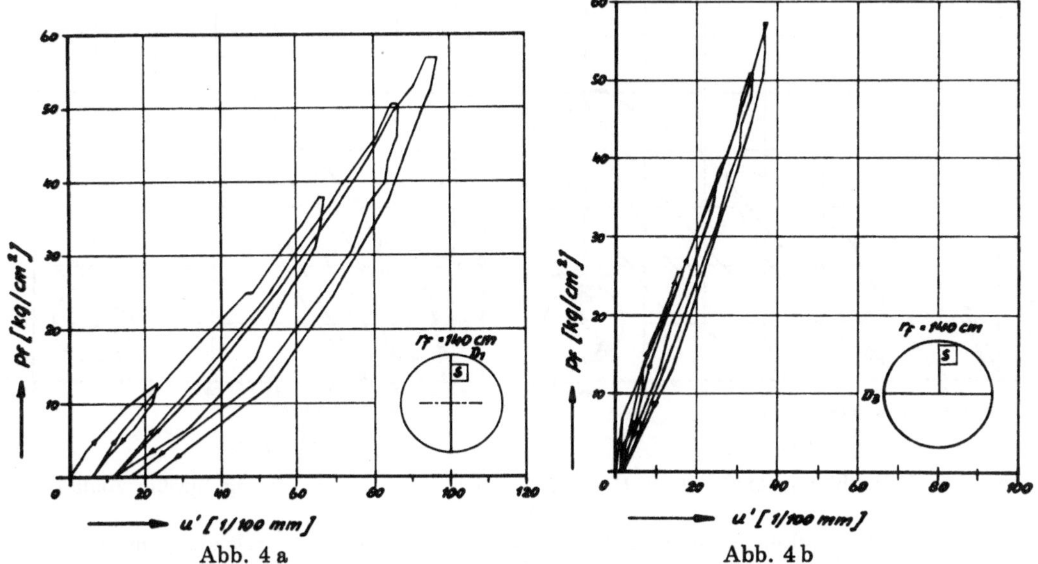

Abb. 4 a   Abb. 4 b

Abb. 4 a. Kalkschiefer, Meßstelle $F\,\text{I}$, nach Vorbelastung injiziert (20 atü), größte Durchmesserverschiebungen

Calcareous schist, measuring point $F\,\text{I}$, with grouting (20 at) after preloading, maximum changes of diameter

Calcschiste — point de mesure $F\,\text{I}$ — avec charge préalable après injection (20 bars) — déformation maximum du diamètre

Abb. 4 b. Desgleichen, kleinste Durchmesserverschiebungen

As per 4 a, minimum changes of diameter

Idem — déformation minimum du diamètre

und darauf nochmals gemessen. Die Schichtung streicht hier quer zur Meßstrecke und fällt fast senkrecht ein.

---

[1] L a u f f e r, H.: Ein Gerät zur Ermittlung der Felsnachgiebigkeit für die Bemessung von Druckstollen- und Druckschachtauskleidungen. Geologie und Bauwesen, Heft 2-3, 1960.

S e e b e r, G.: Auswertung von statischen Felsdehnungsmessungen. Geologie und Bauwesen, Heft 3, 1961.

L a u f f e r, H. und G. S e e b e r: Die Bemessung von Druckstollen- und Druckschachtauskleidungen für Innendruck auf Grund von Felsdehnungsmessungen. Österr. Ingenieur-Zeitschrift, Heft 2, 1962.

Abb. 3 a und 4 a zeigt die Verschiebung des Durchmessers D_3 mit der größten Verschiebung, Abb. 3 b und 4 b die Verschiebung des Durchmessers D_1 mit der kleinsten Verschiebung; Abb. 3 a und b vor, Abb. 4 a und b nach der Injektion und nach der Vorbelastung. Die Verbesserung der Felsqualität, vor allem die Verminderung der bleibenden Verformung, ist deutlich zu sehen. Es läßt sich aber nicht unterscheiden, welchen Anteil an der Verbesserung die Vorbelastung und welchen die Injektionen herbeigeführt haben.

Die folgenden Meßstrecken wurden deshalb als Doppelmeßstrecken eingerichtet; d. h. nach dem Abpressen der 1. Meßstrecke wurde die unbeeinflußte benachbarte 2. Meßstrecke injiziert und darauf abgepreßt. Dadurch ist aus der Messung in der Meßstrecke 2 der Einfluß der Injektion auf einen nicht mehr vorbelasteten

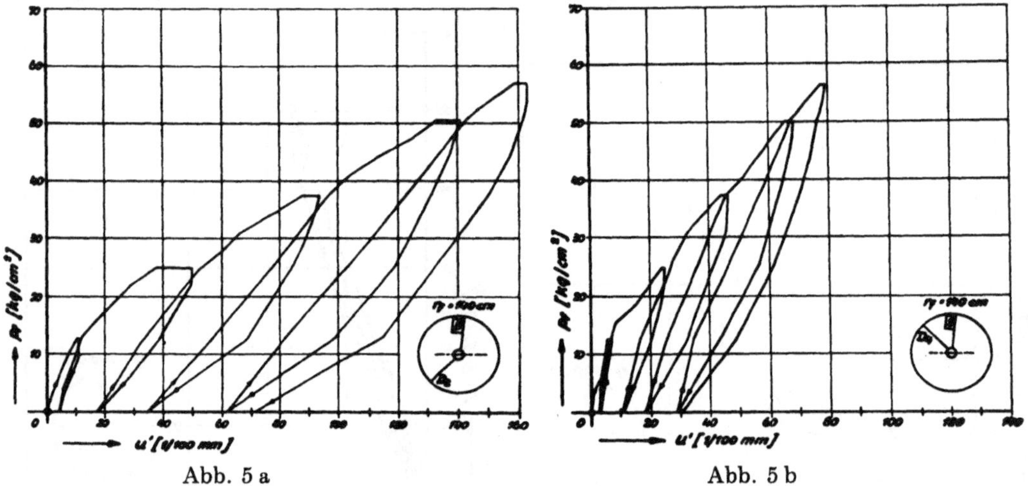

Abb. 5 a
Abb. 5 b

Abb. 5 a. Kalkschiefer, Meßstelle *F* II, nicht injiziert, ohne Vorbelastung, größte Radialverschiebungen

Calcareous schist, measuring point *F* II, without grouting, without preloading, maximum changes of radius

Calcschiste — point de mesure *F* II — avant injection sans charge préalable, déformation maximum du rayon

Abb. 5 b. Desgleichen, kleinste Radialverschiebungen

As per 5 a, minimum changes of radius

Idem — déformation minimum du rayon

Fels ersichtlich; der Einfluß der Vorbelastung läßt sich ohnehin leicht aus den Aufstiegslinien im vorbelasteten Bereich der Messung 1 erkennen. In Meßstrecke *F* II streicht die Schichtung unter etwa 45° über die Meßstrecke.

Die Abb. 5—6 zeigen wieder die Verformungslinien aus der Doppelmeßstrecke *F* II, für die größten (*a*) und kleinsten (*b*) Verschiebungen, jeweils für den rohen, nicht vorbelasteten Fels (5) und den ohne Vorbelastung mit 30 atü injizierten Fels (6). Die Verbesserung der Felsqualität durch die Injektion allein kommt sehr deutlich zum Ausdruck (Abb. 5 a, 6 a : $u_{\mathrm{max}} = 1{,}63$ gegen 0,93 mm). Bis ungefähr zu $^2/_3$ des Injektionsdruckes verhält sich der Fels viel elastischer, die bleibenden Verformungen sind wesentlich geringer. Aber auch oberhalb des Injektionsdruckes ist noch eine wesentliche Verbesserung feststellbar; so ist oberhalb 30 atü bis zum Enddruck die Verschiebung von 1,00 mm in Abb. 5 a auf 0,55 mm in Abb. 6 a gesunken, was eine Erhöhung des Verformungsmoduls auf fast den doppelten Wert bedeutet. Der steile Verlauf der Verformungslinie des rohen Felsens bis etwa 4 atü in Abb. 5 ist in erster Linie auf die Zugfestigkeit der Betonauskleidung zurückzuführen.

Es muß hier noch erwähnt werden, daß die gezeigten Verformungslinien direkt im Mittelquerschnitt C oder E gemessene darstellen, und deshalb nur etwa 90 %

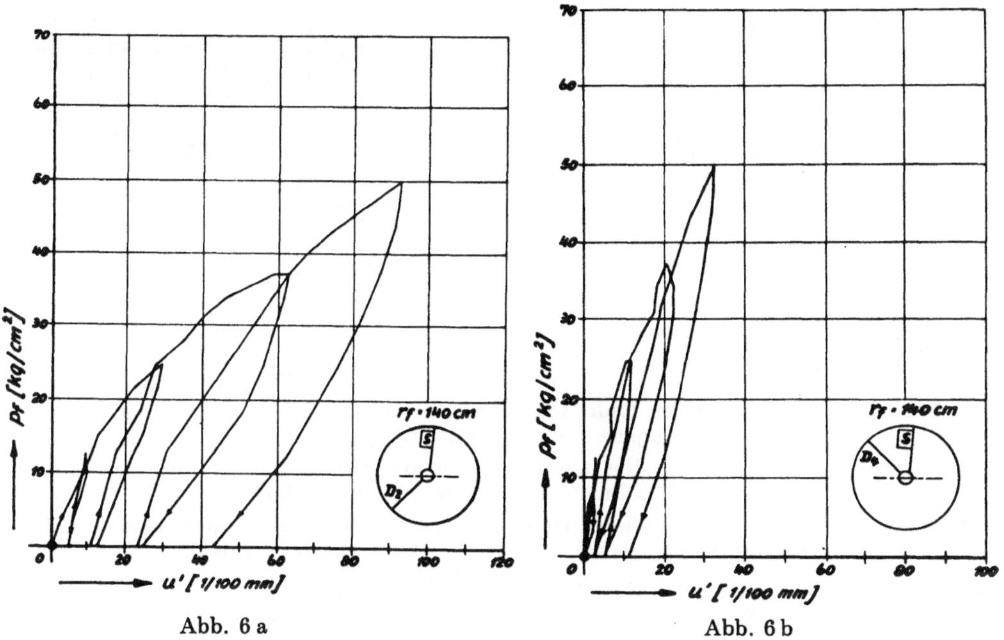

Abb. 6 a
Abb. 6 b

Abb. 6 a. Kalkschiefer, Meßstelle F II, ohne Vorbelastung injiziert (30 atü), größte Radialverschiebungen

Calcareous schist, measuring point F II, with grouting (30 at) without preloading, maximum changes of radius

Calcschiste — point de mesure F II — sans charge préalable après injection 30 bars — déformation maximum du rayon

Abb. 6 b. Desgleichen, kleinste Radialverschiebungen

As per 6 a, minimum changes of radius

Idem — déformation minimum du rayon

der Verschiebungen am unendlich langen Zylinder liefern. Den vollen Wert bringt erst die Überlagerung mit den Verschiebungswerten des A-Querschnittes.

Die Abb. 7 a und b geben ein Bild über das anisotrope Verhalten des Felsens entlang des Umfanges, doch ist dieses in dem kompakten Kalkschiefer nicht sehr stark ausgeprägt. Interessant ist, daß auch durch die Injektion die Anisotropie in den meisten Fällen nicht ausgeglichener, sondern eher ausgeprägter wird. Dies mag daher kommen, daß sehr oft die in der Schichtung liegenden Klüfte von nicht injizierbaren Kluftfüllungen erfüllt sind, und nur Klüfte mehr oder weniger senkrecht zur Schichtung offen und injizierbar sind.

Den Abschluß jeder Messung bildet der Dauerversuch. Wenn auch der Fels sehr fest und elastisch aussieht und bei jeder Druckstufe etwa 20—30 Minuten bis zur Ablesung gewartet wird, ist doch der Zeiteinfluß noch nicht ganz ausgeschaltet. Abb. 8 zeigt das Ergebnis des Dauerversuches im Kalkschiefer (Meßstecke F II).

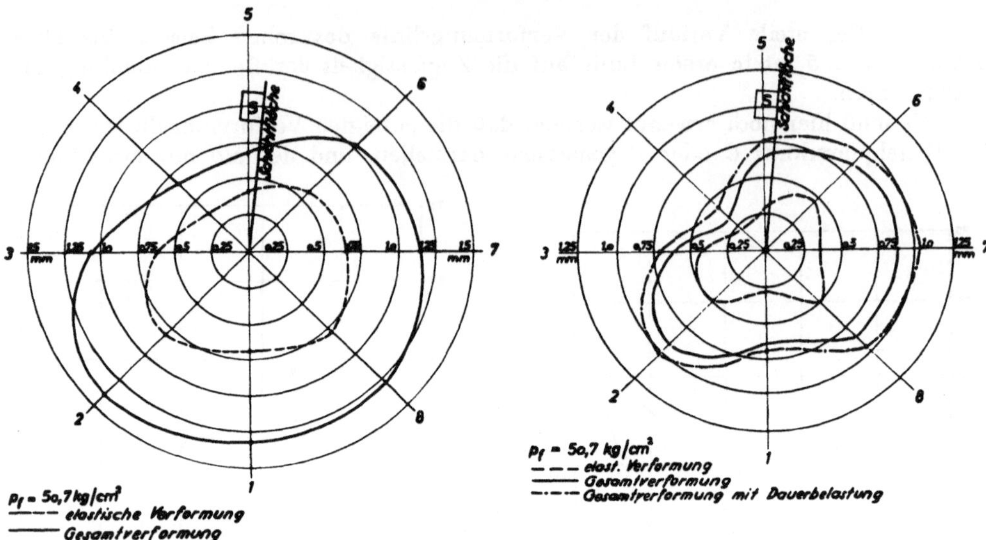

Abb. 7 a
Abb. 7 b

Abb. 7 a. Umfangsverteilung der Radialverschiebungen, Kalkschiefer, Meßstelle *F* II, nicht injiziert, ohne Vorbelastung

Distribution of the changes in radius along the circumference, calcareous schist, measuring point *F* II, without grouting, without preloading

Distribution des déformations radiales le long du périmètre dans le calcschiste au point de mesure *F* II — avant injection — sans charge préalable

Abb. 7 b. Desgleichen, ohne Vorbelastung injiziert

As per 7 a, with grouting, without preloading

Idem sans charge préalable mais avec injection

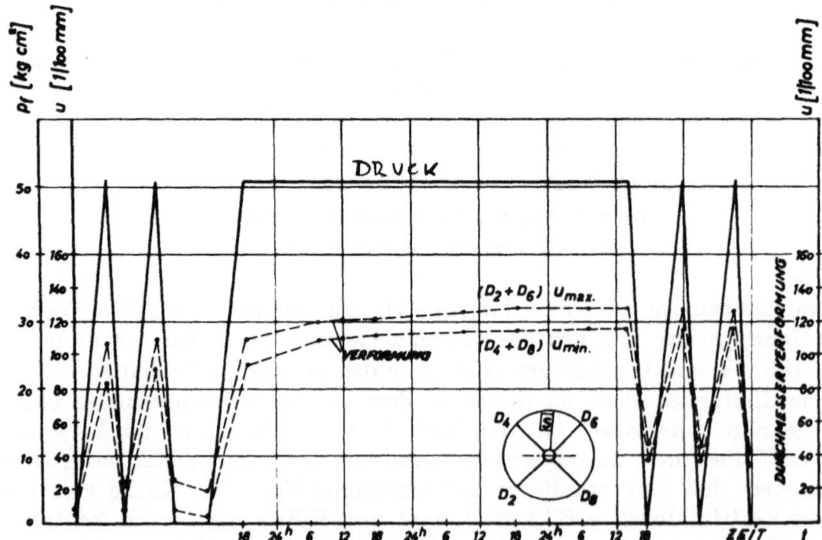

Abb. 8. Dauerbelastung im Kalkschiefer, Meßstelle *F* II, ohne Vorbelastung injiziert

Permanent load in the calcareous schist, measuring pont *F* II, with grouting, without preloading

Charge de longue durée sur calcschiste — point de mesure *F* II — sans charge préalable — après injection

Man sieht, daß nach Erreichen des Druckes die max. Verschiebungen noch um rd. 12 %, die minimalen dagegen um 32 % zunehmen. Der Absolutbetrag der Zunahme ist über den Umfang annähernd konstant (siehe Abb. 7 b). Dadurch wird die Anisotropie in gewissem Maße ausgeglichen.

Die Arbeitslinie für den nicht injizierten Kalkschiefer zeigt Abb. 9 a, die für den injizierten Abb. 9 b. Wenn auch in der Umfangsverteilung die Anisotropie nicht

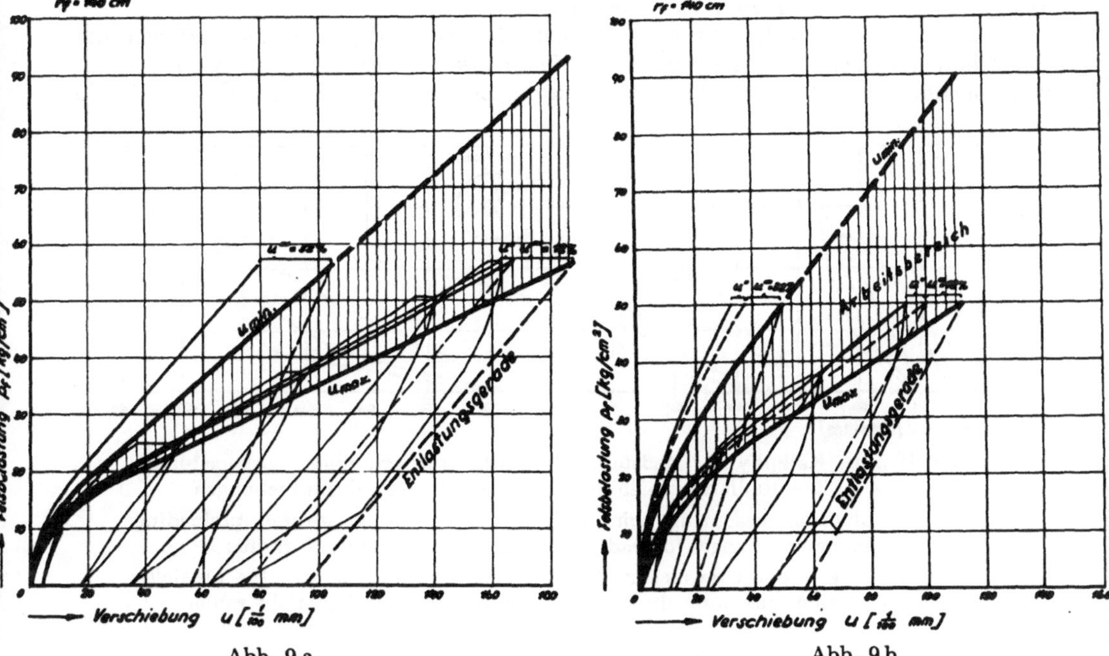

Abb. 9 a   Abb. 9 b

Abb. 9 a. Ermittlung der Arbeitslinien aus Meßergebnissen im Kalkschiefer, Meßstelle *F* II, nicht injiziert

Determination of the deformation lines from the results of measurement in the calcareous schist, measuring point *F* II, without grouting

Détermination des lignes de déformation à partir des résultats de mesure dans le calcschiste — point de mesure *F* II — avant injection

Abb. 9 b. Desgleichen, ohne Vorbelastung injiziert

As per 9 a, with grouting, without preloading

Idem sans charge préalable — après injection

stark auffiel, zeigen die Arbeitslinien für die minimale und maximale Verschiebung doch eine beträchtliche Abweichung.

B. Kalkphyllite der bunten Bündner Schiefer

In der Kalkphyllitzone wurden 4 Meßstellen im Bereich des Druckstollens (*D* II) bzw. der Druckschacht-Schrägstrecke (*D* I, *S* IV, *S* V) ausgeführt. Das Gestein ist ein dünnbankiger schwarzgrauer Kalkphyllit mit stark ausgeprägter Anisotropie. Die Schichtung liegt sehr flach.

Die Abb. 10—11 zeigen wieder die Verformungslinien der größten und kleinsten Verschiebungen, und zwar im rohen und im unvorbelastet injizierten Zustand. Durch Zementmilchinjektion von etwa 30 atü Druck wird eine spürbare Verbesserung erzielt; Abb. 10b, 11b: die maximale Verschiebung sinkt von 5,50 mm auf rd. 1,80 mm bei 31,8 atü Druck, d. h. der Verformungsmodul steigt etwa auf den dreifachen Wert.

Die große Verschiebung in der Sohle in Abb. 11c ist auf Mängel in der baulichen Ausführung der Auskleidung der Meßstrecke zurückzuführen.

Man erkennt hier schon aus dem Vergleich der Verformungslinien, wie stark die Anisotropie ausgeprägt ist und wie sehr das Gebirge nachkriecht, was in Abb. 12 bzw. 13 noch deutlicher zum Ausdruck kommt.

Ein interessantes Ergebnis zeigte der Dauerversuch; zunächst nahmen die Verformungen nach der Erreichung des Enddruckes rasch um etwa 30 % zu; in den folgenden zehn Tagen verkleinerte sich die Zunahme, bis der annähernd stabile

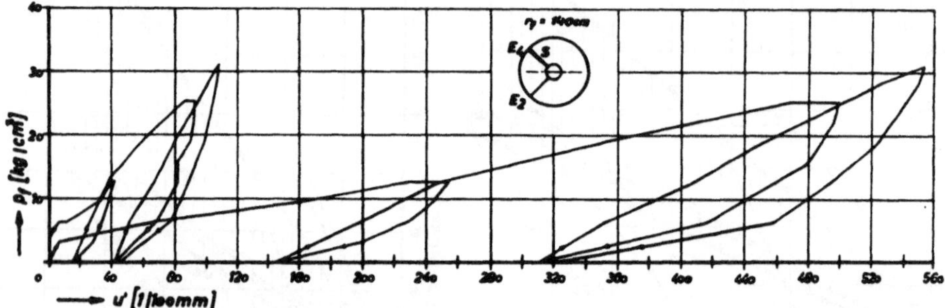

Abb. 10 a, b. Kalkphyllit, Meßstelle *D* I, nicht injiziert, ohne Vorbelastung
a größte Radialverschiebungen; *b* kleinste Radialverschiebungen

Calcareous phyllite, measuring point *D* I, without grouting, without preloading
a maximum changes of radius; *b* minimum changes of radius

Phyllite calcaire — point de mesure *D* I — avant injection, sans charge préalable
a déformation maximum du rayon; *b* déformation minimum du rayon

Endwert von etwa 40 % erreicht war. Nach einer Entlastung und Wiederbelastung wurde aber vorerst nicht mehr der Endwert der Verschiebung gemessen, sondern jener um etwa 30 % erhöhte Wert, der sich nach einigen Stunden einstellte. Der Endwert würde wieder erst nach rd. 10 Tagen erreicht. Hier wird das Verhalten

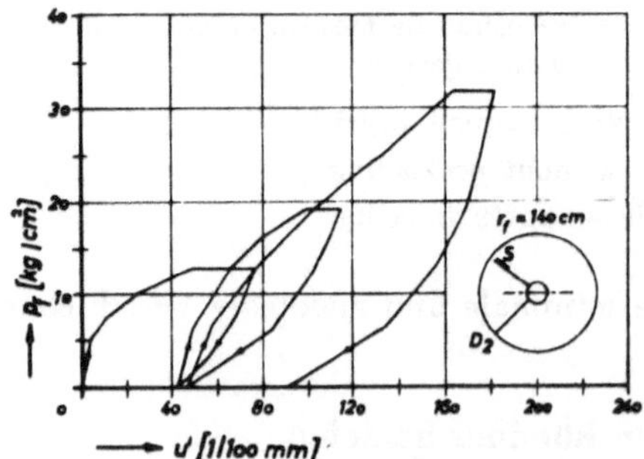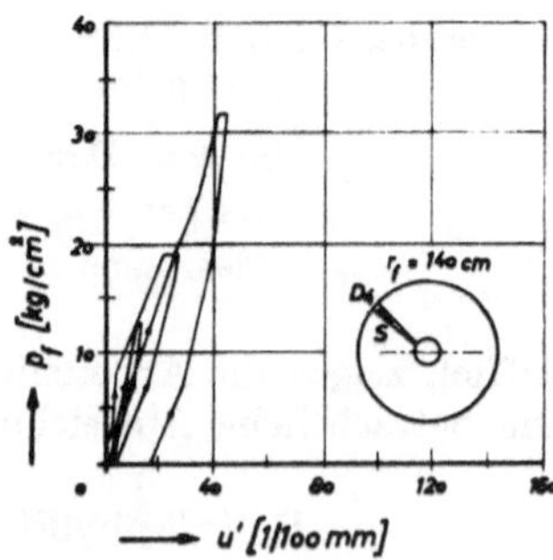

Abb. 11 a, b. Kalkphyllit, Meßstelle *D* I, ohne Vorbelastung injiziert (30 atü)
a Radialverschiebungen entsprechend dem Radius D_2; *b* kleinste Radialverschiebungen

Calcareous phyllite, measuring point *D* I, with grouting (30 at), without preloading
a changes of radius corresponding to radius D_2; *b* minimum changes of radius

Phyllite calcaire — point de mesure *D* I — sans charge préalable — après injection à 30 bars
a déformation radiale pour le rayon D_2; *b* déformation minimum du rayon

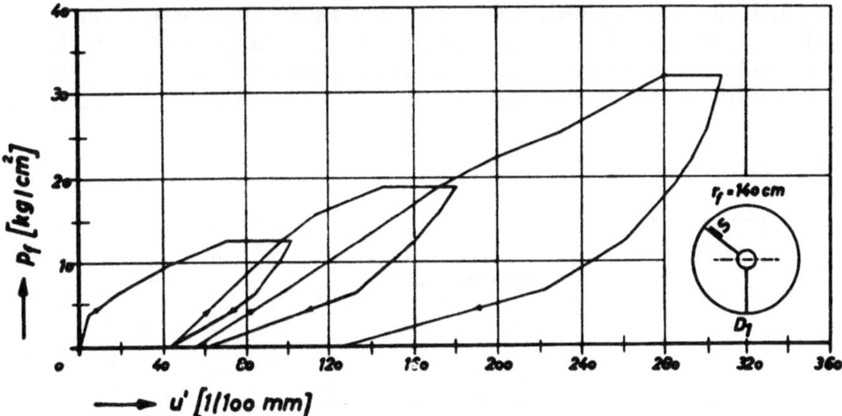

Abb. 11 c. Kalkphyllit, Meßstelle D I, ohne Vorbelastung injiziert (30 atü), absolute größte Radialverschiebungen

Calcareous phyllite, measuring point D I, with grouting (30 at) without preloading, absolute maximum changes of radius

Phyllite calcaire, point de mesure D I — sans charge préalable —, avec injection à 30 bars, déformation maximum absolue du rayon

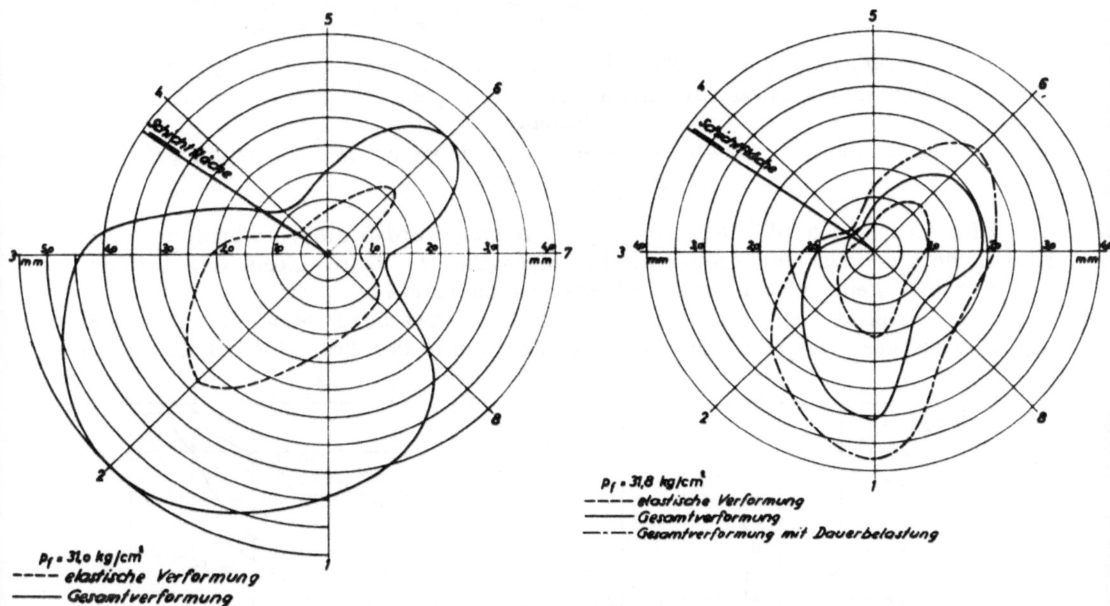

Abb. 12 a   Abb. 12 b

Abb. 12 a. Umfangsverteilung der Radialverschiebungen im Kalkphyllit, Meßstelle D I, nicht injiziert, ohne Vorbelastung

Distribution of the changes of radius along the circumference in the calcareous phyllite, measuring point D I, without grouting, without preloading

Distribution des déformations radiales le long du périmètre dans le phyllite calcaire au point de mesure D I avant injection, sans charge préalable

Abb. 12 b. Desgleichen, injiziert, ohne Vorbelastung

As per 12 a, with grouting, without preloading

Idem après injection, sans charge préalable

der meisten Gebirgsarten, nämlich einer kurzdauernden Belastung einen größeren
Widerstand entgegenzusetzen als einer langdauernden, besonders deutlich.

Die Meßstelle D II wurde so angelegt, daß eine etwa 30 cm starke, mit Mylonit
und Letten gefüllte Kluft (Störungszone) im Abstand von einigen Metern hinter

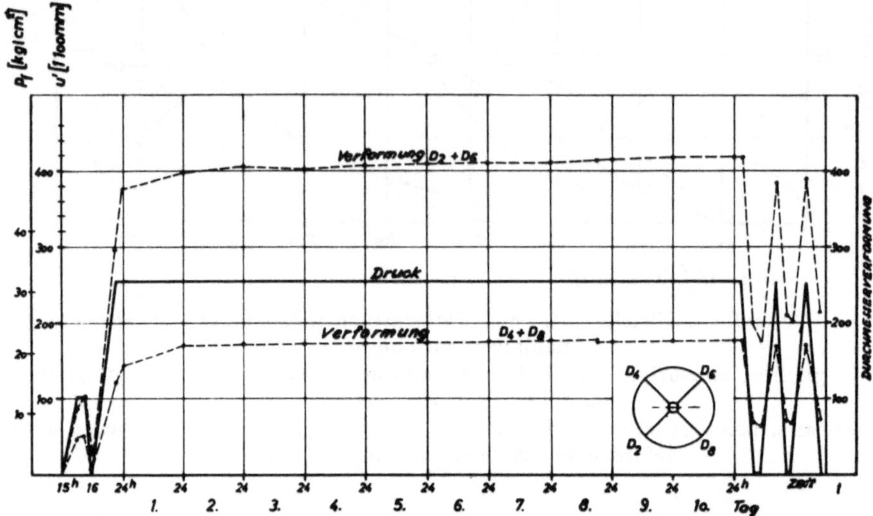

Abb. 13. Dauerbelastung im Kalkphyllit, Meßstelle D I, ohne Vorbelastung injiziert
Permanent load in the calcareous phyllite, measuring point D I, with grouting without
preloading
Charge de longue durée sur phyllite calcaire — point de mesure D I sans charge préalable,
après injection

dem Ulm vorbeistrich. Wie Abb. 14 und 15 zeigen, wurden die Verschiebungen zur
Kluftebene hin bedeutend größer als bei Meßstelle D I. Durch Zementmilch-Injek-
tion war hier keine wesentliche Verbesserung zu erreichen.

Am Aussehen des Felsens war von dieser Schwächung seiner Tragfähigkeit
nichts zu merken. Das bedeutet, daß eine Beurteilung der Tragfähigkeit durch den

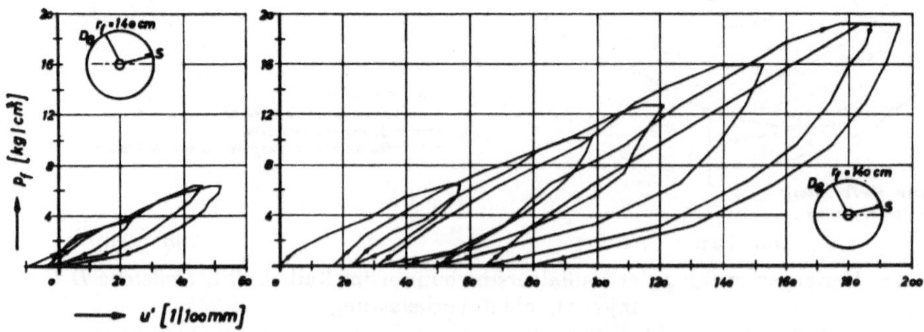

Abb. 14 a, b. Kalkphyllit, Meßstelle D II, größte Radialverschiebungen
a nicht injiziert, ohne Vorbelastung; b nach Vorbelastung injiziert (30 atü)
Calcareous phyllite, measuring point D II, maximum changes of radius
a without grouting, without preloading; b with grouting (30 at) after preloading
Phyllite calcaire — point de mesure D II — déformations maximales du rayon
a avant injection, sans charge préalable; b avec charge préalable après injection à 30 bars

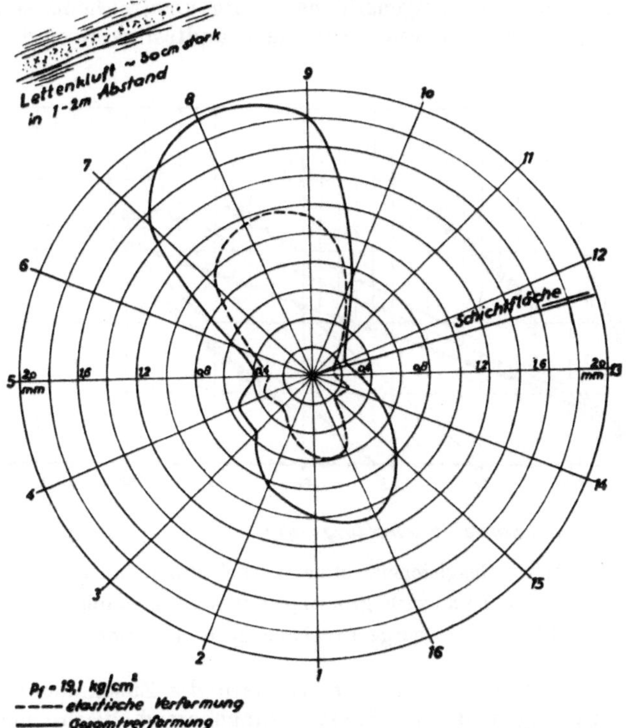

Abb. 15. Umfangsverteilung der Radialverschiebungen im Kalkphyllit, Meßstelle *D* II, nach Vorbelastung injiziert

Distribution of the changes of radius along the circumference in the calcareous phyllite, measuring point *D* II, with grouting after preloading

Distribution des déformations radiales le long du périmètre dans les phyllites calcaires point de mesure *D* II avec charge préalable, après injection

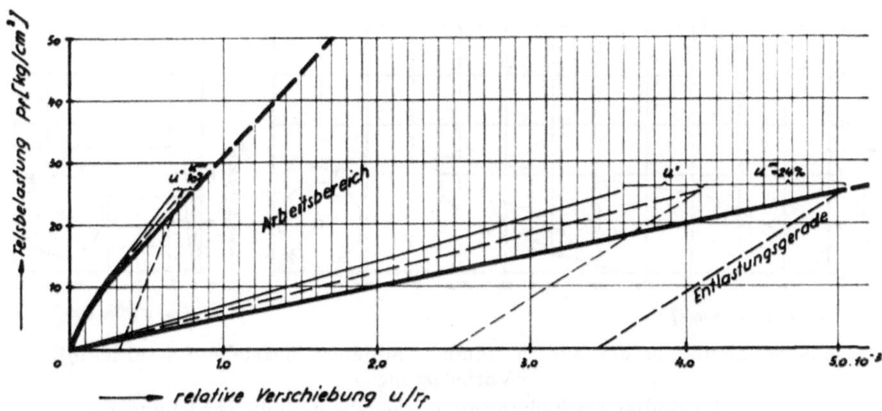

Abb. 16 a. Ermittlung der Arbeitslinien für den Kalkphyllit, Meßstelle *D* I, nicht injiziert, ohne Vorbelastung

Determination of the deformation lines for the calcareous phyllite, measuring point *D* I, without grouting, without preloading

Détermination des lignes de déformation pour le phyllite calcaire, point de mesure *D* I avant injection, sans charge préalable

Geologen, nur auf Grund des Augenscheins, völlig unzureichend sein kann, wenn
für das Tragvermögen des Gebirges nicht die Standfestigkeit und auch nicht die

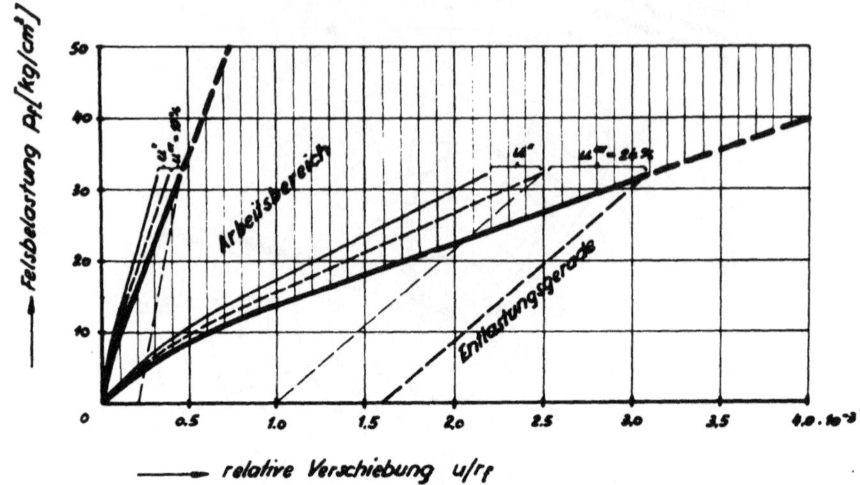

Abb. 16 b. Desgleichen, ohne Vorbelastung injiziert
As per 16 a, with grouting without preloading
Idem sans charge préalable après injection

Widerstandsfähigkeit des Felsens an sich, sondern die Zusammendrückbarkeit von
Klüften und Kluftfüllungen, die im Kraftwirkungsbereich liegen, maßgebend wird.
Die Ermittlung der Arbeitslinie des injizierten Kalkphyllites aus Meßstelle D I
(ohne Klufteinfluß!) zeigt Abb. 16.

C. Serizitschiefer der bunten Bündner Schiefer

Diese Gebirgsformation besteht aus etwa cm-dicken Lagen etwas festerer, teilweise auch kalkiger Serizitschiefer mit ebenfalls i. M. etwa cm-dicken Zwischen-

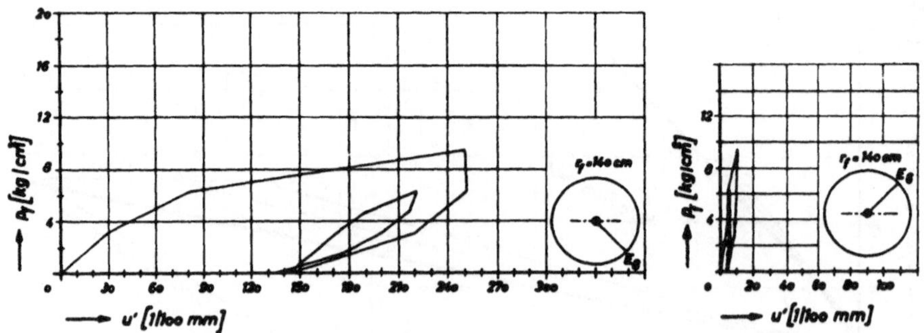

Abb. 17 a, b. Serizitschiefer der bunten Bündner Schiefer, Meßstelle S I, nicht injiziert, mit
Vorbelastung
a größte Radialverschiebungen; *b* kleinste Radialverschiebungen
Sericite-schist of the variegated Bündner schists, measuring point S I, without grouting, with
preloading
a maximum changes of radius; *b* minimum changes of radius
Schistes sériciteux des schistes bariolés de Bunden, point de mesure S I, avant injection,
avec charge préalable
a déformation maximum du rayon; *b* déformation minimum du rayon

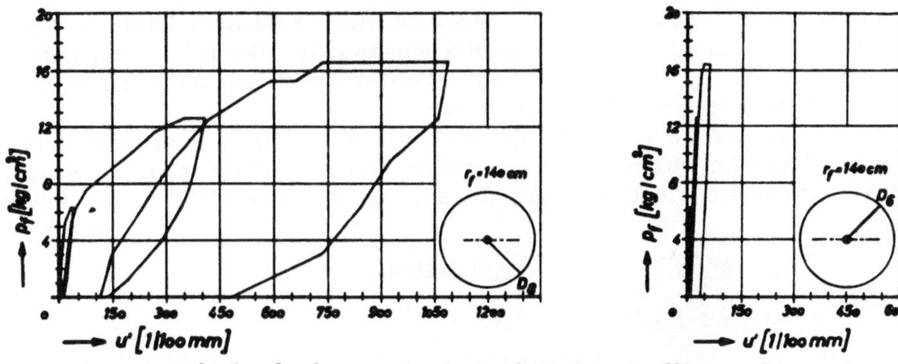

Abb. 18 a, b. Serizitschiefer der bunten Bündner Schiefer, Meßstelle S I, ohne Vorbelastung injiziert (30 atü)

a größte Radialverschiebungen; *b* kleinste Radialverschiebungen

Sericite-schist of the variegated Bündner schists, measuring point S I, with grouting (30 at) without preloading

a maximum changes of radius; *b* minimum changes of radius

Schistes sériciteux des schistes bariolés de Bünden, point de mesure S I, sans charge préalable, apres injection à 30 bars

a déformation maximum du rayon; *b* déformation minimum du rayon

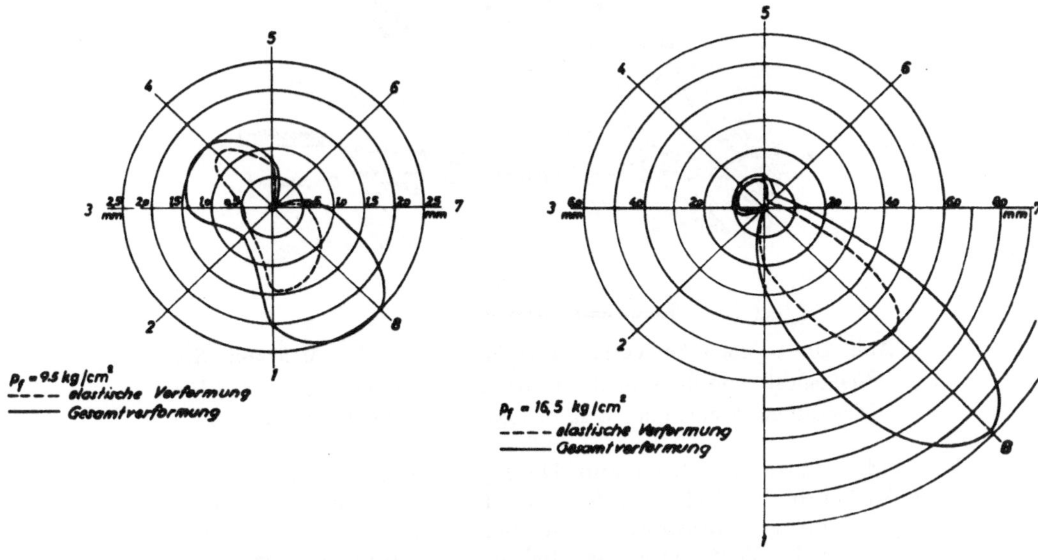

Abb. 19 a   Abb. 19 b

Abb. 19 a. Umfangsverteilung der Radialverschiebungen im Serizitschiefer, Meßstelle S I, nicht injiziert, mit Vorbelastung

Distribution of the changes of radius in the sericite-schist, measuring point S I, without grouting, with preloading

Distribution des déformations radiales le long du périmètre dans le schistes sériciteux au point de mesure S I, avant injection, avec charge préalable

Abb. 19 b. Desgleichen, ohne Vorbelastung injiziert

As per 19 a, with grouting, without preloading

Idem sans charge préalable, après injection

lagen aus weichem, tonigem Material. Verschiedentlich sind auch Bänke von gelb-
lichen, sandigen Rauhwacken und Gipsbänke eingeschaltet. Bei Fehlen von Gebirgswasser ist das Gebirge über einige Zeit (Stunden bis Tage) standfest, doch bei
Wasserzutritt wird es teilweise breiig weich.

    Die Abb. 17 und 18 zeigen die großen Unterschiede der Radialverschiebungen
senkrecht und parallel zur Schichtung. Senkrecht zur Schichtung traten erwartungsgemäß sehr große Deformationen auf, die auf das Zusammendrücken der weichen

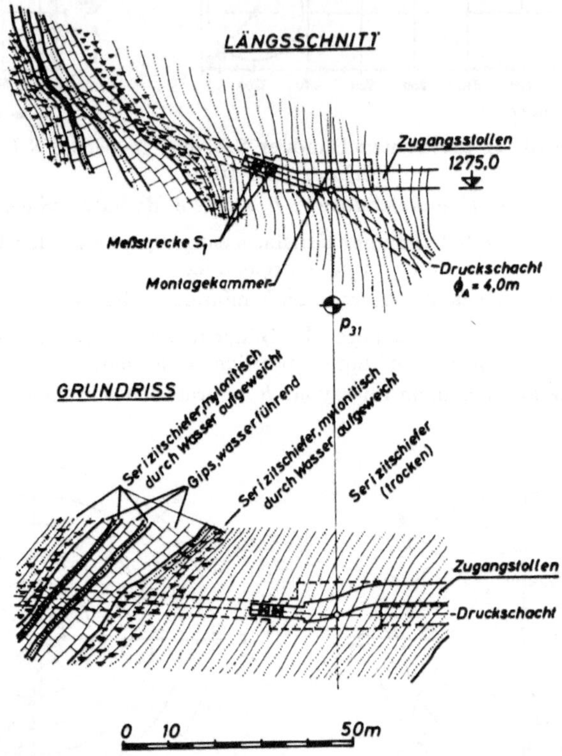

Abb. 20. Geologische Verhältnisse im Bereich der Meßstelle *S* I
Geological conditions in the region of the measuring point *S* I
Conditions géologiques dans la zone du point de mesure *S* I

Zwischenschichten zurückzuführen sind. Die Injektion hatte fast keine Wirkung. Wie
nach dem Ausbruch der Meßstelle festgestellt wurde, drang die Zementmilch nur
in Klüfte senkrecht zur Schichtung, nicht aber in die Zwischenlagen ein. Es wurde
deshalb auch nur die Verformung parallel zur Schichtung verbessert.

    Ein anderes Bild gibt die Umfangsverteilung; während bei einem Druck bis
6 kg/cm² die Anisotropie noch annähernd symmetrisch ist (Abb. 19 a), zeigt sich
bei höherem Druck (größerer Kraftwirkungsbereich) eine ganz ausgeprägte Asym-
metrie (Abb. 19 b). Diese ließ auf eine besonders weiche Zone in Richtung des
Radius 15 schließen. Im Zuge des inzwischen erfolgten Schachtaufbruches (Abb. 20)
wurde eine wasserführende Gipszone angefahren, die selbst sehr fest war, in deren
beiden Randbereichen aber eine, durch das unter Druck stehende Gebirgswasser
völlig aufgeweichte Zone von ausgelaugtem Gips, Serizitschiefer und Rauhwacken
von einigen Metern Mächtigkeit festgestellt wurde. Diese Zone streicht im Abstand
von etwa 20 m an der Meßstelle vorbei. Der Radius 15 zeigt senkrecht darauf.

Der Dauerversuch (Abb. 21) zeigt ein starkes Nachkriechen der Verschiebungen in den ersten $1^{1}/_{2}$ Tagen und darauf noch ein langsames Zunehmen in den folgenden

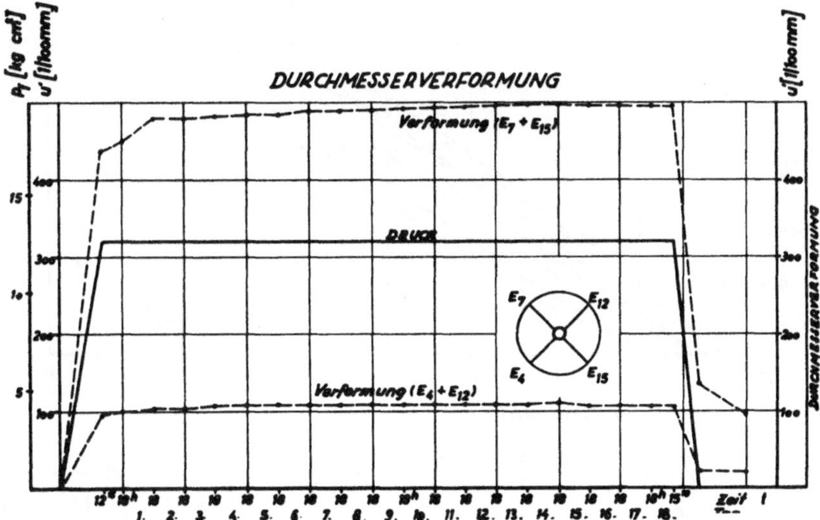

Abb. 21. Dauerbelastung Serizitschiefer der bunten Bündner Schiefer, Meßstelle S I, ohne Vorbelastung injiziert

Permanent load, measuring point S I, with grouting, without preloading in the sericite schists of the variegated Bündner schists

Charge de longue durée dans les schistes sériciteux des schistes bariolés de Bünden au point de mesure S I sans charge préalable après injection

14 Tagen. Danach stellt sich offensichtlich ein Gleichgewichtszustand ein, der wahrscheinlich durch den Überlagerungsdruck bedingt ist.

Da in diesem Abschnitt dem Fels keine Mitwirkung an der Aufnahme des Innendruckes zugemutet werden kann, erübrigt sich die Aufstellung einer Arbeitslinie.

D. Augengneis

Im vorweg ausgebrochenen Umlaufstollen des Dammes wurden dieselben Gesteinsformationen durchfahren, wie sie später im größten Teil des Druckstollens angetroffen werden. Es sind dies Augen- und Schiefergneise. Im Augengneis wurde eine Doppel-Meßstrecke eingerichtet. Das Gestein ist sehr fest und wenig geklüftet.

Die Verschiebungen (Abb. 22 und 23) waren erwartungsgemäß klein. Bei der Injektion mit 30 atü nahm der Fels nur wenig auf und zeigte nachher auch keine wesentliche Verbesserung seiner Eigenschaften.

Die Umfangsverteilungen (Abb. 24) zeigen auch keine ausgesprochene Anisotropie, sondern es ist so, daß örtliche Zerrüttungsstellen, die wohl von der Sprengung herrühren dürften, für die Größe der Verschiebung maßgebend sind. So zeigt in Abb. 24 b der Radius 5 (Scheitel), in Abb. 24 a der Radius 2 (Sohle) die größte Verschiebung, obwohl die beiden Profile nur 3,0 m entfernt sind.

Interessant war die Feststellung, daß trotz des hohen Innendruckes an der Innenleibung von 65 kg/cm² der verhältnismäßig schwache Auskleidungs-Spritzbeton (15—20 cm) keine sichtbaren Risse zeigte. Die größte Dehnung errechnet sich dabei (Abb. 23 a, E 5) zu $\varepsilon = \dfrac{u}{r_f} = \dfrac{0.36}{1400} = 0,26 \ ^{0}/_{00}\,,$

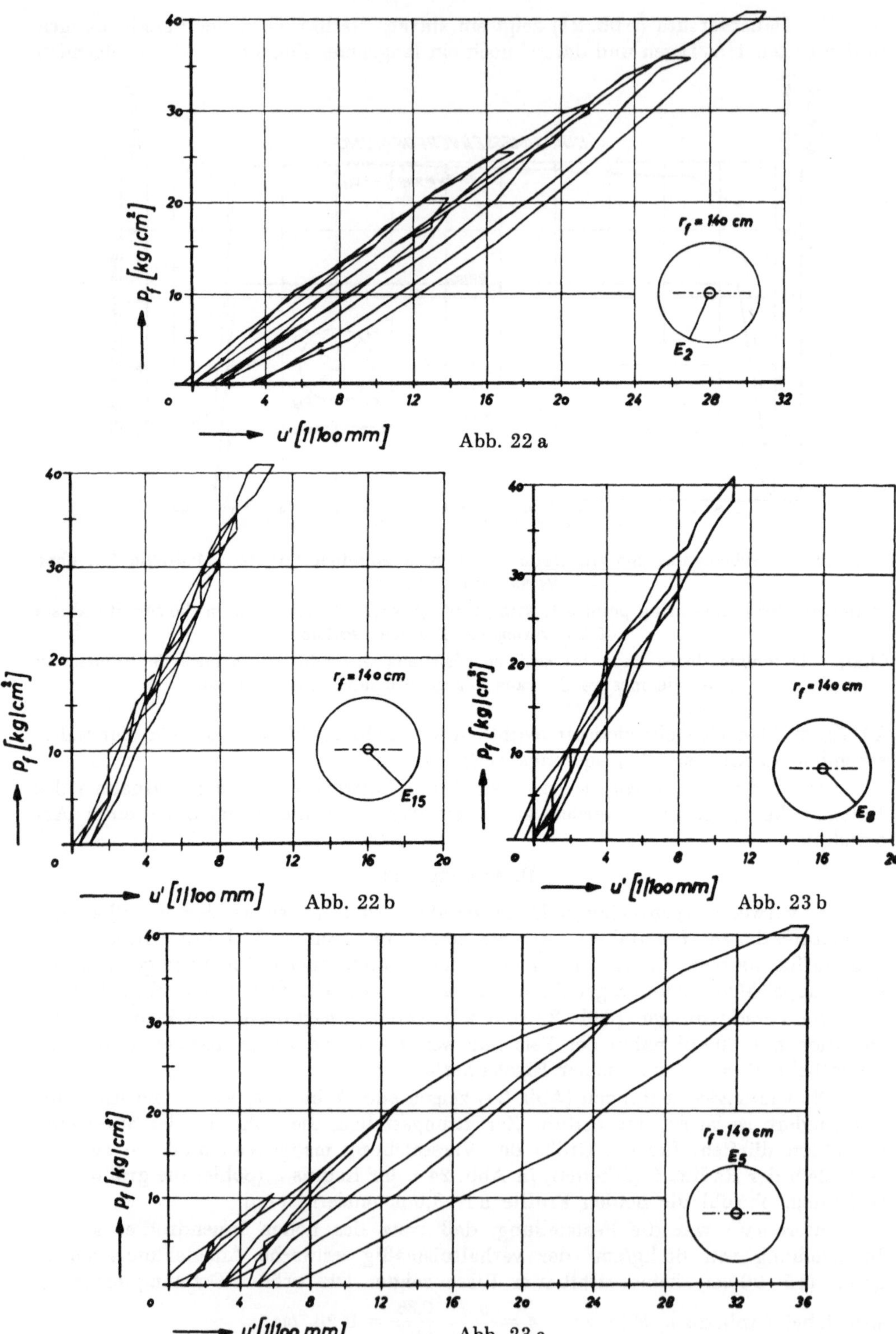

Abb. 22 a

Abb. 22 b

Abb. 23 b

Abb. 23 a

doch ist sie am übrigen Umfang nur etwa halb so groß. Die Verformungen waren größtenteils fast rein elastisch. Im Dauerversuch zeigte der Fels keine Neigung zum Kriechen.

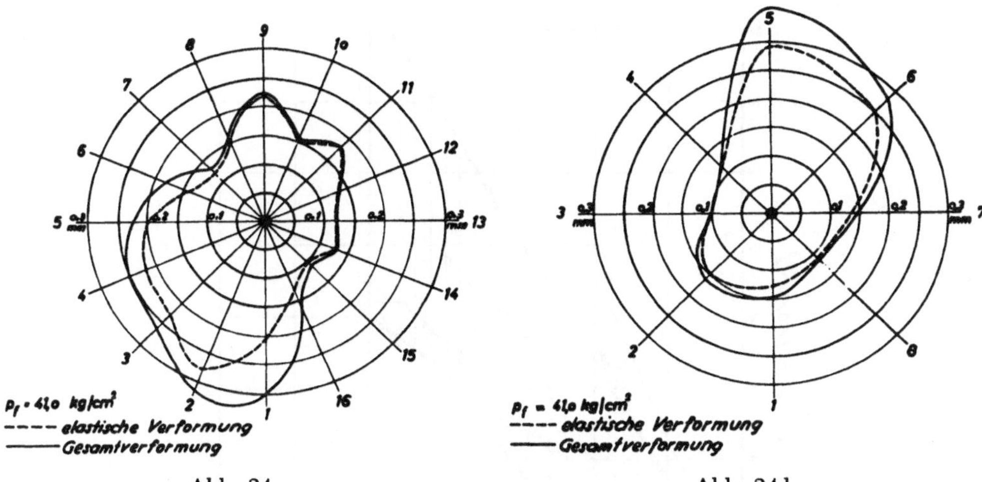

Abb. 24 a   Abb. 24 b

Abb. 22 a. Augengneis, Meßstelle U, nicht injiziert, ohne Vorbelastung, größte Radialverschiebungen

Eye-gneiss, measuring point U, without grouting, without preloading, maximum changes
of radius

Gneiss oeillés — point de mesure U — avant injections sans charge préalable — déformation
maximum du rayon

Abb. 22 b. Desgleichen, kleinste Radialverschiebungen

As per 22 a, minimum changes of radius

Idem — déformation minimum du rayon

Abb. 23 a. Augengneis, Meßstelle U, ohne Vorbelastung injiziert (30 atü), größte Radialverschiebungen

Eye-gneiss, measuring point U, with grouting (30 at), without preloading, maximum
changes of radius

Gneiss oeillés — point de mesure U — sans charge préalable après injection à 30 bars —
déformation maximum du rayon

Abb. 23 b. Desgleichen, kleinste Radialverschiebungen

As per 23 a, minimum changes of radius

Idem — déformation minimum du rayon

Abb. 24 a. Umfangsverteilung der Radialverschiebungen, Meßstelle U, nicht injiziert, ohne
Vorbelastung

Distribution of the changes of radius along the circumference, measuring point U, without
grouting, without preloading

Distribution des déformations radiales le long du périmètre. Point de mesure U avant
injection sans charge préalable

Abb. 24 b. Desgleichen, ohne Vorbelastung injiziert

As per 24 a, with grouting without preloading

Idem — sans charge préalable après injection

Die Arbeitslinie (Abb. 25) wurde für die größte Verschiebung bei nicht injiziertem Fels aufgestellt, da nicht sicher ist, ob die kleinere Verschiebung in der injizierten Strecke durch die Injektion verursacht wurde.

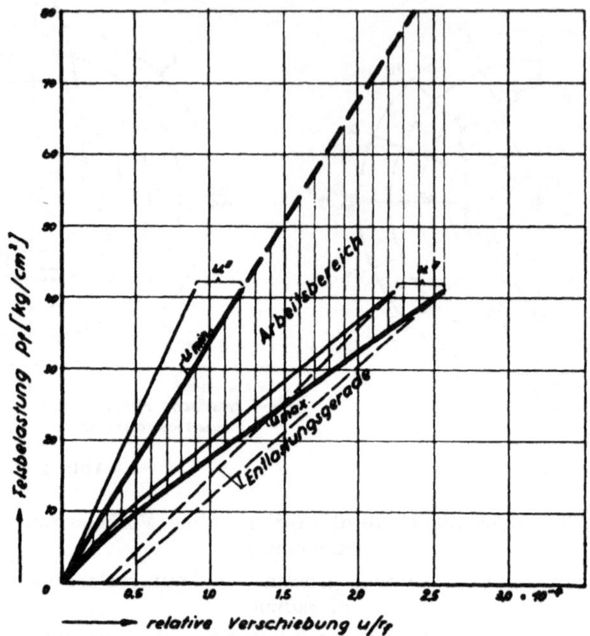

Abb. 25. Ermittlung der Arbeitslinien für den Augengneis, Meßstelle *U*, nicht injiziert, ohne Vorbelastung

Determination of the deformation lines for the Eye-gneiss, measuring point *U*, without grouting, without preloading

Détermination des lignes de déformation pour les gneiss oeillés, au point de mesure *U* — sans injection, sans charge préalable

Abschließend zeigt die Tab. 1 zusammenfassend die aus den vorhin geschilderten Messungen ermittelten Elastizitäts- und Verformungsmoduli.

Tabelle I. Ideelle Verformungsmoduli und E-Moduli nach Messungen am Druckschacht des Kaunertalkraftwerkes (m = 6)

Gestein	nicht injiziert		injiziert ohne Vorbelastung	
	$\max \dfrac{V_{0/p}}{E_{0/p}}$	$\min \dfrac{V_{0/p}}{E_{0/p}}$	$\max \dfrac{V_{0/p}}{E_{0/p}}$	$\min \dfrac{V_{0/p}}{E_{0/p}}$
	kg/cm²	kg/cm²	kg/cm²	kg/cm²
Kalkschiefer	$V_{0/60} = 90\,000$ $E_{0/60} = 190\,000$	$V_{0/60} = 50\,000$ $E_{0/60} = 100\,000$	$V_{0/50} = 163\,000$ $E_{0/50} = 260\,000$	$V_{0/50} = 73\,000$ $E_{0/50} = 148\,000$
Phyllit	$V_{0/25} = 37\,000$ $E_{0/25} = 73\,000$	$V_{0/25} = 5\,900$ $E_{0/25} = 18\,000$	$V_{0/30} = 77\,000$ $E_{0/30} = 150\,000$	$V_{0/30} = 12\,000$ $E_{0/30} = 25\,000$
Gneis	$V_{0/40} = 400\,000$ $E_{0/40} = 425\,000$	$V_{0/40} = 188\,000$ $E_{0/40} = 228\,000$	praktisch gleich	

Deformationsmessungen im Versuchsstollen als Mittel zur Erforschung des Gebirgsverhaltens und zur Bemessung des Ausbaues

Von

F. Pacher, Salzburg*

Mit 11 Textabbildungen

Zusammenfassung — Summary — Résumé

Deformationsmessungen im Versuchsstollen als Mittel zur Erforschung des Gebirgsverhaltens und zur Bemessung des Ausbaues. Beim Anschlag eines Hohlraumes im verspannten Gebirge werden Spannungen gelöst und Materialwanderungen eingeleitet. Dieser Vorgang hält eine gewisse Zeit an bzw. klingt aus, sobald sich ein neuer Gleichgewichtszustand eingestellt hat. Die Formänderungen des Öffnungsquerschnittes und seiner Umgebung hängen vom Festigkeits- und Formänderungsverhalten des Gebirges, dem Spannungszustand im unverritzten Gebirge, vom Querschnitt, von der Zeit, nicht zuletzt vom Ausbau und dessen Eigenschaften selbst ab.

Es liegt nahe, Äußerungen aller Art des Gebirges zu beobachten und systematisch zu messen nicht nur, um daraus Schlüsse hinsichtlich des Gebirgsverhaltens usw. zu ziehen, sondern um darüber hinaus die ermittelten Lastdeformationskurven zur Bemessung der Ausbauquerschnitte heranzuziehen.

Jeder Gebirgsart kommt eine charakteristische Kennlinie zu.

Die zahlreichen beeinflussenden Faktoren, die zu trennen unmöglich wäre, sind im Ergebnis bei richtiger Anlage solcher Versuchsstrecken zwar implizit, aber doch im richtigen Verhältnis vertreten.

Weitere Details werden an Hand ausgeführter und in Ausführung begriffener Beispiele erläutert.

Measurements of Deformation in a Test Gallery as a Means of Investigating the Behaviour of the Rock Mass and of Specifying Lining Requirements. If an excavation is made in a stressed rock mass stresses are released and material migration is initiated. This process lasts for some time; it ceases as soon as a new state of equilibrium is reached. The deformation of the opening cross section and surrounding area depends upon the strength and deformation behaviour of the rock mass, the state of stress in the unworked rock mass, the cross section, the time, and also upon the lining and its properties.

It is important to make observations and systematic measuremens of all the effects exhibited by the rock mass. This enables conclusions to be drawn not only as to the behaviour of the rock mass etc.; the measured load-deformation-curves can also be used to determine the dimensions of the lining cross sections. Every kind of rock mass has a characteristic load-deformation-curve.

The numerous effective factors are impossible to be separated, but exert their combined influence in test galleries. Provided the galleries are correctly laid out, they are combined in the correct relationship to each other.

Further details are explained by means of practical examples, some of which are already completed and others are now in progress.

---

* Dipl.-Ing. Franz P a c h e r, Ingenieurbüro für Geologie und Bauwesen, Salzburg.

Mesure des déformations dans les galeries d'essai afin d'étudier le comportement du massif et de calculer le revêtement. L'ouverture d'une cavité dans un massif où règnent des contraintes libères ces contraintes et provoque des déplacements du matériau. Ce processus dure un certain temps puis s'amortit lorsqu'un nouvel état d'équilibre est atteint. Les déformations de la section du tunnel et de la roche encaissante dépendent de la résistance et de la déformabilité du massif rocheux, du profil, du temps, et plus encore du revêtement et de ses propriétés.

On recommande d'observer toutes les manifestations du massif rocheux et de les mesurer systématiquement, non seulement pour en tirer des conclusions sur le comportement du massif, mais aussi pour utiliser les courbes obtenues au calcul de la section du revêtement.

A chaque type de massif rocheux correspond une courbe caractéristique.

Il serait impossible d'énumérer ici les nombreux facteurs qui jouent un rôle. En vérité pour que l'installation de telles galeries d'essais soit correcte, il faut savoir quelle est la part de chaque facteur.

Des détails plus approfondis sont fournis par les exemples de travaux achevés ou en cours.

Einleitung

Die Belastung und damit die Stärke eines Tunnelgewölbes hängen nicht nur vom Gebirge und vom Tunnelquerschnitt, sondern auch von der Art und dem Zeitpunkt des Ausbaues ab. Das Beobachten des Gebirgsverhaltens beim Öffnen des Hohlraumes, insbesondere das genaue Messen der Deformationen, gewinnt an Bedeutung, seit der Ausbau um vieles rascher als mit den klassischen Baumethoden eingebracht werden kann und den infolge unvermeidlicher Firstsenkung zur Auslösung kommenden „Auflockerungsdruck" zu vermeiden weiß. Bei den modernen Bauweisen, insbesondere der Spritzbetonbauweise (nach Rabcewicz „Neue Österreichische Bauweise") hat der Ausbau, abgesehen vom tektonischen Gebirgsdruck, vor allem den Gebirdruck aus Spannungsumlagerungen, den sogenannten „Umlagerungsdruck", aufzunehmen, während bei den früher üblichen langsamen Ausbauarten der vor allem in der Firste entstehende Auflockerungsdruck die maßgebende Belastung des Ausbaues darstellt.

Um die Klarstellung der Zusammenhänge zwischen der Überlagerungshöhe, dem Gebirgsdruck und dem Ausbauwiderstand hat sich L. Rabcewicz[1] sehr bemüht.

Da das Gebirgsverhalten bzw. die Gesamtwirkung auf den Ausbau äußerst schwer vorauszuberechnen ist, drängt es immer mehr, dieses empirisch zu erfassen (L. Müller[2]). Der Gedanke liegt nahe, bereits mit dem Auffahren des Probestollens Erfahrungen über den zu erwartenden Gebirgsdruck zu sammeln.

So sucht man, auf den Erkenntnissen der Boden- und Felsmechanik aufbauend und die Vorteile des Spritzbetons nützend, neue theoretische Grundlagen für eine Bemessung zu erhalten. Als wesentlichste Faktoren sind der Schichtaufbau, die Festigkeits- und Formänderungseigenschaften des Gebirges und der Spannungszustand in demselben zu nennen. Aber auch die Auswirkungen des Gefüges und der gefügebedingten Anisotropie der Bergarten sind bekanntlich sehr groß; ihren Einfluß auf den Gebirgsdruck theoretisch zu erfassen ist schwierig. Ein Versuchsstollen jedoch zeigt alle diese Auswirkungen und verspricht selbst dann eine Umrechnung auf einen größeren Querschnitt, wenn diese Anisotropie nicht in ihren einzelnen Komponenten, sondern nur in ihrer Gesamtwirkung erkennbar ist.

Aus den gemessenen Deformationen und ihrem zeitlichen Ablauf werden — in Einklang mit den theoretischen Erkenntnissen — die Rückschlüsse auf die Ausbaubelastung gezogen.

Zweck und Grundlagen der Versuche

Beim Durchörtern des Gebirges mit einem Stollen entstehen vor allem Formänderungen in Richtung zum Hohlraum hin. Dabei sind zu unterscheiden:

a) Formänderungen, welche der Spannungsumlagerung (Entlastung) im elastischen Bereich zuzuschreiben sind;

b) Formänderungen, welche durch Bruchfließen infolge Überschreitens der Gesteinsfestigkeit entstehen und

c) Formänderungen, welche bei fortschreitender Auflockerung durch die Schwerkraft hervorgerufen werden.

Angenommen, es bestünde die Möglichkeit, die Deformation bestimmter Punkte des Stollenumfanges unter verschieden hohem, aber vorgegebenem Innen-Gegendruck (Ausbauwiderstand) zu ermitteln und dafür die Arbeitslinien (1, 2 usw.) zu zeichnen, so könnte man daraus unmittelbar die Beziehungen zwischen Verformungsweg und Ausbauwiderstand in Abhängigkeit von der Zeit ableiten.

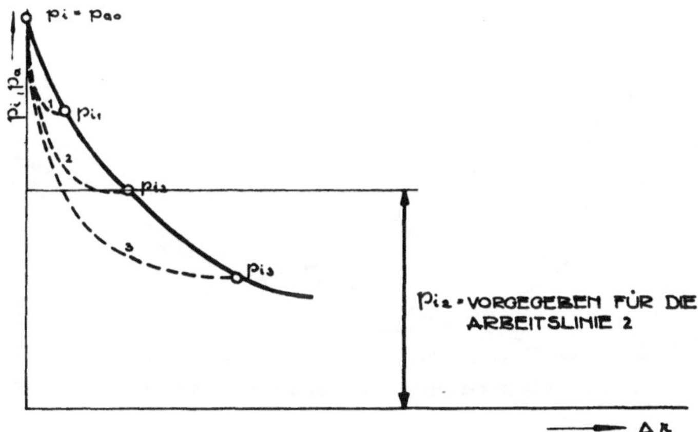

Abb. 1. Arbeitslinie des Gebirges bei verschieden starker Entlastung infolge vorgewählten Ausbauwiderstandes

Load-deformation-curve of the rock mass for different cases of stress relief, depending on the preselected lining resistance

Courbe de déformation du massif rocheux à la libération de contraintes variées pour un revêtement dont la résistance est donnée à l'avance

Man könnte auch stufenweise die End-Deformationen $\Delta r_{1,2\ldots n}$ durch einen starr gedachten Ausbau festhalten und den sich einstellenden Gebirgsdruck (Außendruck, wirksame Ausbaubelastung) ermitteln. Solange dabei das Gebirge im wesentlich elastischen Verhaltensbereich bleibt, wären Deformation und Ausbauwiderstand etwa verkehrt proportional zueinander. Nach Überschreiten der Gebirgsfestigkeit — vom Stollenumfang ausgehend — bildet sich die sogenannte „Schutzzone" aus, welche einen Übergang vom Hohlraum zum intakten Gebirge herstellt, einen nach außen zunehmenden Widerstand aufbaut und wie ein Puffer lastausgleichend wirkt. Unter Umständen kann sich infolge zunehmender Auflockerung beim Fließvorgang ein Bereich der Schutzzone aus der allgemeinen Gebirgsverspannung lösen. Das entfestigte Material dieser Zone beginnt unter dem Einfluß der Schwerkraft den gefürchteten „Auflockerungsdruck" zu erzeugen, dessen Anstieg schlagartig erfolgen kann. Mit Auslösen des Auflockerungsdruckes wird ein Wiederansteigen der außerdem einseitigen Belastung eintreten.

Verbindet man die Punkte der so erhaltenen Linienzüge, so bekommt man charakteristische Linien für das Verhalten verschiedener Gebirgsarten bei gleichem Stollendurchmesser.

Ein standfestes Gebirge (I) würde sich deutlich vom nachbrüchigen Gebirge (II), erst recht von einem zu rascher Entfestigung neigenden, von einem treibenden oder blähenden (druckhaften) Gebirge (III) unterscheiden.

Bei allen Kennlinien muß auch der Einfluß der Zeit mit berücksichtigt werden, da dieser bei der Wahl des *Ausbaues* sehr entscheidend ist. Gebirge und Ausbau suchen im Zusammenspiel von Spannungsumlagerung und Deformation einen neuen

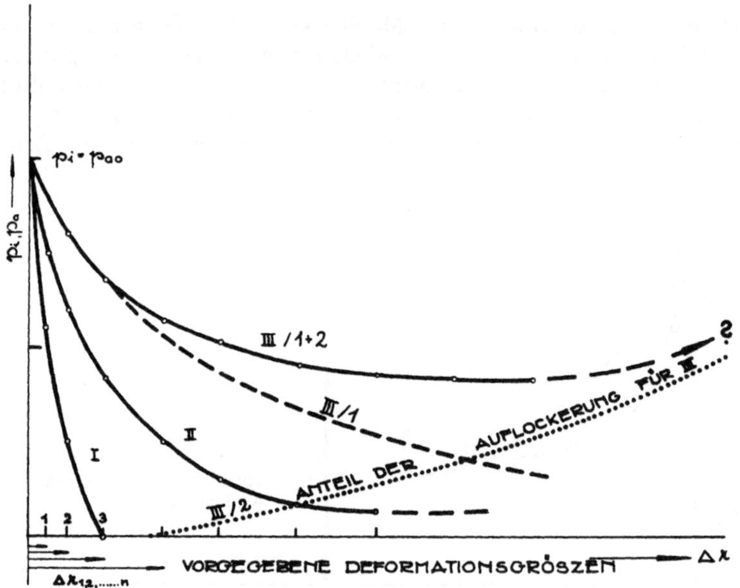

Abb. 2. Schematische Kennlinien verschiedener Gebirgsarten auf Grund stufenweise vorgegebener Deformation

Schematic load-deformation-curves for different kinds of rock mass with gradually increasing deformation

Courbes caractéristiques schématiques de différents types de massifs rocheux pour une déformation initiale donnée par degrés

Gleichgewichtszustand aufzunehmen: zunächst steht der Deformation des Gebirges kein Widerstand gegenüber. Nach dem Einbringen des Ausbaues kann das umgebende Gebirge seine Deformation dem Ausbau nur so weit aufzwingen als es der entstehende Ausbauwiderstand zuläßt. Man muß also sowohl die Last-Deformationskurve des Gebirges wie jene des Ausbaues in Abhängigkeit von der Zeit kennen.

Für einen Ausbau in Stahl oder mit Betonfertigteilen, die sofort tragfähig sind, können die Last-Deformationskurven des Ausbaues ohne weiteres gezeichnet werden. Die Zunahme der Festigkeit und des E-Moduls bei Beton und Spritzbeton muß durch Versuche festgestellt und bei der Ermittlung der Arbeitslinien berücksichtigt werden.

Betrachtet man die Wechselwirkungen an einem Gedankenmodell, so laufen die Spannungsumlagerungen etwa folgendermaßen ab: Nach einer, durch die Vortriebsweise gegebenen unbehinderten Anfangsdeformation ($A—B$ in der unteren Hälfte des Diagrammes) nimmt die Zeit-Deformationskurve einen ausbauabhängigen

Verlauf (*B—C*). In dieser Zeit wird der Ausbauwiderstand von 0 (*D*) ansteigen, bis er mit der wirksamen Ausbaubelastung im Punkt G_1 im Gleichgewicht steht (obere Hälfte des Diagrammes). Aus einem Versuch erhalten wir demnach die Deformationsabhängigkeit des Gebirges von der Zeit (Kurve *A—B—C*) und die Ausbaubelastung p_a im Punkt G_1, rückgerechnet aus der Ausbaubeanspruchung ($p_{i1} = p_{a1}$).

Um den Verlauf der Kennlinie *F—G—H* des Gebirges kennenzulernen, müßte ein zweiter bzw. dritter Versuch im gleichen Gebirge durchgeführt werden, welche die Punkte G_2 bzw. G_3 liefern würden. (Der Einfachheit halber wird nur *eine* Kennlinie für einen bestimmten Umfangspunkt betrachtet.)

Den Anfangspunkt *F* der Kurve kann man entweder durch unmittelbares Messen der Primärspannung im Gebirge oder auch aus Überlegungen, z. B. nach den Formeln von Fenner[3] und Kastner[4], erhalten. Diese Formeln geben die Größe des wirksamen Druckes in Abhängigkeit von Kennziffern des Gebirges, von der Überlagerung und vom Verhältnis des Hohlraumradius zum Halbmesser der Schutzzone an. Der Halbmesser der Schutzzone kann nach Rabcewicz aus seismischen Messungen ermittelt werden.

Bei der Bemessung der Druckschachtpanzerung, für welche Lauffer und Seeber[5] ein Verfahren entwickelt haben, liegen die Verhältnisse einfacher, weil

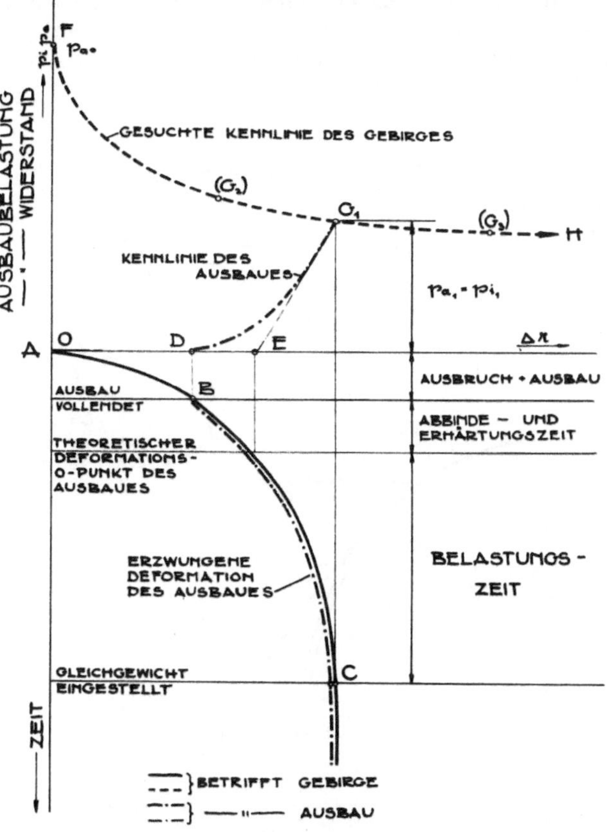

Abb. 3. Zusammenspiel von Gebirgsdruck und Ausbauwiderstand im Versuch

Relationship between rock mass pressure and lining resistance in the test

Influences simultanées du poids du massif rocheux et de la résistance du massif rocheux sur l'essai

jederzeit Innendruck und Deformation bekannt sind und die Arbeitslinien gleich für den endgültigen Durchmesser des fertigen Ausbaues erhalten werden.

Im Falle der Bemessung einer Tunnelauskleidung auf Grund von Deformationsmessungen am Versuchsstollen kleinen Durchmessers hingegen ist es noch notwendig, die Ergebnisse auf das größere Profil umzurechnen.

Bei den im Ingenieurbau vorkommenden Fällen ist es durchwegs vorteilhaft, einen Ausbau zu wählen, der rasch einen ausreichend hohen Ausbauwiderstand erreicht, der hohlraumlos gegen das Gebirge abschließt und eine hohe Biegezug- und Scherfestigkeit aufweist. Besonders die sofortige Entwicklung des Ausbauwiderstandes, wobei in diesem Stadium meist geringe Rückstellkräfte genügen, verhindert die weitere Auflockerung. Gebirge und Ausbau sollen eine innige Verbin-

154 F. Pacher:

dung eingehen. Spritzbeton und Spritzbetontechnik eignen sich hiefür ausgezeichnet; nach Müller gleicht die Spritzbetonschicht einer Sofortversiegelung, welche nahezu vollkommen intakt gebliebene Kluftkörpergruppen in die Gewölbewirkung einbezieht.

Bei den tiefliegenden Tunneln und Schächten des Bergbaues dagegen muß oft erst die Entspannung des umgebenden Gebirges und die Bildung einer Schutzzone abgewartet werden, um dem sonst enormen Gebirgsdruck auszuweichen. Dies steht mit dem zuvor Gesagten nicht in Widerspruch. Art, Stärke und Zeitpunkt des Ausbaues sind entsprechend der charakteristischen Linie des Gebirges zu bestimmen.

Allgemein gesprochen, soll der Ausbau steif genug sein, um eine unerwünschte Auflockerung zu verhindern, andererseits aber nachgiebig genug, um die Bildung der Schutzzonen, welche eine Verringerung des notwendigen Ausbauwiderstandes zur Folge haben, zu ermöglichen.

Hinsichtlich der Zeitspanne, welche zwischen der Öffnung des Hohlraumes und dem Eintritt der Tragfunktion des Ausbaues verstreichen darf, sind natürlich stark nachbrüchige bis druckhafte Gebirge besonders empfindlich, wie nebenstehende Abbildung zu verdeutlichen sucht. In dieser stark schematisierten Skizze sind drei Fälle dargestellt, welche sich hinsichtlich der Größe der ungehinderten Anfangsdeformation Δr unterscheiden. In der oberen Hälfte des Diagrammes der Abb. 4 sind für alle drei Fälle die Arbeitslinien der elastischen Verformung $(E{-}G)$ verschiedener Ausbaustärken $(d$ bis $5\,d)$ eingetragen.

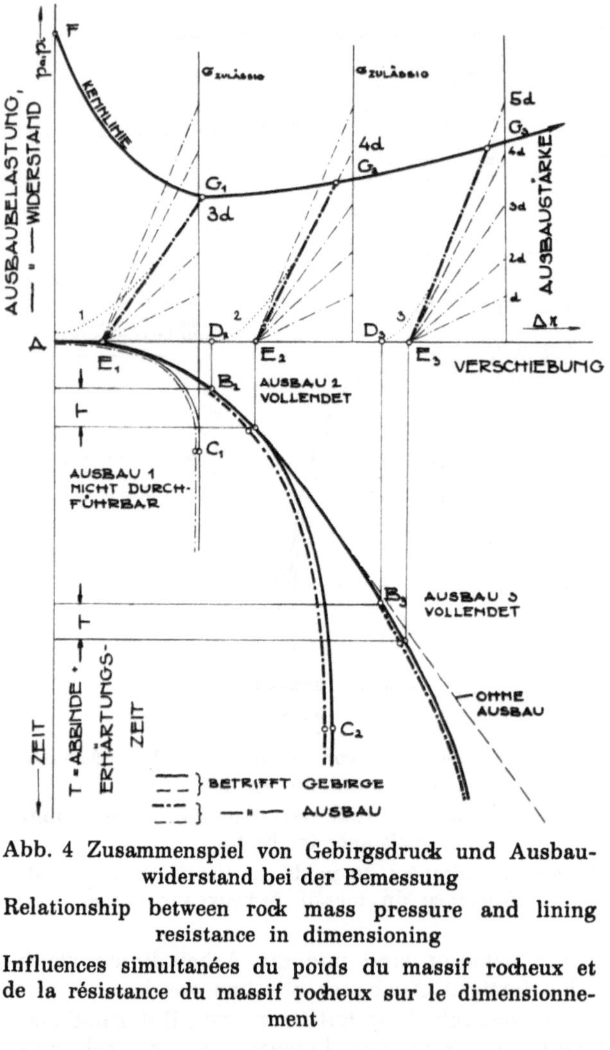

Alle Ausbau-Arbeitslinien, welche die Kennlinie innerhalb der zulässigen Beanspruchung schneiden, vermögen einen Gleichgewichtszustand herzustellen. Im Falle 1 $(E_1{-}G_1)$ würde eine Ausbaustärke von $3\,d$ genügen, wenn es gelänge, den Ausbau rechtzeitig einzubringen. Wegen des steilen Anstieges der Zeit-Deformations-Kurve $(A{-}B_2)$ fehlt jedoch leider die für Abbinden und Erhärten nötige Zeit T.

Der Fall 2 $(D_2{-}E_2{-}G_2)$ würde eine Ausbaustärke von $4\,d$, der Fall 3 $(D_3{-}E_3{-}G_3)$ schon eine solche von $5\,d$ erfordern.

Abb. 4 Zusammenspiel von Gebirgsdruck und Ausbauwiderstand bei der Bemessung

Relationship between rock mass pressure and lining resistance in dimensioning

Influences simultanées du poids du massif rocheux et de la résistance du massif rocheux sur le dimensionnement

Die richtige Einschätzung des Gebirgsdruckes bzw. der zu erwartenden Ausbaubelastung und -deformation ist demnach technisch und wirtschaftlich von größter Bedeutung. In allen Fällen bringen Stollendeformationsmessungen neue Erkenntnisse.

Ein Versuchsprogramm ist so anzulegen, daß es Auskunft gibt über:

a) die Deformation des den Hohlraum umgebenden Gebirges in Abhängigkeit von Zeit und Ort (Abstand von der Tunnelachse),

b) den Verformungsmodul des Gebirges in Abhängigkeit von der Entfernung vom Hohlraum.

c) das Festigkeits- und Formänderungsverhalten des Ausbaues,

d) die Beanspruchung des Ausbaues nach Einstellen des Gleichgewichtes und

e) die Festigkeits- und Formänderungseigenschaften des verwendeten Betons oder Spritzbetons, ermittelt an Betonproben unter den Bedingungen des Stollens.

Im Zusammenhang damit vermögen richtig angesetzte Modellversuche wertvolle Erkenntnisbeiträge zu liefern. Letzten Endes bietet eine Versuchsstrecke im Maßstab 1 : 1 die allerbesten Beobachtungsmöglichkeiten.

Im folgenden sollen die Anlage eines solchen Versuchsstollens, die Meßausrüstung und die Meßergebnisse an Hand von Beispielen besprochen werden.

Beispiel eines ausgeführten Versuchsstollens
(Generelle Anordnung siehe Abb. 5)

Den Abschluß des Zugangsstollens bildet eine Meßkammer, von welcher aus der eigentliche Versuchsstollen vorgetrieben wird. Die Kammer enthält den Standpunkt des Präzisionsnivellierinstrumentes, welcher gleichzeitig ein Fixpunkt der

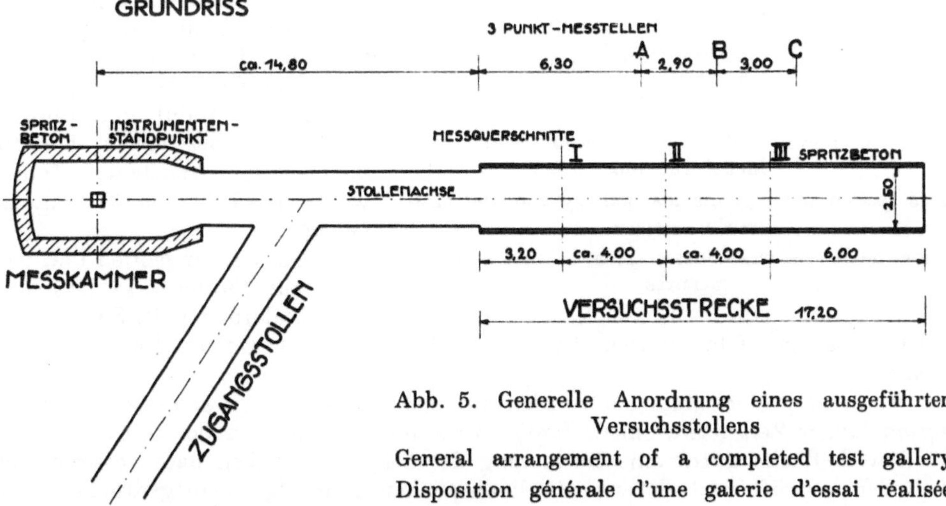

Abb. 5. Generelle Anordnung eines ausgeführten Versuchsstollens

General arrangement of a completed test gallery

Disposition générale d'une galerie d'essai réalisée

optischen Achse, der Bezugslinie sämtlicher Messungen, ist. Um eventuelle Bewegungen der Meßkammer auszuschalten, wurde die Verkleidung sehr stark ausgebildet und eine entsprechend lange Beruhigungszeit vor Beginn des Versuchsstollenvortriebes abgewartet.

Die Versuchsstrecke in den dort anstehenden festgelagerten, schwach bis mäßig verkitteten Feinst- und Mittelsanden wurde in langsam-gleichmäßigem Vortrieb ausgebrochen und sofort durch eine Spritzbetonauskleidung von 5 bis 7 cm Stärke gesichert. Der lichte Durchmesser des kreisrunden Versuchsstollens betrug 2,5 m.

Rechtwinkelig zur Achse des Stollens waren drei Meßquerschnitte angeordnet, in welchen die radialen wie polygonalen Verformungen eines Achteckes gemessen wurden.

Für die Beobachtung der Radienveränderung wurde ein optisch zentrierter Bock verwendet, dessen Schwenkarm die Entfernung zu den im Spritzbeton eingelassenen Bolzen abtastete. Die polygonalen Entfernungen der Bolzen wurden mit einer Lehre ausgemessen. Die Aufzeichnung beider Messungen erfolgte elektrisch.

Die Arbeiten liefen so ab, daß der Stollen jeweils nur in 1-m-Abschnitten voraus ausgebrochen wurde und hierauf sofort die Auskleidung mit Spritzbeton erfolgte. Die Meßbolzen sowie die Flurplatte für die Aufstellung des Gerätes wurden noch während der Spritzbetonarbeiten versetzt.

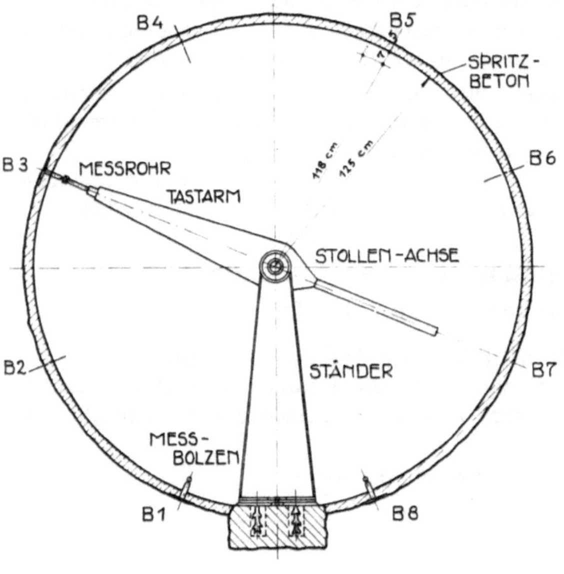

Abb. 6. Querschnitt durch den Versuchsstollen mit Meßeinrichtung
Cross section through the test gallery showing measuring device
Coupe de la galerie d'essai avec le dispositif de mesures

Die Messungen selbst begannen 12 bis 24 Stunden nach dem Einbau, sozusagen am noch nicht erhärteten Beton, um tunlichst früh ein Anfangsmaß zu erhalten. Sie erfolgten während der Versuchsdurchführung laufend, mit Wiederholungen im Abstand von mehreren Monaten über ein ganzes weiteres Jahr. Danach konnte man annehmen, daß der zu erwartende Gebirgsdruck sich voll ausgebildet und ein Gleichgewicht gefunden habe. Nebenher wurden noch zusätzliche Untersuchungen, sowohl am Beton als auch am anstehenden Gebirge, durchgeführt, um den *E*-Modul und die Festigkeit zu bestimmen. Die für das Gebirge ermittelten Moduln waren mit 1000 bis 2000 kg/cm^2 gering.

Um die tangentiell und parallel zur Stollenachse laufenden Formänderungen des Spritzbetonausbaues zu ermitteln, wurden Dreipunktdeformeterstrecken nach Talobre[6] eingebaut.

Auch diese Verformungen wurden auf elektrischem Wege gemessen. Nach genügend langer Zeit (etwa einem Jahr) wurde die Deformation des Spritzbetons in der Gewölbefläche durch eine Entlastung der 3-Punkt-Strecken nach verschiedenen Methoden geprüft und daraus die Beanspruchung infolge Gebirgsdruckes, bzw. dieser selbst, zurückgerechnet.

Die Kleinheit der Meßwerte brachte teilweise eine Streuung in die Ergebnisse, so daß diese ausgeglichen werden mußten. Trotzdem zeigte die Durchmesserverformung bei allen drei Meßquerschnitten dasselbe typische Bild. Am deutlichsten war die Verformungstendenz bei der Radienverformung zu erkennen. Die vertikalen Durchmesser haben sich stärker verkürzt als die horizontalen und das ganze Stollenprofil wurde etwas in den rechten unteren Ulm hineingedrückt. Diese Beobachtung wurde auch durch die Verkürzung der Polygonseiten bestätigt. Das mittlere Verkürzungsmaß für einen Radius betrug 0,22 bis 0,37 mm.

Unter Annahme eines allseitig gleichen Druckes wurde aus den Betonspannungen eine angenäherte Belastung des Ausbaues zwischen 0,8 und 1,5 kg/cm² ermittelt. Aus der unterschiedlichen Durchmesserverkürzung kann geschlossen werden, daß die Verteilung über den Umfang nicht gleichmäßig ist, sondern daß die Spannungen in vertikaler Richtung ca. um 10 bis 30 % größer, in horizontaler Richtung um 10 bis 30 % kleiner sind. Die Druckverteilung entspricht also nicht dem hydrostatischen Zustand, sondern eher dem in der Bodenmechanik angenommenen Seitendruckverhältnis.

Diese in den Jahren 1960/61 durchgeführte Arbeit brachte wertvolle Erfahrungen, aber auch die Erkenntnis, daß es notwendig ist, die gesamte Deformation — vom Beginn des Ausbruches bis zum Erreichen des Gleichgewichtszustandes — zu erfassen. Bei einem zur Zeit in Ausführung befindlichen Versuchsstollen wird daher auf den schon seinerzeit von Müller[2] gemachten

Abb. 7. Meßeinrichtung in Funktion
Measuring device in action
Dispositif de mesure en service

Vorschlag zurückgegriffen, wonach die Bewegungen des umgebenden Gebirges mittels voreilend versetzter Meßmittel beobachtet werden.

Abb. 8. Dreipunktdeformeter
Three-point deformation meter
Triangle extensomètrique

Beispiele eines in Ausführung befindlichen Versuchsstollens mit langarmigen Bewegungsanzeigern
(Deformationsindikatoren)

Dieser Versuchsstollen liegt in mergeligem Gebirge mit geringem *E*-Modul und niedriger Festigkeit. Der verhältnismäßig homogene Aufbau des Gebirges läßt auswertbare Versuchsergebnisse erwarten. Die Stollenleibung wird ebenfalls in Spritzbetonbauweise, jedoch mit Einlage von Baustahlgewebe, gesichert. Der Vortrieb wird unter möglichster Schonung der Wandung, jedoch sonst unter denselben Bedingungen wie im Vollprofil, d. h. Lösen durch Sprengen, erfolgen. Meßkammer, Ver-

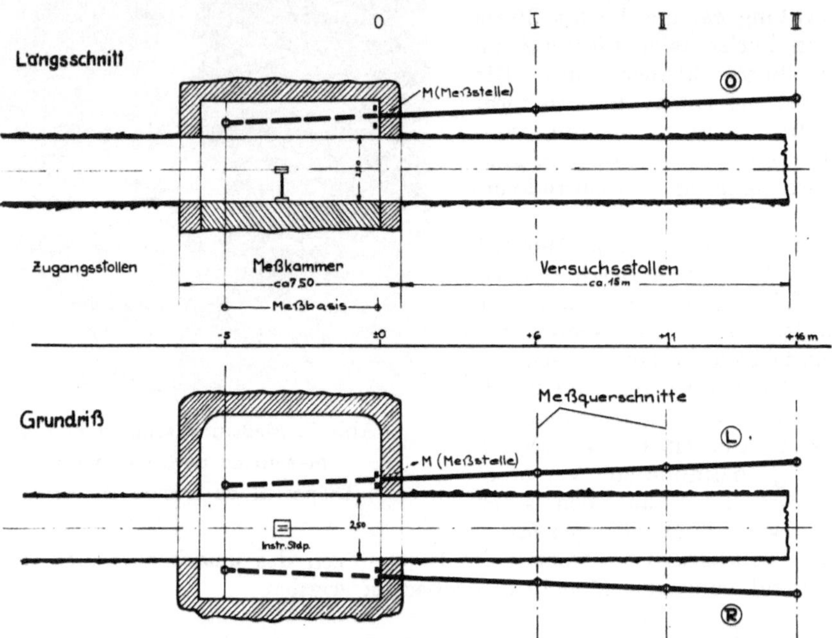

Abb. 9. Generelle Anordnung eines geplanten Versuchsstollens
General arrangement of a proposed test gallery
Disposition générale d'une galerie d'essai en projet

suchsstrecke und Meßquerschnitte sind ähnlich wie im vorigen Beispiel angeordnet. Die Meßkammer dient als Basis für die Beobachtung der Gebirgsbewegungen infolge Öffnen und Ausbau des Hohlraumes und erhält deshalb einen kräftigen Ausbau. Die Versuchsstrecke wird eine Länge von 20 m und einen lichten Durchmesser von 2,5 m aufweisen, ihre durchschnittliche Ausbaustärke in Spritzbeton soll etwa 6 cm betragen. Die Versuchsstrecke begleitend werden die Bohrlöcher *O*, *L* und *R*, welche die Deformationsindikatoren aufzunehmen haben, angeordnet und voreilend von der Meßkammer aus gebohrt. Bevor der Vortrieb der eigentlichen Versuchsstrecke beginnt, müssen alle Arbeiten in der Meßkammer sowie die Montage und Prüfung der Deformationsindikatoren beendet sein. Ferner muß sichergestellt sein, daß sich das Gebirge wieder vollkommen beruhigt hat. Die gleich Fühlern vorgeschobenen Meßgeräte gestatten eine kontinuierliche Bewegungsbeobachtung ausgewählter Punkte im Umkreis des Stollens mit hoher Genauigkeit, ohne Behinderung durch den Baubetrieb. 2 Fixpunkte in der Meßkammer bilden die Ausgangsbasis und das im ungestörten Gebirge steckende Ende einen Kontrollpunkt. Die Meßbasis selbst wir ständig durch ein Präzisionsnivellement kontrolliert. Vorbeschriebene Messungen werden durch ein zweites System ergänzt, welches, wie im vorigen Beispiel, die Veränderungen des horizontalen und des vertikalen Durchmessers über Bolzen festhält. Die Betonbeanspruchungen sollen mittels Dreipunktdeformeterstrecken oder Dehnmeßstreifen ermittelt werden.

Die Ausschläge der Deformationsindikatoren wie der Dreipunktdeformeterstrecken werden elektrisch gemessen, die Durchmesser dagegen mechanisch.

Zum Unterschied vom vorerwähnten Beispiel tritt hier keinerlei Beeinträchtigung des Vortriebes ein, da die Messungen fast ausschließlich in der Meßkammer anfallen. Nur beim Durchörtern der Meßquerschnitte wird ein gewisser Zeitraum zur Montage der Bolzen usw. benötigt.

Die Messungen erfolgen zunächst während des Vortriebes und ein bis zwei Wochen nach Beendigung desselben. Kontrollen im Abstand von etwa 4 Wochen über einen genügend langen Zeitraum sind vorgesehen.

Die Verfeinerung der Meßtechnik gegenüber dem ersten Entwurf war nur möglich, da in der Zwischenzeit entsprechend genau arbeitende Deformationsindikatoren für diese und andere Zwecke entwickelt werden konnten.

Messungen im Vollprofil

Sind die Gebirgsverhältnisse stark wechselnd, so daß die Umrechnung vom Versuchsstollen auf das Vollprofil mit einem großen Risiko behaftet wäre, ist es

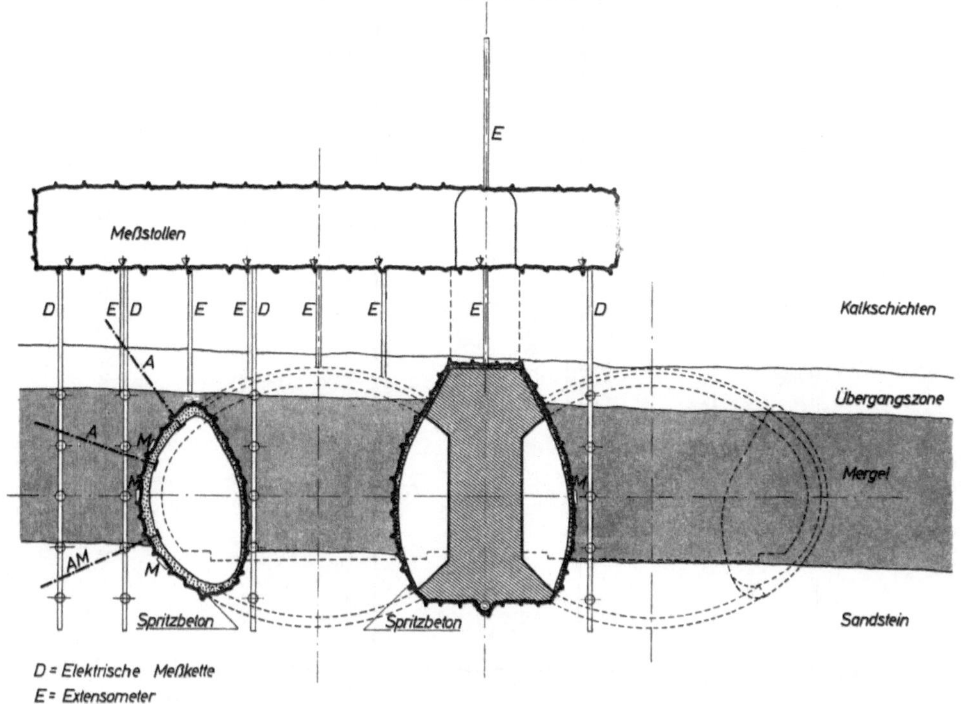

Abb. 10. Beispiel einer Meßanordnung im Vollprofil
Example of a measuring device in full profile
Example d'un appareil de mesures en pleine section

zweckmäßig und notwendig, die Deformation im Vollprofil zu erfassen. Das grundlegende Meßsystem bleibt im wesentlichen dasselbe. Die Deformationsindikatoren

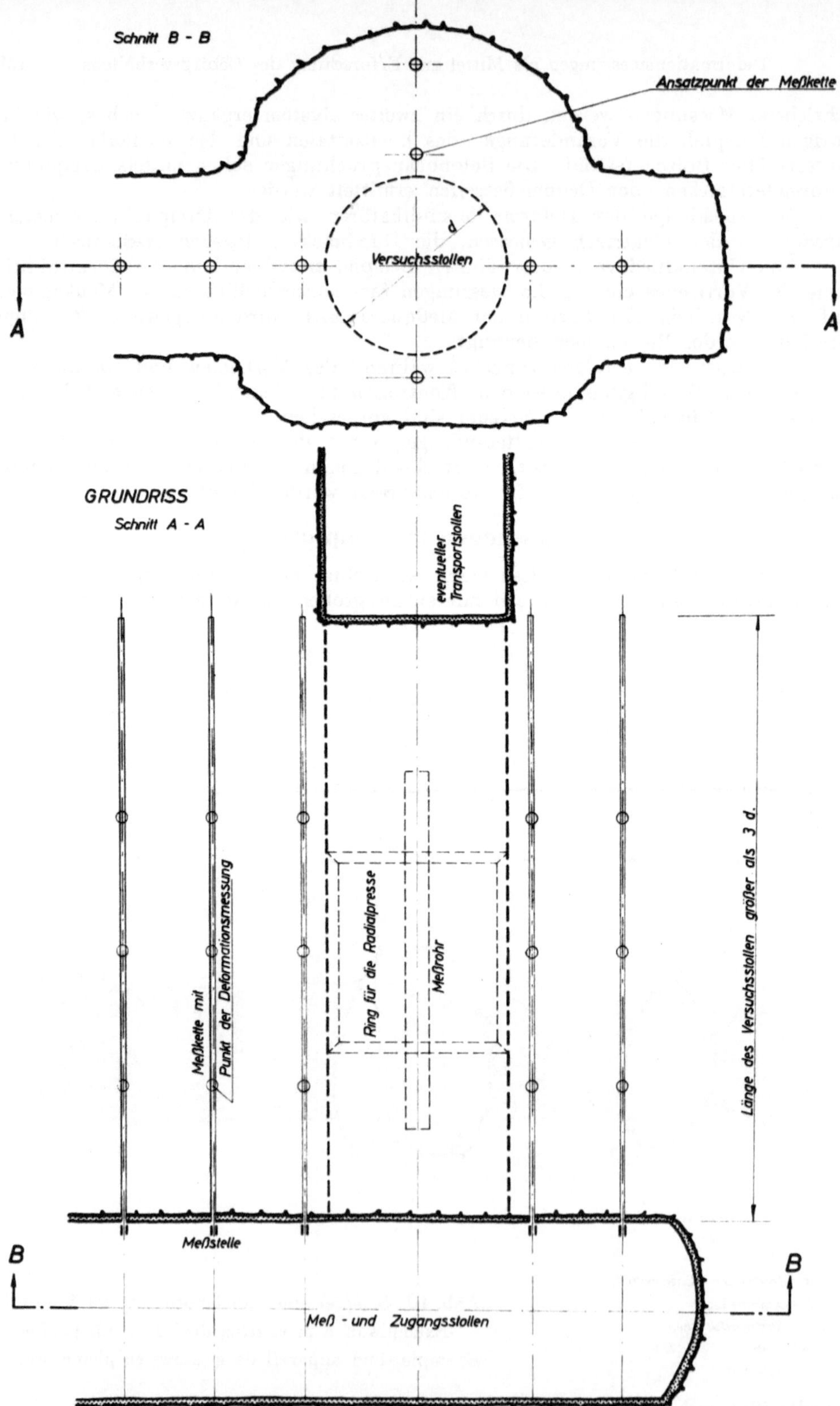

Abb. 11. Meßanordnung beim kombinierten Ent- und Belastungsversuch
Arrangement of the measuring device for a combined load-unload-test
Appareil de mesure pour un essai combiné de libération des contraintes et de remise en charge

werden vermehrt und durch Meßgeräte ergänzt, welche von einem darüberliegenden Meßstollen aus angesetzt sind. Im Prinzip wird aber derselbe Gedanke verfolgt wie im vorigen Beispiel, nämlich, ungestört die gesamten Formänderungen beobachten zu können. Zusätzliche Kontrollgeräte im Stolleninneren beobachten die Längenänderung einzelner ausgezeichneter Strecken. Druckmeßdosen, Meßanker und Einrichtungen zur Bestimmung der Betonbeanspruchung vervollständigen die Ausrüstung.

Ausblick

Die bisherigen Ergebnisse haben die Richtigkeit der Verfolgung dieses Weges bestätigt. Erst die Kenntnis und der Vergleich von Kennlinien verschiedener Gebirgsarten werden Unterlagen für ein Bemessungsverfahren liefern. Deformationsmessungen an der fertiggestellten Tunnelröhre, unmittelbar im Anschluß an ihre Ausführung vorgenommen, bilden eine wertvolle Ergänzung der Messungen im Versuchsstollen. Sie dienen der Beobachtung des Verhaltens des definitiven Ausbaues und der Sicherstellung, daß die Auskleidung keine unzulässige Deformation bzw. keine Überbeanspruchung erleidet.

Darüber hinaus beabsichtigen M ü l l e r und P a c h e r , den Entlastungsversuch mit dem Stollenabpreßversuch zu vereinen und daraus einen Standard-Versuch der Felsprüfung zu entwickeln. Ein solcher Versuch würde die bekannten Vorteile des Stollenabpreßversuches, insbesondere die große Wirkungstiefe bei einfacher Lastaufbringung, nützen, das Feld der Ergebnisse aber wesentlich erweitern. Aus ihm werden wertvolle Erkenntnisse

über Gebirgsbewegungen in der Ent- wie Belastungsphase, aber auch bei wiederholter Ent- und Belastung,

bei der Bestimmung des E- und Verformungsmoduls des Gebirges in Abhängigkeit von Zeit und Ort,

bei der Bestimmung der Poissonschen Konstanten,

für die Erfassung der Anisotropie,

über die Entwicklung und Ausdehnung der Schutzzone,

über die Injektionswirkung und

über den Einfluß der Zeit bei allen Vorgängen

erwartet.

Literatur

[1] v. R a b c e w i c z , L.: Aus der Praxis des Tunnelbaues. Geologie und Bauwesen, Jg. 27, H. 3-4, 1962.

v. R a b c e w i c z , L.: Bemessung von Hohlraumbauten. Felsm. u. Ing. Geol., Vol. I/3-4, 1963.

[2] M ü l l e r , L.: Setzungen von Bauwerken auf Felsuntergrund. Baugrundtagung Hannover 1953.

M ü l l e r , L.: Der Felsbau. Ferd.-Enke-Verlag, Stuttgart 1963.

[3] F e n n e r , R.: Untersuchungen zur Erkenntnis des Gebirgsdruckes. Glückauf, Jg. 74, 1938.

[4] K a s t n e r , H.: Statik des Tunnel- und Stollenbaues. Springer-Verlag, Berlin 1962.

[5] L a u f f e r , H. und G. S e e b e r : Die Bemessung von Druckstollen- und Druckschachtauskleidungen für Innendruck auf Grund von Felsdehnungsmessungen. Österr. Ing. Zeitschr., Jg. 5, H. 2, 1962.

[6] T a l o b r e , J.: La Mécanique des Roches. Dunod-Paris 1957.

Beeinflussung der Gebirgsfestigkeit durch Sprengarbeiten

Von

Leopold Müller*

Mit 12 Textabbildungen

Zusammenfassung — Summary — Résumé

Beeinflussung der Gebirgsfestigkeit durch Sprengarbeiten. Neuere technologische Untersuchungen in Fels haben gezeigt, daß schon geringfügige Auflockerung des Gebirges dessen Festigkeiten ganz entscheidend tief herabsetzt. Dieser Umstand läßt manche der modernen Methoden zur Lösung des Gesteins mittels Sprengungen in einem neuen Lichte erscheinen und zwingt zu einer kritischeren Einstellung.

Die Empfindlichkeit des Gebirges gegen Sprengerschütterungen ist durch Spannungskonzentrationen an Klüften (Kerbspannungen) sowie durch Zugspannungen in deren Nachbarschaft auch in einem sonst zugspannungsfreien Bereich bedingt und ist daher von der Beschaffenheit und Orientierung der Klüfte, insbesondere vom Durchtrennungsgrad der einzelnen Kluftscharen, abhängig. Die Stellung der Kluftscharen zu den bei einer Sprengung auftretenden Beanspruchungen spielt eine ebenso große Rolle wie die Herabsetzung der Verbandfestigkeit durch „Herausklopfen" der Reibung (F ö p p l) aus dem Kluftkörperverband.

Untersuchungen am anstehenden Gebirge sowie im Laboratorium haben gezeigt, daß die Tiefenwirkung vieler Sprengungen weit größer ist als allgemein angenommen wird, daß sie aber durch künstliche Maßnahmen wirkungsvoll gemildert werden kann.

Felsbereiche, denen eine tragende oder stützende Funktion im Bauwerk zukommt, erfordern deshalb eine ganz besondere Sorgfalt hinsichtlich der Auswahl unter den Möglichkeiten der Sprengtechnik. Die Vielfalt dieser Möglichkeiten erlaubt bei richtiger Wahl in jedem einzelnen Fall eine hinreichend schonende Behandlung des Gebirges, doch wird in der Praxis leider die Auswahl unter den sprengtechnischen Möglichkeiten zum Nachteil der Bauwerke und ihrer Sicherheit sehr oft nach Gesichtspunkten getroffen, welche nicht oder zumindest nicht allein maßgebend sind.

The Effect of Blasting on the Strength of a Rock Mass. Although the fact that blasting concussions tend to loosen a rock mass has always been known, the enormous reduction in strength associated with such loosening is not generally recognised. Recent investigations of the effect of blasting on a rock mass have shown that even slight loosening may markedly reduce the strength. This factor introduces a new slant on modern blasting methods some of which will have to be considered more critically.

The sensitivity of the mass to impacts depends on stress concentrations at the end of joints (Kerbspannungen) and on tensile stresses developed in the immediate vicinity of a joint, even in regions where the field stresses are not tensile. Thus, the sensitivity depends on the shape and orientation of the joints and particularly on the degree of separation of the individual joint families. The orientation of the joint families to the strains produced by blasting is just as important as the reduction in overall strength arising from the drop in friction due to the effects of the "knocking out" process (F ö p p l) on the joint body complex. Concussions may critically increase the degree of separation of the joint network thus reducing the tensile and shear strengths of the rock mass.

* Baurat h. c., Dipl.-Ing. Dr. techn. Leopold M ü l l e r, Ingenieurbüro für Geologie und Bauwesen, Salzburg, Franz-Josef-Straße 3.

Investigations both in the laboratory and on the in-situ rock mass have shown that, in many blasts, the effects penetrate to a much greater depth than generally assumed, but measures may be taken which effectively alleviate them.

The choice between various possible blasting techniques must be made very carefully, therefore, in areas where the rock mass is to have a bearing or supporting function particularly abutments of rock dams, near surface areas of high rock slopes, and the walls of underground excavations. Considering the great variety of blasting techniques available, it should be possible to select a method which, if correctly chosen, would ensure that the strength of the rock mass is adequately preserved. Unfortunately, however, in practice the choice of blasting technique is frequently decided, at least partially, on other considerations which result in unfavorable effects on the construction and its safety.

Effets des travaux à l'explosif sur la résistance du massif rocheux. On sait depuis toujours que les ébranlements dus à l'explosif affaiblissent le massif rocheux; mais on méconnait l'énorme diminution de résistance qui accompagne cet affaiblissement. Des études récentes sur les techniques du tir ont montré que même un affaiblissement insignifiant en apparence peut abaisser cette résistance d'une façon tout à fait décisive. Ceci éclaire d'un jour nouveau plusieurs méthodes modernes d'abattage des roches à l'explosif et oblige à une mise au point critique sur ces méthodes.

C'est par la concentration des contraintes aux limites des fissures (effet d'entaille), ainsi que par l'apparation de tractions dans le voisinage des fissures — et dans tout domaine où les tractions n'existaient pas — que le massif rocheux se montre sensible aux ébranlements dus à l'explosif. C'est pourquoi cette "susceptibilité" dépend de la nature et de l'orientation des fissures et tout particulièrement du degré de séparation de chaque famille. La position de ces familles, par rapport aux sollicitations dues à l'explosif, joue un rôle important, et aussi l'abaissement de la résistance globale du milieu discontinu, produit par la suppression brutale de l'adhérence (F ö p p l). Les ébranlements augmentent le degré de séparation des réseaux de fissures jusqu'à une valeur critique, ce qui diminue donc les résistances à la traction et au cisaillement du massif rocheux.

Les recherches enterprises, tant sur le massif rocheux qu'au laboratoire, ont montré que l'effet de beaucoup d'explosifs est plus profond que l'on ne le croit généralement, mais qu'il peut être atténué d'une manière efficace par des mesures appropriées.

Il faut donc apporter un soin tout particulier au choix des techniques de tir parmi les possibilités existantes, toutes les fois que le rocher a une fonction de support ou de soutien dans les ouvrages, et spécialement dans les fondations des barrages, les parties superficielles des versants de grande hauteur, et l'environnement des cavernes artificielles. Comme il y a un grand nombre de possibilités, on peut choisir dans chaque cas particulier un traitement qui ménage suffisamment le massif rocheux. Malheureusement, ce choix se réfère très souvent en pratique à des critères qui ne sont pas — ou du moins pas exclusivement — les critères décisifs. Il peut alors s'avérer désavantageux pour les ouvrages et leur sécurité.

Einleitung

Das Lösen großer Gesteinsmassen geschieht heute bedeutend rascher, zum Teil auch billiger, vielfach aber auch auf brutalere Weise als vor 10 oder 15 Jahren. Während der Felsaushub einer Talsperre von 100 bis 130 m Höhe noch vor 15 Jahren im Durchschnitt 10 bis 15 Monate dauerte, wird ein Aushub gleichen Ausmaßes heute mitunter in 4 bis 6 Monaten vollendet; die Schichtleistungen großer Felsbaustellen sind innerhalb dieses Zeitraumes von etwa 250 bis 350 m³ vielfach auf das Doppelte, ja in extremen Fällen sogar auf das Vierfache angewachsen und Tagebaubetriebe bewältigen heute bis zu 250 000 m³ pro Tag, was einer Leistung von etwa 10 000 bis 15 000 m³ pro Schicht und Abbauort entspricht.

Von einer derartigen Leistungssteigerung sind naturgemäß die Abbaubetriebe und Bauunternehmungen höchst befriedigt, weil sie die Wirtschaftlichkeit der Felsgewinnung erhöht; die Bauherren und Finanzkreise loben an ihr die Verkürzung der Ausführungs- und Amortisationsfristen, und noch vielen anderen Beteiligten bringen diese Leistungssteigerungen Vorteile. Für das Bauwerk selbst aber, im besonderen für diejenigen Bauteile, welche über die Sicherheit einer Fundierung oder die Standsicherheit einer Böschung entscheiden, ist ein derart beschleunigter Abbau mitunter sehr nachteilig; er setzt die Standfestigkeit und damit die Sicherheit und Gesamtwirtschaftlichkeit des Bauwerkes oder Abbaubetriebes herab und steht, wie im folgenden gezeigt werden soll, in mancherlei Hinsicht mit den neueren Erkenntnissen der Geomechanik nicht in Einklang, auch nicht mit Grundsätzen, wie sie erfahrene Ingenieurgeologen, z. B. unser unvergeßlicher Professor S t i n i, schon vor Jahrzehnten aufgestellt haben.

Das muß, auch wenn es nicht gerne gehört wird, mit aller Deutlichkeit gesagt und erläutert werden, damit Mittel und Wege gefunden werden, welche einerseits die Möglichkeiten der modernen Bohr- und Sprengtechnik ausnützen, andererseits aber das Gebirge schonend zu behandeln gestatten. Wie weit man ohne Schaden für das Bauwerk und seine Sicherheit mit der Beschleunigung eines Felsabbaues gehen darf, muß im Einzelfall entschieden werden.

Massenfestigkeit und Auflockerung

Je mehr wir darüber aufgeklärt werden, wie gering oftmals die Gebirgsfestigkeiten — die Massenfestigkeiten zerklüfteten Gesteins — sind, desto sorgsamer sind wir darauf bedacht, oder sollten es doch sein, diese Restfestigkeiten zu erhalten und nicht durch den Abbauvorgang weiter abzumindern. Jeder Praktiker weiß, daß Sprengarbeit fast unvermeidlich das Gebirge auflockert; im Tunnelbau sprechen wir von einem Auflockerungshof, im Talsperrenbau führen wir Kontanktinjektionen aus, um diese Auflockerung wieder wettzumachen, und M a k o v e c[1] hat aus Durchlässigkeitsuntersuchungen am Kraftwerk Ybbs-Persenbeug das Maß und die beträchtliche Tiefe der Auflockerung in einem Baugrubenaushub nachgewiesen. Nun haben uns bereits die ersten technologischen Großversuche am Gebirgskörper gezeigt, wie sehr und wie rasch mit einer Gesteinsauflockerung die Gebirgsfestigkeiten abnehmen. Das Diagramm (Abb. 1) gibt einen solchen Zusammenhang zwischen der Gebirgs-Beanspruchung σ_3, der Volumenänderung $\frac{\Delta V}{V}$ und der Querdehnungszahl m quer zur Hauptdruckrichtung, welcher in einem Großversuch in Japan ermittelt wurde. Dieser Zusammenhang stammt zwar aus einem Triaxialversuch, bei welchem die Kraftaufbringung das Primäre, die Auflockerung deren Folge war. Wir haben aber allen Grund, anzunehmen, daß eine ähnliche Beziehung auch bei umgekehrtem Kausalnexus, nämlich bei Entfestigung durch Auflockerung, gegeben ist (Abb. 2). Schon eine Aufdehnung von 1 %, 2 % oder 3 % (Raumteile) kann bei vielen Gefügetypen wahrscheinlich Entfestigungen von 50 %, 80 % bzw. 90 % im Gefolge haben. Sprenglockerungen von 1 %, ja selbst von 5 % sind in Oberflächennähe keine Seltenheit, wie man durch Messungen der Kluftöffnungen vor und nach einer Sprengung unschwer feststellen kann. Aber auch noch 5 m unter der Ausbruchbegrenzung sind häufig noch Auflockerungen von 2 bis 3 % zu beobachten.

Man muß daher stets mit einer Entfestigung durch Sprenglösung rechnen, welche erheblich größer ist, als die meisten wohl erwarten. Wie tief reichen solche Entfestigungen? Welches sind ihre Ursachen, welches ihre Folgen? Wie kann man sie verstehen und wie ihnen begegnen?

Daß Sprengungen das Gebirge auflockern, ist nichts neues; das haben die Praktiker schon immer gewußt. Neu ist aber die Erkenntnis, *wie groß* die daraus entspringenden Festigkeitsverluste sein können.

Die mechanischen Wirkungen von Sprengerschütterungen

Die besondere Empfindlichkeit des Gebirges gegen Sprengerschütterungen wird verständlich, wenn man sich daran erinnert, daß das Gebirge in den meisten Fällen ein bereits durch und durch zerbrochener, klüftiger Körper ist. Ein Material, bei

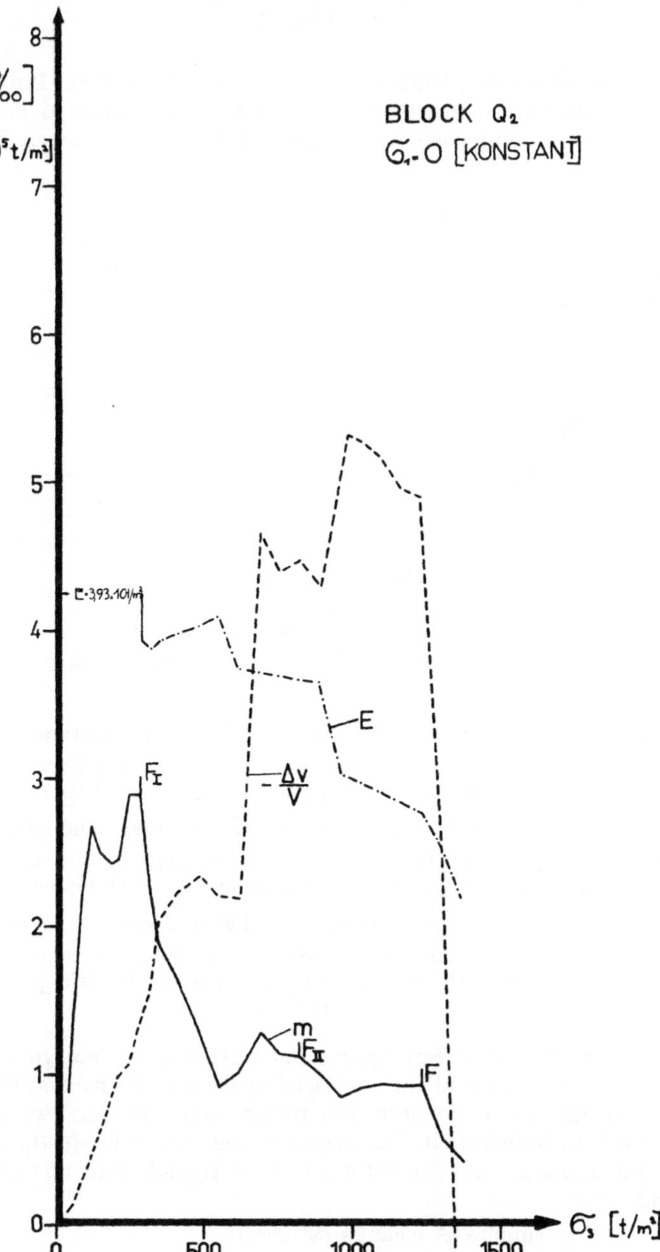

Abb. 1. Volumen-Zu- und Abnahme $\frac{\Delta V}{V}$ (‰) im Verlaufe eines Triaxialversuches in situ Biotit-Granit vom Kurobe-Tal, Japan

σ_3 negativ größte Hauptnormalspannung; F_I, F_{II} Fließgrenzen; m Poisson-Zahl

Volumetric change $\frac{\Delta V}{V}$ (‰) during a triaxial in-situ-test of biotite granite in the Kurobe valley in Japan

σ_3 negative major principal stress; F_I, F_{II} flow limits; m Poisson's ratio

Variation du volume $\frac{\Delta V}{V}$ (en 10⁻³) au cours d'un essai triaxial in situ sur le granite à mica noir de la vallée de Kurobe au Japan

σ_3 est la contrainte principale négative la plus élevée; F_I et F_{II} sont les limites d'écoulement plastique; m est l'inverse du coefficient de Poisson

welchem schon geringe Energiemengen — bei freilich großem Impuls — genügen, um beträchtliche Massen zu lösen und abzuwerfen, *muß* auch in denjenigen Teilen, welche nach der Sprengung eben noch stehen bleiben, erheblichen Schaden leiden*.

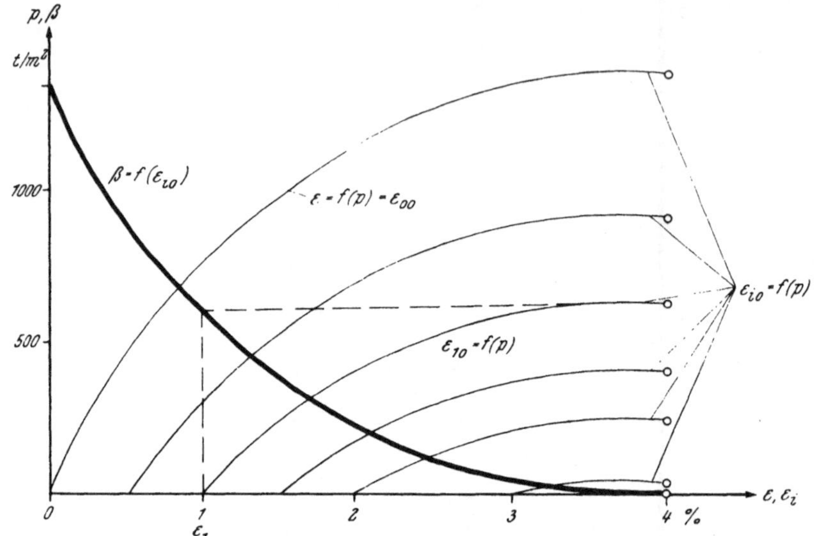

Abb. 2. Zusammenhang zwischen Querdehnung bzw. Auflockerung und Festigkeit

p Beanspruchung; β Festigkeit; ε Querdehnung; ε_i Anfangsauflockerung; ε_{00} Querdehnung bei Anfangsauflockerung bei $\varepsilon_i = 0$; ε_{i0} Querdehnung bei Anfangsauflockerung ε_i

Relation between lateral expansion or loosening, and strength

p stress; β strength; ε lateral expansion; ε_i initial loosening; ε_{00} lateral expansion at initial loosening $\varepsilon_i = 0$; ε_{i0} lateral expansion at initial loosening ε_i

Relation entre l'extension transversale (ou l'affaiblissement) et la résistance

p contrainte; β resistance; ε extension transversale; ε_i affaiblissement initial; ε_{00} extension latérale pour l'affaiblissement initial $\varepsilon_i = 0$; ε_{i0} extension latérale pour l'affaiblissement initial ε_i

Wie tief die Wirkung einer Sprengung ins anstehende Gebirge reicht, haben für kontinuierliche Massen vor allem Langefors[2] und Rinehart[3] untersucht. Im zerklüfteten Gebirge kann sie noch wesentlich tiefer reichen, weil eine Reihe von Umständen die Empfindlichkeit des Gebirges gegenüber Entfestigung durch Sprengungen, Erschütterungen und Vibrationen im Vergleich zum unzerklüfteten Gestein erhöhen, und zwar:

a) seine viel geringere Ausgangsfestigkeit;

b) die hohe Kerbwirkung bzw. Anweisungswirkung (Spannungskonzentration) am blinden Ende von Klüften;

c) das Auftreten von Zugspannungen an den Klufträndern;

d) der plötzliche Verlust der Reibung auf den Kluftflächen;

e) Schockwellen-Interferenzen.

a) Gebirgsfestigkeit. Über den großen Unterschied zwischen Gebirgs- und Gesteinsfestigkeit, welche zueinander in einem Größenverhältnis von 1 : 1 bis 1 : 200 stehen können, wurde im Kreise der Salzburger Kolloquien schon so viel gesprochen,

---

* Die Energiemengen einer Sprengung sind ja im Durchschnitt so gering, daß mit ihnen bei Umsetzung in Wärme-Energie die zu lösende Felsmasse kaum um einen Grad Celsius erwärmt würde.

daß es sich erübrigt, darauf näher einzugehen. Es leuchtet ein, daß ein Material von so geringer Massenfestigkeit durch eine Stoßbeanspruchung viel mehr geschädigt wird als ein unversehrtes Gestein; wie eben ein kranker Mann weniger aushält als ein gesunder.

b) **Kerbwirkungen.** Die Klüfte, welche die Ursachen der Festigkeitsminderung in der Gebirgsmasse sind, verursachen örtlich hohe Spannungskonzentrationen. Die Mitglieder dieses Kolloquienkreises erinnern sich der spannungsoptischen Versuche,

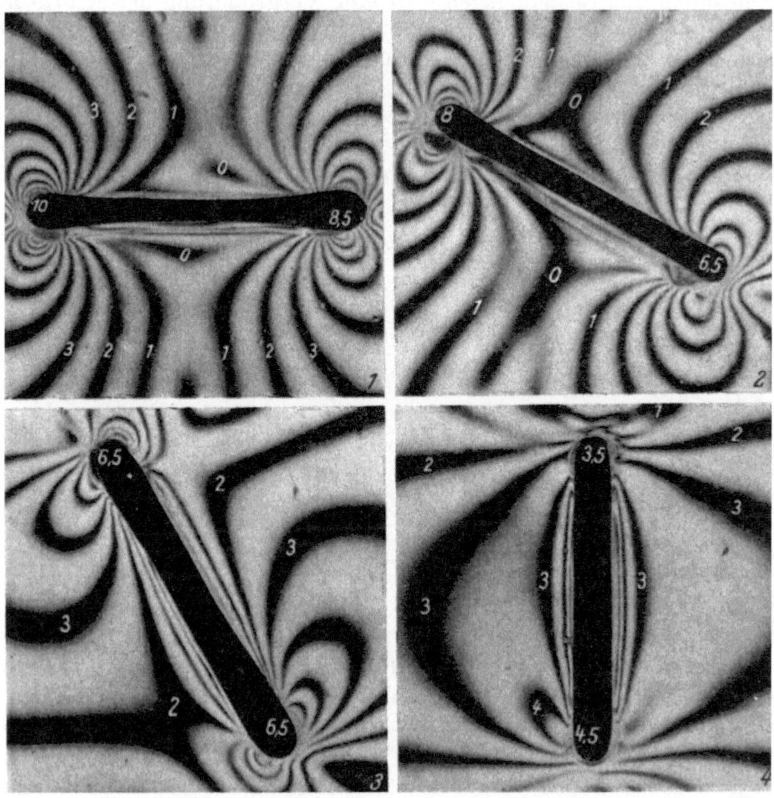

Abb. 3. Beanspruchung des Gebirges in der Umgebung einer Kluft in Abhängigkeit ihrer Stellung zur (lotrechten) Hauptspannung σ_3 (aus Sonntag[5])

Variation of rock stresses in the vicinity of a joint for varying orientation to the direction of the (vertical) principal stress σ_3 (from Sonntag[5])

Répartition des contraintes au voisinage d'une fissure suivant son orientation par rapport à la contrainte principale (verticale) (d'après Sonntag[5])

mit welchen Föppl[4] und Sonntag[5] den Spannungszustand des Gebirges in der Umgebung von Klüften demonstriert haben. Die Abb. 3 soll die Zeit der ersten Kolloquien, in denen manche grundlegende Erkenntnisse erarbeitet wurden, in Erinnerung bringen; sie zeigt, daß am blinden Ende einer Kluft insbesondere dann eine Spannungskonzentration auftritt, wenn die Kluft quer zum Kraftfluß streicht.

c) **Zugspannungen.** War vor der Sprengung schon eine Zugspannung quer zu Klüften vorhanden, so läßt sich leicht einsehen, daß eine Sprengerschütterung vorhandene Kerbspannungen noch erhöhen kann; denn die Longitudinalwellen, welche von einem Erschütterungsherd ausgehen, bestehen ja aus einer abwechselnden

Stauchung und Dehnung des Materials. Und wenn die Formänderungsbeträge einer
Stoß-Fortpflanzung auch nicht groß sind, so erfolgen andererseits die Formänderungen so rasch, daß das Material, welches ohnehin spröde ist, nur noch spröder
reagiert, so daß sich der Riß verlängert.

Aber auch wenn ein anderer, allgemeinerer Spannungszustand im Fels herrscht,
sind Zugspannungen an den Klufträndern, bzw. an einem Teil derselben, vorhanden,
wie Griffith[6] nachgewiesen hat. In einem solchen Falle werden nicht nur die
Longitudinalwellen, sondern auch die Transversalwellen, welche ja Schubbean-
spruchungen durch das Gestein jagen, auf eine Vergrößerung der Klüfte hinwirken;
sie werden die Zugspannungen an den Klufträndern vergrößern, auch wenn, groß-
räumig betrachtet, im Gestein ein Spannungszustand herrscht, welcher keine Zug-
spannungsanteile aufweist.

Daß die Bruchtheorie von Griffith (Abb. 4) der in der Natur vor sich gehen-
den Bruchbildung tatsächlich entspricht, läßt sich experimentell demonstrieren.
Der Verfasser konnte ein dem Gedankenmodell entsprechendes Beispiel aber auch

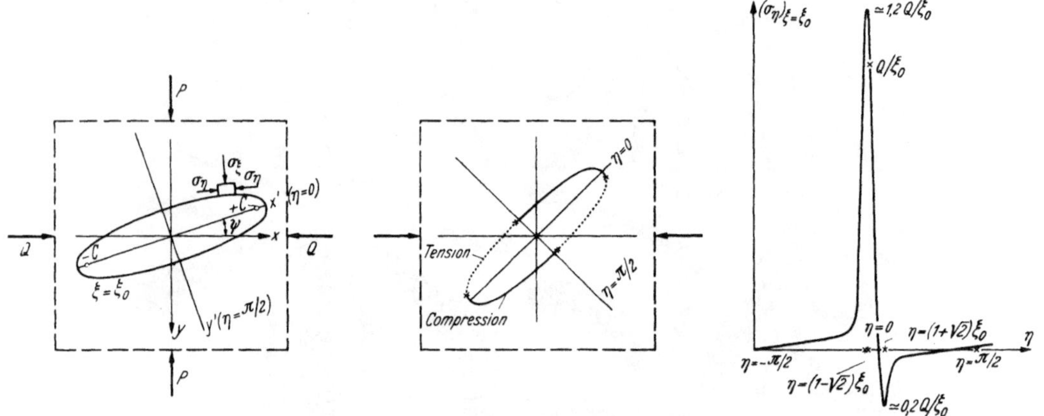

Abb. 4. Spannungsverlauf am Rande einer elliptischen Kluft nach Griffiths[6] Bruchtheorie
a allgemeiner Ansatz; *b* Verteilung von Druck- und Zugspannungen an einem unter 45°
zur Hauptdruckrichtung geneigten Anriß von elliptischer Gestalt; *c* Größe der Tangential-
spannung $(\sigma_n)\,\xi = \xi_0$ am Umfang dieses Risses (nach Odé, aus Müller[19])

Stress distribution around an elliptical joint according to Griffith's[6] failure theory
a diagram; *b* distribution of compression and tensile stresses around an elliptical fracture,
inclined at 45° to the direction of the principal stresses *P* and *Q*; *c* Magnitude of the
tangential stress $(\sigma_n)\,\xi = \xi_0$ on the wall of the fracture (after Odé, from Müller[19])

Distribution des contraintes autour d'une fissure à section elliptique d'après la théorie
de Griffith sur la rupture[6]
a schéma général; *b* répartition des compressions et des tractions autour d'une fissure de
forme elliptique inclinée à 45° sur la direction des contraintes principales de compression;
c grandeur des contraintes tangentielles sur la paroi de cette fissure $(\sigma_n)\,\xi = \xi_0$
(d'après Odé, dans Müller[19])

in der Natur, und zwar in der Skandinavischen Schieferhülle, beobachten (Abb. 5):
An einer 120 cm langen Kluft erlaubte die im spitzen Winkel zu dieser streichende
Schieferung eine genaue Ablesung der Beträge, um die sich die Kluftränder gegen-
einander verschoben hatten. Die kinematische Analyse dieser Relativverschiebungen
ergab eindeutig das Vorhandensein zweier Zugzonen und zweier Druckzonen an den
Klufträndern. Diese Zonen unterscheiden sich allerdings von der Spannungsverteilung, welche Griffith annimmt, durch ihre unterschiedliche Lage und Intensität.

In diesem Unterschied können wir den Einfluß der Kluftreibung und des rheologischen Verhaltens erkennen, welche die G r i f f i t h - Theorie bekanntlich vernachlässigt. B r a c e[7] ist um die Erforschung dieses Einflusses bemüht, wie wir auf der Felsmechanik-Tagung in Santa Monica gehört haben.

d) **Reibungseinbuße.**

Aus den Untersuchungen über Haft- und Gleitreibung, welche F ö p p l[8—11] vor neun Jahren hier vortrug, wurde bekannt, daß Erschütterungen die Reibung, welche auf Kluftflächen herrscht, sozusagen „herausklopfen". Diese Reibung ist aber, wie P a c h e r s[12] Theorie der Verbandwiderstandsziffern und Versuche der Internationalen Versuchsanstalt für Fels in Salzburg lehren, ein wesentlicher Teilfaktor der Festigkeit zerklüfteter Massen.

Auch der vorübergehende Verlust der Reibung infolge der Sprengerschütterungen mindert die Gebirgsfestigkeiten ab. Dies ist insbesondere in solchen Gebirgsverbänden bedeutungsvoll, deren Festigkeit nicht auf der Festigkeit der Materialbrücken, sondern mehr oder weniger auf Reibungsschluß beruhen. Es sind dies der Bausteinkasten- und der Mauerwerksverband, welche gar keine Materialbrücken zwischen den Kluftkörpern besitzen. T i n c e l i n und S i n o u[13] haben zwar durch Messungen in Lothringischen Gruben nachgewiesen, daß die durch Sprengungen ausgelöste Entfestigung

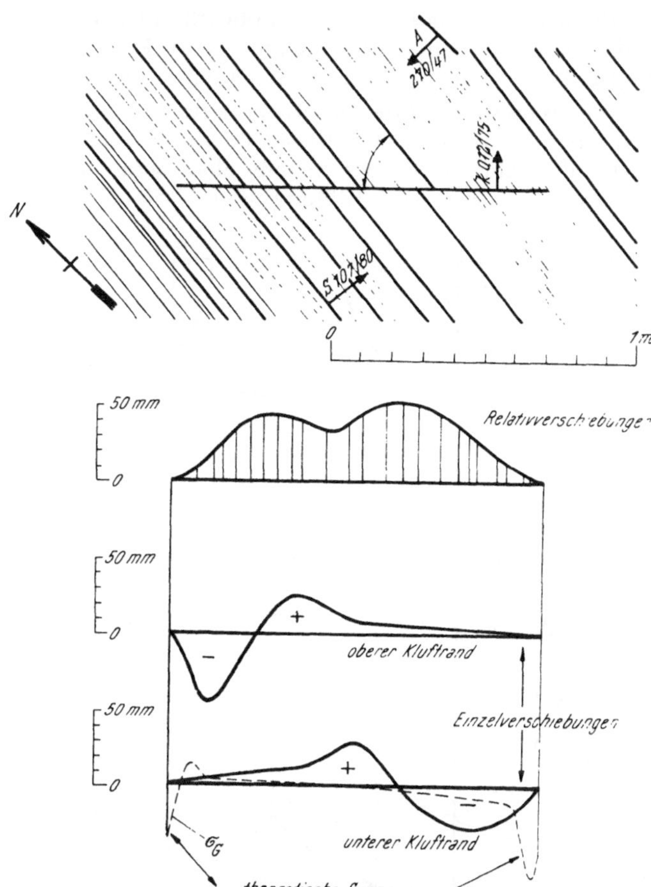

Abb. 5. Relativ-Verschiebungen der Schnittränder einer Verschiebungskluft in der skandinavischen Schieferhülle

a Ansicht (*A* Gesteinsoberfläche, *S* Schieferung, *k* Kluftfläche); *b* Relativ-Verschiebungen der Schnittränder; *c* rekonstruierte Einzelverschiebungen des oberen Schnittrandes, *d* des unteren Schnittrandes; σ_G theoretische Spannungen nach G r i f f i t h (siehe Abb. 4)

Relative displacements of the edges of a joint in Scandinavian schist cover

a situation (*A* rock surface, *S* planes of schistosity, *k* joint); *b* relative displacements of the joint edges; *c* individual displacements of the upper edge, reconstructed; *d* same for the lower edge; σ_G theoretical stress distribution according to G r i f f i t h's theory (see fig. 4)

Déplacements relatifs des parois d'une fissure dans un schiste scandinave

a schéma (*A* surface du rocher, *S* schistosité, *k* fissure); *b* déplacement relatif des parois de la fissure; *c* reconstitution du déplacement de la seule paroi supérieure; *d* idem pour la seule paroi inférieure; *e* distribution théorique des contraintes d'après G r i f f i t h (voir fig. 4)

der Stollenfirsten nach 3000 bis 5000 Stunden wieder zurückgeht, doch kommt der Praktiker nur selten in den Genuß dieser Festigkeitserholung; denn die Auflockerungsfolgen treten in den meisten Fällen früher ein.

e) **Schockwellen-Interferenzen.** Durch alle diese Umstände kommt es zu einer Vergrößerung vorhandener Risse, damit zu einer Erhöhung des Durchtrennungsgrades des Kluftnetzes. Es kommt aber auch zur Bildung von Neuklüften. Nach Rinehart[3] lassen sich Bildung und Verlauf von Neuklüften als geometrischer

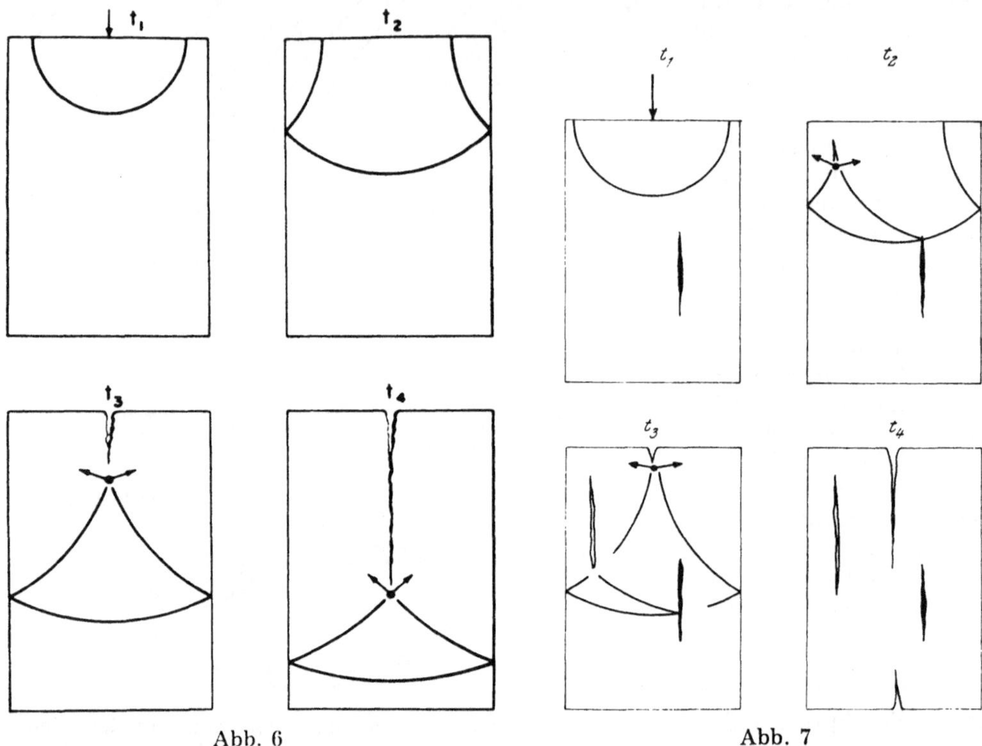

Abb. 6   Abb. 7

Abb. 6. Entstehung eines Bruches durch eine Stoßwelle und deren Reflexion
(nach Rinehart[3])
Generation of a fracture under impulsive loading, by reflection of the longitudinal wave
(after Rinehart[3])
Naissance d'une fracture par réflexion d'une onde de choc (d'après Rinehart[3])

Abb. 7. Entstehung mehrerer Brüche durch Stoßwellen-Reflexion bei Anwesenheit einer Kluft
Generation of several fractures by reflection of longitudinal waves influenced by a pre-existing joint
Naissance de plusieurs fractures par réflexion d'une onde de choc en présence d'une fissure initiale

Ort der Schnittpunkte zusammentreffender Longitudinalwellen konstruieren. Die Abb. 6 zeigt den Verlauf einer solchen Kluftbildung in vier Zeitabschnitten t_1, t_2, t_3 und t_4. Abb. 7, auf die gleiche Weise konstruiert, zeigt den Einfluß eines in einem sonst gleichgeformten Kluftkörper schon vor der Stoßwelle vorhandenen Risses: das Endergebnis zur Zeit t_4 ist eine bedeutend intensivere Zerteilung des Kluftkörpers, als sie sich bei dem Kluftkörper gleichen Formates ohne Risse in seinem Inneren ergab. Je nach der Lage und Richtung eines vorhandenen Risses orientieren sich die Neuklüfte verschiedenartig (Abb. 8).

Dieser Effekt der Stoßrefraktion an den Grenzen der Kluftkörper schafft also Neuklüfte und erhöht demnach die Klüftigkeitsziffer wie den Durchtrennungsgrad des Kluftnetzes ganz erheblich.

Der Einfluß des Durchtrennungsgrades

Der Durchtrennungsgrad, der durch alle die genannten Einflüsse erhöht wird, ist aber, wie in früheren Kolloquien des öfteren gezeigt wurde, derjenige Faktor, welcher wie kein anderer die Abminderung der Gesteinsfestigkeit zur Gebirgsfestigkeit bestimmt (s. Diagramm Abb. 9). Wo noch Materialbrücken vorhanden sind,

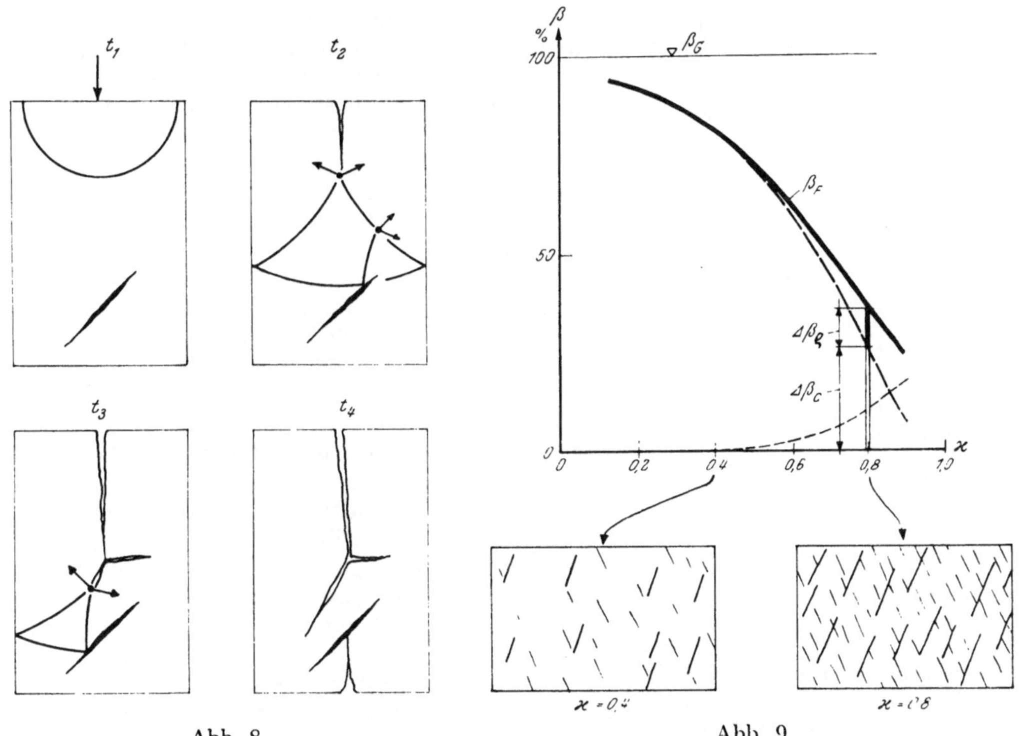

Abb. 8   Abb. 9

Abb. 8. Wie Abb. 7, bei allgemeiner Kluftstellung
Same as fig. 7, for general orientation of the pre-existing joint
Comme fig. 7, pour une orientation quelconque de la fissure initiale

Abb. 9. Abhängigkeit der Gebirgsfestigkeit vom Durchtrennungsgrad des Kluftnetzes
β Druckfestigkeit; $\varkappa$ Durchtrennungsgrad; β_G Gesteinsfestigkeit (Substanzfestigkeit); β_F Gebirgsfestigkeit (Massenfestigkeit); $\Delta\beta_\varrho$ reibungsbedingter Festigkeitsanteil; $\Delta\beta_C$ Materialfestigkeitsanteil

Relation between strength of rock mass and degree of separation of the joint network
β compression strength; $\varkappa$ degree of separation; β_G rock strength (strength of the substance); β_F strength of the rock mass; $\Delta\beta_\varrho$ strength component due to joint friction; $\Delta\beta_C$ strength component due to shear strength

Relation entre la résistance du massif rocheux et le degré de séparation du réseau de fissures
β résistance à la compression; $\varkappa$ degré de séparation; β_G résistance du materiau rocheux; β_F résistance du massif rocheux; $\Delta\beta_\varrho$ partie de la résistance due au frottement; $\Delta\beta_C$ partie de la résistance due à la résistance propre du matériau

werden diese durch die infolge der Sprengung hervorgerufene Verlängerung und Vermehrung der Klüfte verringert; wo die Materialbrücken aber bereits verloren gingen und Kluft an Kluft grenzt, so daß die Verbandfestigkeit nur noch aus Reibungswiderständen besteht, geht sie durch das „Herausklopfen" der Reibung — wenigstens vorübergehend — zum großen Teil verloren.

Eine Beeinträchtigung der Gebirgsfestigkeit durch die geschilderten Umstände findet bei jedem Sprengstoß statt. Je mehr Stöße, desto größer der Festigkeitsverlust. Darauf beruht der Gedanke der Millisekunden-Sprengtechnik: Der Stoß, der beim Millisekunden-Schießen durch das Gebirge geht, ist an sich nicht geringer als der der einzelnen Sprengladungen einer Abschlag-Serie; aber es ist praktisch nur *ein* Stoß anstelle einer ganzen Folge von Erschütterungen, bei welchen jede einzelne die vom vorangegangenen Sprengstoß geschaffenen oder erweiterten Risse im Gebirge abermals progressiv vergrößert.

Jeder Sprengstoß erhöht den Durchtrennungsgrad und macht das Gebirge empfindlicher gegen Erschütterungen, so daß jeder folgende Stoß größere Auflockerungen erzeugt als der vorangegangene.

Die Vergrößerung des Durchtrennungsgrades sowie der Verlust der Kluftreibung setzt die Zugfestigkeit und die Scherfestigkeit des Gebirges herab. Dies und die geringere Ausgangsfestigkeit sind der Hauptgrund, weshalb zerklüftetes Gebirge um so viel empfindlicher gegen Sprengerschütterungen ist als unzerklüftetes Gestein. Wir haben aber noch andere Faktoren zu betrachten.

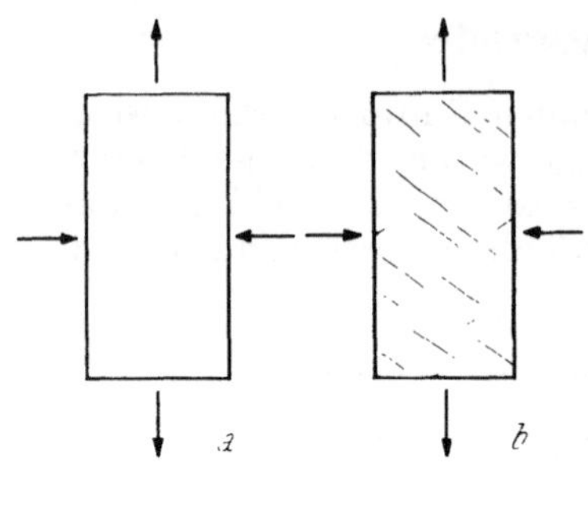

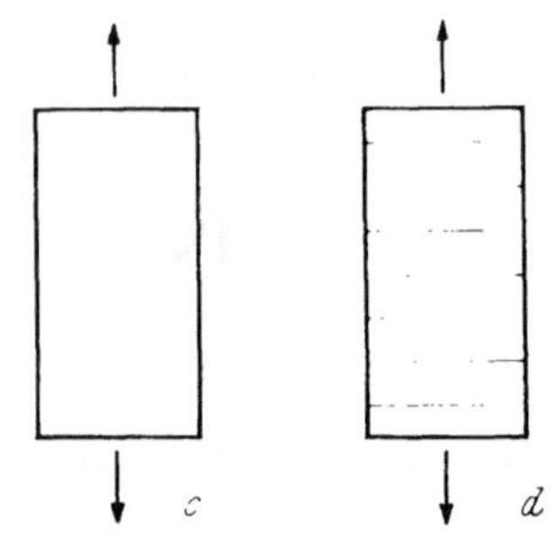

Abb. 10. Änderung der elastischen Eigenschaften durch Zerklüftung

a ungeklüfteter, *b* geklüfteter Felsblock unter Scherbeanspruchung; *c* und *d* Vergleich mit einem kompakten und einem eingeschnittenen Federstab unter einachsiger Beanspruchung

Alteration of the elastic properties by jointing

a unjointed, *b* jointed rock under shear stress; *c* and *d* comparison with a compact and a cut in spring byr under uniaxial load

Changement des propriétés élastiques par la fissuration

a milieu continu; *b* bloc rocheux fissuré soumis au cisaillement; *c* et *d* comparaison avec le comportement d'une tige élastique en compression simple, intacte puis avec une entaille

Der Einfluß der elastischen Eigenschaften des Gebirges

Die Fortleitungsgeschwindigkeit von Schall- und Stoßwellen ist im zerklüfteten Gebirge bekanntlich wesentlich (zwei- bis viermal) kleiner als im gleichen, aber unzerbrochenen Gestein. Das hat zwei Ursachen: Einmal zwingt der Umweg um gewisse Klüfte die Wellen, einen längeren Weg zu nehmen; zum anderen aber gleicht das elastische System sowohl hinsichtlich des Durchganges von Longitudinal- wie auch von Transversalwellen dem einer Feder (Abb. 10), in welcher der Kraftweg verlängert ist und die Schwingungen nicht mehr geradlinig, sondern als Biege- und Torsionswirkungen durchgeleitet werden. Dies macht das Gebirge gegenüber dem Gestein weicher und befähigt es zu größeren Formänderungen ohne eigentlichen

Bruch. Dadurch wird die Erschütterungsempfindlichkeit des Gebirges, welche durch die vorher betrachteten Wirkungen so sehr erhöht wird, etwas gemindert; allerdings nur in Verbänden mit nicht allzu hohem Durchtrennungsgrad, vorwiegend im „Verschränkten Kluftkörperverband".

Ein anderer Faktor, den wir betrachten müssen, ist die Dämpfung der Stoß-wellen, welche umso größer ist, je stärker das Gebirge zerklüftet ist, wie u. a. Buchheim[14] erläutert hat. Auch die Dämpfung begrenzt glücklicherweise die Tiefenreichweite der Erschütterungen. Sie besteht darin, daß auf dem Wege der Stoßfortleitung Energie aufgezehrt wird. Aber eben dieses Aufzehren von Energie bewirkt erhöhte Zerbrechung des Gebirges bis zur Zerstückelung, ja Pulverisierung; die Dämpfung bewirkt also Begrenzung der Auflockerungszone auf Kosten stär-kerer Auflockerung.

Der Auflockerungshof

Dies ist der Grund, weshalb wir oft recht deutlich einen mehr oder minder klar begrenzten Auflockerungshof beobachten können. Wir dürfen nur aus solchen Beobachtungen nicht auf allzu geringe Maße der Auflockerungszone schließen, wie

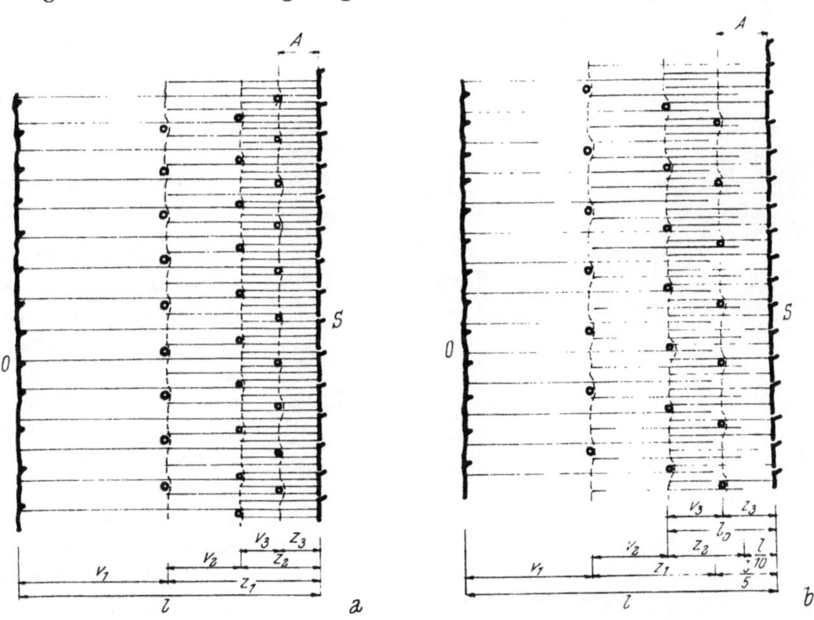

Abb. 11. Zweckmäßiger Abstand der Sprenglöcher von der Ausbruchgrenze bei Annäherung an dieselbe unter der Voraussetzung einer Auflockerungstiefe gleich der Vorgabe
a theoretisch, b praktisch; O Oberfläche; S Sollausbruch; l Ausbruchtiefe; v Vorgabe; z Auflockerungstiefe; A Ausbruch ohne Sprengung

Optimum distance of blasting holes from the excavation limit approaching this limit
Assumed depth of loosening equal to overburden
a theoretical, b practical; O surface; S pay-line; l depth of excavation; v overburden; z depth of loosening; A excavation without blasting

Distance convenable entre les trous de mine et la limite des fouilles, lorsqu'on s'en approche, en supposant une profondeur d'affaiblissement égale à la couverture
a theorique, b pratique; O terrain naturel; S limite à atteindre; l profondeur des fouilles; v couverture; z épaisseur de l'affaiblissement; A zone à excaver sans explosif

es aus Zweckoptimismus oft geschieht. Nach Langefors[2] reichen Rißbildungen im Kontinuum bei spröden Körpern etwa so weit ins Innere, wie die Vorgabe tief ist. In zerklüfteten Felsmassen müßten wir eine viel größere Tiefenwirkung an-nehmen, weil diese bedeutend erschütterungsempfindlicher sind.

Makovec[1] hat nachgewiesen, daß die Auflockerung unter einer Wehrbaugrube, welche gewiß mit bescheidenen Vorgaben von 1 bis 1,5 m ausgesprengt wurde, 8 bis 9 m tief reichte.

Beobachtungen in Japan haben interessante Aufschlüsse gegeben. 5 m hinter einer Sprengkammer von 13 m Vorgabe und 315 kg maßgebender Ladung wurde ein Stollen zum Einsturz gebracht; 15 m tiefer im Gebirge wurde ein zur Beobachtung aufgebrachter Mörtelverputz rissig und platzte ab. Bis 15 m hinter die Sprengkammer waren die Klüfte nachweislich teilweise bis zu 25 mm weit geöffnet worden. Wir müssen annehmen, daß die mit freiem Auge nicht mehr feststellbare Auflockerung des Gebirges, die Erhöhung seiner Rissigkeit, viel tiefer, etwa 30 m, in den Berg hinein reichte.

Kriterien der Sprenglockerung

Versuchen wir zu folgern, worauf es bei der Beurteilung der Entfestigung durch Sprengungen ankommt, so können wir, solange keine weiteren Versuchsergebnisse vorliegen, leider noch wenig quantitative Angaben machen, aber doch etliche wichtige Schlüsse ziehen:

a) Große Vorgaben erhöhen die Tiefenreichweiten der Gesteinslockerung. Sie müssen daher überall dort vermieden werden, wo das Gebirge geschont werden soll, also besonders im Hohlraum- und im Staumauerbau. Bei großer Tiefenabmessung der zu lösen-

Abb. 12. Schonender Felsabbau bei Anordnung enggestellter Großlochbohrungen (Werkphoto Atlas-Copco)
Careful rock excavation using closely spaced boreholes (Atlas-Copco industrial photograph)
Abatage ménagé par trous de mines rapprochés (photographie Atlas-Copco)

den Felsmasse müssen daher die Vorgaben mit Annäherung an die beabsichtigte Ausbruchgrenze etwa entsprechend Abb. 11 schrittweise vermindert werden.

b) Parallel zur Felsoberfläche angesetzte Bohrlöcher sind bei sonst gleicher Lademenge je Volumeneinheit des zu lösenden Gebirges günstiger als stumpfwinkelig zur Oberfläche gebohrte.

c) Hohe örtliche Konzentration von Sprengladungen, wie sie Kammersprengungen, aber auch Großlochsprengungen, mit sich bringen, muß als besonders verbandlockernd und daher entfestigend und in ihrer Wirkung tiefreichend angesehen werden. Sie sollte besonders beim Ausbruch von Felswiderlagern gewölbter Talsperren nur unter besonderen Vorkehrungen und beschränkt angewendet werden. Mit Verstand eingesetzt, können Großlochbohrmaschinen durchaus gesteinsschonend verwendet werden (Abb.12).

d) Verteilte Ladungen, wie sie bei Kleinkaliber-Sprengungen, besonders aber beim Smooth-Blasting, angewendet werden, schonen das Gebirge, erhalten seine Festigkeit, und ihre Auflockerungswirkung reicht nur wenig tief. Nicht oder nur teilweise geladene Bohrlöcher gewähren, wie die Experimente von Langefors[2] beweisen, den besten Schutz gegen tiefreichende Auflockerung des Berginneren.

e) Hohe Brisanz der Sprengmittel scheint, soweit dies heute schon beurteilt werden kann, das Gebirge bei geringer Vorgabe (bzw. Verspannung) zu schonen, bei großer Vorgabe (bzw. Verspannung) die Auflockerung und Entfestigung zu vergrößern.

f) Die Millisekunden-Sprengtechnik vermindert die Stärke der Auflockerung wesentlich, die Tiefenreichweite der Auflockerung jedoch in geringem Maße.

g) Wasserfüllung der Klüfte vermehrt die Stoßwirkung und deren Tiefenreichweite sehr wahrscheinlich bedeutend.

h) Je höher der Durchtrennungsgrad des Kluftnetzes, desto größer ist der Einfluß der unter a) bis f) aufgezählten Einflüsse.

i) Die Stellung der Kluftscharen zur Felsoberfläche ist von wesentlichem Belang. Diesbezüglich geben die alten Bergmanns- und Steinbruchregeln Anhalt.

Schlußfolgerungen

Wenn man erlebt, wie brutal der Fels mit den Mitteln der neueren Technik auf vielen Baustellen mißhandelt wird, dann könnte man zu der Ansicht gelangen, die letzte Entwicklung der Sprengtechnik entspräche nicht in allen Dingen den Erkenntnissen der Geomechanik. Ein solches Urteil wäre voreilig: Nicht die Sprengtechnik ist schuld, wenn man sie vielfach aus Unkenntnis geomechanischer Zusammenhänge fehlerhaft anwendet.

Und doch ist ein peinlicher Anachronismus festzustellen: In eben den Jahren, in denen die Praxis zu immer robusteren, um nicht zu sagen brutaleren Methoden geschritten ist, um die Leistungen zu erhöhen und die Kosten der Gesteinslösung herabzusetzen, hat die Geomechanik aufgezeigt, wie gering oftmals die Gebirgsfestigkeiten sind; in welchem Maße die Standfestgkeit von Felsböschungen von den Festigkeitseigenschaften der äußeren, meist aufgelockerten Schicht abhängt; wie sehr es bei Felswiderlagern von Talsperren auf die Erhaltung der Festigkeit ankommt; und in welchem Maße der Gebirgsdruck in Tunneln und Kavernen von der Auflockerungsentfestigung abhängt; mit einem Wort: wie sehr das Gebirge, welches ein Teil unserer Baukonstruktionen ist, äußerster Schonung und pfleglicher Behandlung bedarf.

Diese Erkenntnisse kommen, obwohl von Stini, Kahler[15] und anderen längst voraus verkündet, für die breitere technische Öffentlichkeit sozusagen zu spät. Niemand wird erwarten, daß die Felsbaupraxis auf dem beschrittenen Wege umkehrt. Die angeschafften Großlochbohrmaschinen verlangen nach Einsätzen, die großen Ladegeräte und Bagger wollen mit entsprechenden Tagesleistungen gefüttert werden, und sei es gegen bessere Einsicht. Nichts anderes wird übrig bleiben, als durch eifriges geomechanisches Studium erlaubte Grenzen der Gebirgsbehandlung für verschiedene Zwecke und Gefügetypen abzustecken und von den Schonung verheißenden Auskunftsmitteln der Millisekunden-Sprengtechnik und des Smooth-Blasting häufiger Gebrauch zu machen. Bei denjenigen Felsteilen aber, auf deren

Standfestigkeit es in erster Linie ankommt, wird man die Anwendung konservativer Abbautechniken fordern und wenn nötig erzwingen müssen. Das wird möglich sein, wenn diese Beschränkung weise und nicht diktatorisch gehandhabt wird. Das allererste aber, womit begonnen werden muß, ist Aufklärung und Vertiefung der Einsicht aller an der Felsarbeit Beteiligten.

Die Sicherheit und die Wirtschaftlichkeit des *Gesamtbauwerkes* muß als maßgebend erachtet werden und nicht die Wirtschaftlichkeit von Einzelleistungen. Die Unternehmungen dürfen nicht allein auf die Kosten der Lösungsarbeit blicken; die Bauherren müssen sich abgewöhnen, übertriebene Terminforderungen zu stellen und eine Baufirma lediglich nach deren Einhaltung oder Nichteinhaltung zu beurteilen. Zu kurze Bautermine, welche Verspätungen der Bauvorbereitung wettmachen sollen, stehen der Sorgfalt der Ausführung im Wege, vermindern die Sicherheit und erhöhen die Kosten. Allein eine rechtzeitige Planung ermöglicht eine solide und sichere Baudurchführung.

Bei forcierter Abbauweise müssen Felsböschungen in vielen Fällen um 10 bis 20^0 flacher geböscht werden, als bei schonender Felsbehandlung erforderlich wäre. Was man an Kosten der Gesteinslösung je Raumeinheit gewinnt, legt man, ohne sich darüber klar zu werden, durch wesentliche Kubaturvermehrung mehrfach zu.

Häufig müssen Talsperrenwiderlager nachträglich vertieft, künstlich gesichert, nachinjiziert, zusätzlich drainiert, verankert oder gar aufgegeben werden, weil man sich rasche Fertigstellungstermine eingebildet und nur die Wirtschaftlichkeit der Felsarbeit, nicht die des Gesamtbauwerkes im Auge hatte. Die damit verbundene Einbuße an Sicherheit kann heute angesichts der Größe derartiger Bauwerke nicht mehr verantwortet werden.

Der Gebirgsdruck, welcher den Ausbau von Stollen, Tunneln, Schächten und Kavernen belastet, kann, wie wir aus den Untersuchungen von Rabcewicz[16, 17] und Pacher[18] gelernt haben, auf einen Bruchteil der bisher gewohnten Maße reduziert werden, wenn es gelingt, Auflockerung und damit Entfestigung des Gebirges weitgehend zu vermeiden. Ausgerechnet jetzt aber, da wir im Tunnelbau am liebsten fräsen und schrämen oder mit dem Rüttelmeißel abbauen möchten anstatt zu sprengen, um das Gebirge nicht unnötig zu lockern, ausgerechnet jetzt sind forcierte Vortriebe mit Tagesleistungen von 10 bis 30 m Mode geworden. Diese lockern, wenn keine Gegenmaßnahmen ergriffen werden, das Gebirge weit mehr auf, als es die alten Vortriebsweisen taten. Das bedeutet stärkere Auskleidungen, bei Wasser- und Druckstollen aber überdies einen erhöhten Verbrauch an Injektionszement und -arbeit, welche wieder notdürftig zusammenkitten müssen, was die nach Wirtschaftlichkeit und Bauterminen einseitig beurteilte Vortriebsarbeit zerstört hat. Am meisten hat sich die Erkenntnis von der Überlegenheit der Schrämarbeit untertage gegenüber der Sprengarbeit im Bergbau durchzusetzen begonnen. Aber auch im Ingenieurbau gibt es schon vereinzelt Fachleute, welche die Zusammenhänge klar erkennen*.

Forcierte Ausbrucharbeit ist nicht nur wegen der mit ihr verbundenen Auflockerungen, welche diesmal betrachtet wurden, sondern auch deshalb von Nachteil, weil sie dem Gebirge nicht Zeit läßt, Spannungsumlagerungen zu vollziehen und sich den veränderten Randbedingungen anzupassen. Sie berücksichtigt nicht den Zeitfaktor der Gebirgseigenschaften, den man nicht außer acht lassen darf, wenn man vor unerwünschten Nachbewegungen, elastischen Nachwirkungen, Nachbrüchen, Nachlockerungen und Nacharbeiten sicher sein möchte. Dieses Thema wird an anderer Stelle behandelt werden.

---

* Bereits vor etlichen Jahren teilte mir Herr Direktor Fischer, Graz, mit, um wieviel höher der Injektionsaufwand bei rasanten Stollenvortrieben ist.

Literatur

[1] **Makovec, F.**: Das Ausmaß der Felsauflockerung bei Sprengarbeiten. Geol. u. Bauw., Jg. 28, H. 1, 1962.

[2] **Langefors, U.**: Calculation of Charge and Scale Model Trials. Quart. Colorado School of Mines, Vol. 54, No. 3, 1959.

[3] **Rinehart, J. S.**: The Role of Stress Waves in Comminuation. Quart. Colorado School of Mines, Vol. 54, No. 3, 1959.

[4] **Föppl, L.**: Störungen des Spannungszustandes in der Umgebung eines Druckstollens durch Spalten. Geol. u. Bauw., Jg. 23, H. 1, 1957 a.

[5] **Sonntag, G.**: Die Beanspruchung des Gebirges in der Umgebung eines Schlitzes in Abhängigkeit seiner Richtung zur Erdoberfläche. Geol. u. Bauw., Jg. 23, H. 1, 1957.

[6] **Griffith, A. A.**: The Theory of Rupture. Proc. 1st Int. Congr. Appl. Mech., Delft 1924.

[7] **Brace, W. F.**: Brittle Fracture of Rocks. Int. Conf. on State of Stress in the Earth's Crust, Santa Monica 1963.

[8] **Föppl, L.**: Ergebnisse spannungsoptischer Versuche an geschichtetem Material. Geol. u. Bauw., Jg. 21, H. 3, 1955 a.

[9] **Föppl, L.**: Der Übergang von der Haftreibung zur Gleitreibung. Geol. u. Bauw., Jg. 21, H. 4, 1955 b.

[10] **Föppl, L.**: Elastische Spannungszustände in Körpern mit ebenen Schnitten. Forschung Bd. 22, H. 2, 1956.

[11] **Föppl, L.**: Elastische Spannungszustände in Körpern mit ebenen Schnitten. Geol. u. Bauw., Jg. 23, H. 1, 1957 b.

[12] **Pacher, F.**: in **Müller, L.**: Geomechanische Auswertung gefügekundlicher Details. Geol. u Bauw., Jg. 24, H. 1, 1958 a.

[13] **Tincelin, E.** und **Sinou, P.**: Schießarbeit und Ausbau in Strecken mit schwierigem Hangenden in einer Lothringischen Eisenerzgrube. III. Int. Bergbaukongr. Salzburg 1963.

[14] **Buchheim, W.**: Geophysikalische Methoden zur Erforschung des Spannungszustandes des Gebirges im Steinkohlen- und Kalisalzbergbau. Int. Gebirgsdrucktagung Leipzig 1958.

[15] **Kahler, F. v.**: Gefahren des Maschineneinsatzes in steilen Halden. Geol. u. Bauw., Jg. 25, H. 4, 1960.

[16] **Rabcewicz, L. v.**: Aus der Praxis des Tunnelbaues. Geol. u. Bauw., Jg. 27. H. 3-4, 1962.

[17] **Rabcewicz, L. v.**: Bemessung von Hohlraumbauten. Felsmech. u. Ing.-Geol., Vol. I/3-4, 1964.

[18] **Pacher, F.**: Deformationsmessungen im Versuchsstollen als Mittel zur Erforschung des Gebirgsverhaltens und zur Bemessung des Ausbaues. XIV. Salzb. Felsmech.-Koll., Springer, Wien 1964.

[19] **Müller, L.**: Der Felsbau. Enke, Stuttgart 1963.

Über die Bestimmung des in-situ-Charakters des Gebirges

Von

R. Richter*

Mit 2 Textabbildungen

Zusammenfassung — Summary — Résumé

Über die Bestimmung des in-situ-Charakters des Gebirges. Die vorliegende Arbeit bezeichnet als in-situ-Charakter jene Kennzeichen des anstehenden Gebirges, welche mit Hilfe eines durch gegebene Einwirkung hervorgerufenen Bewegungsbildes der Kruste beschreibbar sind. Im Falle der Linearelastizität ergeben sich als Kennzeichen der Elastizitätsmodul, die Poissonzahl sowie der Quotient der vertikalen und horizontalen Primär-Normalspannungen. Diese Studie gibt die Werte der Charakteristiken als Funktion der erforderlichen Messungsergebnisse an.

On the Determination of the in-situ-Characteristics of Rock Masses. The term 'in-situ-characteristics' is used to describe those properties of rock masses which help to define the stress and displacement fields in the earth's crust. Poisson's number as well as the ratio of the primitiv (vertical and horizontal) principal stresses, have been taken to be characteristic of the elastic moduli for linearly elastic material.

The paper gives expressions for the characteristics as functions of the necessary experimental data.

Sur la détermination des caractéristiques in situ du massif rocheux. On entend par caractéristiques in situ du massif rocheux les critères à l'aide desquels on peut définir la répartition des déformations du matériau sous une influence donnée. Dans le cas de l'élasticité linéaire, ces critères sont le module d'élasticité, le coefficient de Poisson, et le rapport des contraintes principales initiales verticales et horizontales. L'article indique les valeurs de ces critères en fonction des résultats des mesures nécessaires.

Das Gleichgewicht der in der Erdkruste errichteten Bauobjekte, wie Böschungen, Hohlräume und deren Einbauten, ist u. a. auch eine Funktion der in der Erdkruste in-situ herrschenden Zustände. Diese in-situ-Zustände kennzeichnen wir als den Primärzustand. Unsere Aufgabe ist es, diesen zu bestimmen.

Der primäre Zustand kann unmittelbar, und auch prinzipiell, nur am Modell bestimmt werden, weil die zur Bestimmung erforderlichen Meßinstrumente mit dem Entstehen der Gesteine zugleich eingebaut werden müßten, was im Falle der Erdkruste nicht möglich ist.

Durch mittelbare Verfahren können als Funktion des Primärzustandes erscheinende Wirkungen gemessen werden. In diesem Falle muß aber die funktionale Beziehung zwischen dem Primärzustand und der Wirkung als bekannt vorausgesetzt werden.

Im Schrifttum über Gebirgsmechanik ist allgemein verbreitet, daß man den Primärzustand mit einem Spannungszustand Φ_p kennzeichnet und als Wirkung die auf bekannte Kraftwirkung erfolgende Verschiebung von materiellen Punkten, bzw.

---

* Professor Dr.-Ing. Richard R i c h t e r, Technische Universität für Schwerindustrie, Miskolc, Ungarn.

die zur Verhinderung deren Verschiebung erforderlichen Kräfte und Spannungen betrachtet. Es sei bemerkt, daß eine derartige Lösung der Aufgabe nicht die einzige Möglichkeit darstellt.

Bei Beobachtungen in einem Spannungs-Formänderungs-Verschiebungssystem sind die Meßergebnisse Formänderungen bzw. Verschiebungen, die außer Φ_p Funktionen der die Formänderung erzeugenden Einwirkung (W) sowie der Materialkonstanten (C) des Gebirges sind:

$$\vec{t} = \varphi \, (\Phi_p, W, C).$$

Selbst im einfachsten Falle kennzeichnet φ allgemein und unter der Annahme einer linear-elastischen Gesteinsumgebung, den primären Spannungszustand: die Spannungskomponenten p_2 und p_x, die Einwirkung: ein unendlich langer, in der Hauptrichtung gelegener, kreisförmiger Hohlraum (Schacht oder Strecke bzw. paralleles Bohrloch) und die Materialkonstante: den Elastizitätsmodul (E) sowie die Poissonsche Zahl (m) des Gesteins.

Laut einschlägigem Schrifttum kann als gute Annäherung die Beziehung $p_z = z \, \gamma$ angenommen werden, und somit sind als Charakteristiken die Werte

$$a = \frac{p_z}{p_x}, \quad E \text{ und } m$$

anzusehen.

Everling[1] befaßt sich ausführlich mit der möglichen Gestaltung von a.

In Übereinstimmung mit ihm kann festgestellt werden, daß die Annahme $a = m - 1$ nur in außergewöhnlichen Fällen angenommen werden kann, und falls auch mit dem Vorhandensein von geologischen Kräften zu rechnen ist — wie es allgemein der Fall ist —, muß a als Charakteristik der gegebenen Stelle aufgefaßt werden, und unsere Messungen müssen sich auf seine Bestimmung beziehen.

Eine besondere Aufmerksamkeit müssen wir dieser Frage deshalb beimessen, weil wir die Ursache der scheinbaren Änderung der als Funktion der Tiefe erscheinenden Charakteristiken von E und m darin erblicken. Mit der Annahme einer elastischen Gesteinsumgebung vollziehen sich die mit der Hohlraumbildung eintretenden Deformationen zum Teil schon bevor man dieselben messen kann. Deshalb könnten die Meßergebnisse der im Laufe der Hohlraumbildung meßbaren Formänderungs- bzw. Verschiebungsänderungen nur dann genützt werden, wenn der Spannungszustand Φ_E um Hohlräume endlichen Kreisquerschnittes bekannt wäre, weil dann mit Hilfe des bekannten Spannungszustandes Φ_u um Hohlräume von Kreisquerschnitt unendlicher Ausdehnung Verschiebungen von der Form

$$\vec{t} = f \, (\Phi_u - \Phi_E)$$

berechnet werden könnten, deren Vergleich mit den Meßergebnissen $(\vec{t}_0)$

$$\vec{t} = f \, (\Phi_u - \Phi_E) = t_0,$$

die Berechnung der als unbekannt aufgefaßten Gesteinscharakteristiken ermöglichen würde. Von dieser Möglichkeit müssen wir mangels der Kenntnis von Φ_E jedoch absehen. Demzufolge müssen wir unsere Meßgeräte in dem bereits hergestellten, kreisförmigen Hohlraum anordnen, dann mit einer neuerlichen Kraftwirkung eine Formänderung herbeiführen, die sowohl berechnet wie auch gemessen werden kann und sich somit zur Bestimmung von Unbekannten eignet. Eine solche neuerliche Einwirkung kann z. B. die Einwirkung eines inneren Gasdruckes auf den kreisförmigen Hohlraum oder die Freilegung von Oberflächenblöcken bzw. die Errichtung eines nahe gelegenen neuen kreisförmigen Hohlraumes mit paralleler Achse sein.

A. In einem kreisförmigen Hohlraum mit einer Länge vom 4- bis 5fachen Durchmesser kann durch dichte Absperrungen ein innerer Gasdruck von q_0 hergestellt werden. Dieser Gasdruck bewirkt am Umfang des Hohlraumes eine von φ unabhängige, spezifische tangentiale Dehnung, die mittels Meßstreifen meßbar bzw. berechenbar ist[2,3].

$$-\varepsilon_q = \frac{q_0}{E} \cdot \frac{(m+1)}{m},$$

somit ist der in-situ-Wert der Poissonschen Zahl:

$$m = \frac{-q_0}{E \cdot \varepsilon_q + q_0}$$

bzw. der des Elastizitätsmoduls:

$$E = \frac{-q_0}{\varepsilon_q} \cdot \frac{(m+1)}{m}.$$

Auch bei veränderlichem Gasdruck q_0 sich ergebende, auf die tangentiale Dehnung sich beziehende Meßergebnisse eignen sich nicht zur gesonderten Bestimmung von m und E. Somit erhält man mit dieser Methode nur dann eine Lösung, wenn man eine Charakteristik am Probekörper im Laboratorium bestimmt.

B. Einzelne Blöcke der Oberfläche von Hohlräumen mit kreisförmigem Querschnitt können durch Herstellen von Schlitzen (Um-

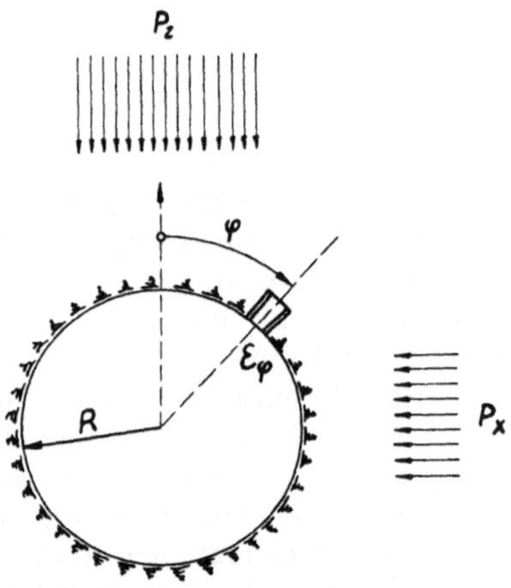

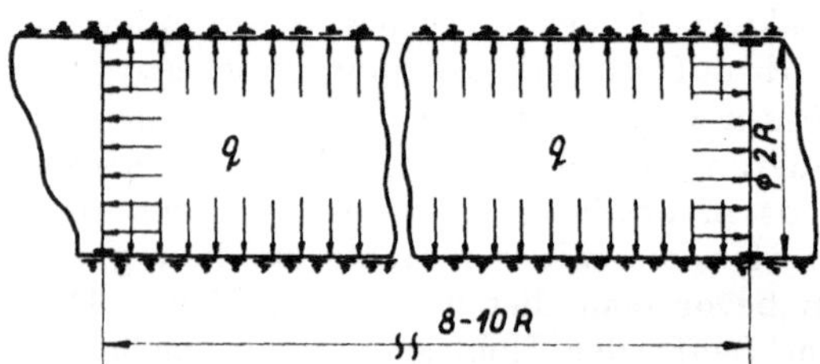

Abb. 1. Druckwirkung in der Versuchsstrecke
Pressure effect in the experiment roadway excavation

Abb. 2. Position des freigelegten Gesteinsblockes
Position of the exposed rock-block

schrämen) entlastet werden, und somit können in Oberflächenpunkten beliebiger Lage die tangentialen Dehnungen gemessen werden.

Allgemein gilt:

$$-\varepsilon = \frac{m^2 - 1}{E \cdot m^2} \cdot \frac{z \cdot \gamma}{a} \cdot [(a+1) - 2(a-1) \cdot \cos 2\varphi],$$

woraus sich die kleinste, im Regelfall die zu $\varphi = 0^0$, 180^0 gehörende Dehnung zu

$$-\varepsilon_0 = z\,\gamma \cdot \frac{m^2 - 1}{m^2 \cdot E} \cdot \frac{3-a}{a}$$

bzw. die größte, im Regelfall die zu $\varphi = 90^0$, 270^0 gehörende Dehnung

$$-\varepsilon_{90} = \gamma\,z \cdot \frac{m^2 - 1}{m^2 \cdot E} \cdot \frac{3a-1}{a}$$

ergeben.

Mit diesen Werten erhält man

$$a = \frac{3\,\varepsilon_{90} + \varepsilon_{0}}{3\,\varepsilon_{0} + \varepsilon_{90}}.$$

Meßergebnisse von dazwischen liegenden Punkten ermöglichen die Bestimmung des Wertes a durch Ausgleich, wobei die allgemeine Form der vermittelnden Gleichungen

$$-\varepsilon = B \cdot \frac{(a+1) - 2\,(a-1)}{a} \cdot \cos 2\,\varphi$$

ist.

C. Bei kombinierten Messungen an Hand von Beobachtungen von $\varepsilon_q, \varepsilon_0, \varepsilon_{90}$ beträgt die primäre Hauptspannung, d. h. die lokale Abweichung gegenüber $z \cdot \gamma$,

$$z\,\gamma = \frac{3\,\varepsilon_{90} - \varepsilon_0}{8} \cdot \frac{q_0^{\,2}}{q_0^{\,2} - (\varepsilon_q E - q_0)^2}$$

bzw. können unter Voraussetzung der Hauptspannung $p_z = z\,\gamma$ die Zusammenhänge nach E aufgelöst werden:

$$E = -\frac{2\,q_0}{\varepsilon_q} + \frac{q_0^{\,2} \cdot (3\,\varepsilon_{90} + \varepsilon_0)}{8\,z \cdot \gamma \cdot \varepsilon_q^{\,2}}.$$

Unter Beachtung der unter A und B erhaltenen Ergebnisse können somit die in-situ-Werte oder die Charakteristiken m, a, E aus den kombinierten Messungen bestimmt werden.

D. Im Laufe der aufeinanderfolgenden Herstellung von parallelen Bohrlöchern wird sich das erste Bohrloch nach seiner Herstellung infolge radialer Verschiebung u_0 seiner Oberflächenpunkte deformieren. In diesem Bohrloch wird man nun ein Meßgerät anordnen, das weitere radiale Verschiebungen einzelner Oberflächenpunkte mißt. Nach der Errichtung des zweiten parallelen Bohrloches beträgt die meßbare Verschiebung[4]

$$\Delta = u_0 - u_{00},$$

worin u_{00} jene radiale Oberflächenverschiebung bedeutet, die als Folge zweier, in der Entfernung D errichteter paralleler, kreisförmiger Hohlräume entsteht. Da beide Bewegungskomponenten sowohl berechnet wie auch gemessen werden können, eignen sie sich zur Bestimmung einer unbekannten Charakteristik.

Mittels der in unendlicher Reihe ausdrückbaren allgemeinen Form der Funktion u_{00}:

$$u_{00} = f\,(m, E)\left[\varphi\left(P_z \cdot \frac{a+1}{a}\right) + \varPhi\left(\sum_{1}^{\infty} p_z \cdot \frac{a-1}{a}\right)\right]$$

können die Charakteristiken zweckmäßigerweise auch aus Messungen in senkrechter oder waagrechter Achsenrichtung berechnet werden[4].

Die in den Punkten A—D zusammengefaßten Messungen ermöglichen die Bestimmung der Charakteristiken des Primärzustandes und liefern somit konkrete Daten zur Kennzeichnung des in-situ-Zustandes von Gesteinsblöcken.

Literatur

[1] E v e r l i n g, G.: Der Spannungszustand im unverritzten Gebirge. Glückauf, H. 19, S. 1199—1202, 1960.

[2] S z é c h y, K.: Beitrag zur Gebirgsdruckverteilung um einen kreisförmigen Tunnelquerschnitt. Öst. Ing. Zschr., H. 4, S. 126—128, 1962.

[3] R i c h t e r, R.: Über in-situ-Messungen der Gebirgsmechanik in kreisförmigen Hohlräumen. Bergakademie, Nr. 5, S. 334—343, 1963.

[4] R i c h t e r, R.: Measurement of rock pressure in situ in elastic rocks, Mitt. Techn. Univ. f. Schwerind. Ungarn, Bd. XXI, S. 25—27.

Einige Betrachtungen betreffend die Meyerhofsche Methode für die Berechnung der Tragfähigkeit von Bogenmauerauflagern

Von

Petar Stojić*

Mit 4 Textabbildungen

Zusammenfassung — Summary — Résumé

Einige Betrachtungen über die Anwendung der Meyerhofschen Methode bei der Berechnung der Tragfähigkeit der Talflanken für Bogenmauern. Der Felsmechanik wurde die Aufgabe gestellt, die für die Flankenstabilität der Felsmasse geeigneten analytischen Methoden zu finden. Die Erscheinung der Bodenbrüche ist in der Bodenmechanik ausführlich bearbeitet worden, was einige Autoren veranlaßte, diese Methoden auch für die Berechnung der Tragfähigkeit der Bogenmauerauflager in zerklüfteter Felsmasse anzuwenden.

Die Meyerhofsche Methode fand eine besonders weitgehende Anwendung auf (mit zentrischer und senkrechter Kraft belasteter Base der schrägen Bandfundamente) schrägen Streifenfundamenten, deren Grundfläche zentrisch mit einer senkrechten Kraft belastet ist.

Die Betrachtungen der praktischen Beispiele zeigen, daß auch im Falle der homogenen Masse die Anwendung dieser Methode für die Nachprüfung der Stabilität der Bogenmauerauflager keine Berechtigung hat, da die Ergebnisse weit von der Wirklichkeit entfernt sein können.

Consideration on Application of Meyerhof's Method for Bearing Capacity Analysis of Arch Dam Foundations. Suitable analytical methods are required in rock mechanics for studying the stability of the rock mass of abutments.

The phenomenon of soil failure has been studied in detail in soil mechanics, and some authors have applied these methods also to analyse the bearing capacity of arch dam foundations in non-homogeneous rock mass.

Meyerhof's method has been applied largely to inclined strip foundations, loaded with a central, normal load acting upon its base.

On considering some practical examples it is realised that application of this method for checking the stability of an arch dam foundation is not justifiable even in the case of homogeneous media since the results can be far from reality.

Considérations sur la méthode de Meyerhof pour calculer la capacité portante des fondations des barrages voûtes. On devrait trouver, à l'aide de la mécanique des roches. des méthodes de calcul permettant d'évaluer la stabilité du rocher au voisinage des appuis des barrages de retenue.

Le développement de la rupture des sols a été largement traité en mécanique des sols. ce qui a conduit certains auteurs à appliquer également ces méthodes au calcul de la capacité portante des fondations de barrages voûtes sur des roches fissurées.

La méthode de Meyerhof a été appliquée à des fondations en bande inclinée, chargées par une force normale centrée.

L'examen d'exemples concrets montre que même dans l'hypothèse d'un massif homogène, l'utilisation de cette méthode pour évaluer la stabilité des appuis d'un barrage voûte n'est pas justifiée car les résultats peuvent être très loin de la réalité.

---

* Dipl.-Ing. Petar Stojić, Energo-invest, J. N. A. 20, Sarajevo, Jugoslawien.

Die Entwicklung der hohen Bogenstaumauern sowie die Errichtung solcher Sperren in Tälern mit weniger günstigen Profilen hat nach dem Krieg das Bedürfnis nach einer Berechnungsmöglichkeit für die Standsicherheit der Widerlager geschaffen.

Die Felsmechanik, eine der jüngeren Bauwissenschaften, wurde vor die Aufgabe gestellt, eine analytische Methode zu entwickeln, die zur Beurteilung der Stabilität der Felsmasse en bloc geeignet ist.

Im Anfangsstadium der Entwicklung dieser Wissenschaft versuchte man, der Analyse der zerklüfteten Felsmasse mit weniger geeigneten Vorstellungen der Bodenmechanik, die für bestimmte Probleme der Felsmechanik abgewandelt wurden, näherzukommen.

Heutzutage aber können uns geologische und hydrogeologische Angaben, Felsprüfungen im Laboratorium und Untersuchungen des Felsens „in situ" eine große Anzahl von Grundlagen zur Behandlung aller Probleme beim Entwurf und für die Ausführung der Konstruktion, welche mit dem Fels eng verbunden ist, geben. Eine Stabilitätsanalyse der Felsmasse auf diesen Grundlagen ist allerdings mühsam und verlangt ziemlich viel Zeit und beträchtliche finanzielle Mittel.

Es ist eine Tatsache, daß die Mechanik der zerklüfteten Felsmasse (Diskontinuum) komplizierter als die Bodenmechanik ist und ihre Möglichkeiten heute noch ziemlich begrenzt sind. Dies berechtigt uns aber nicht zu einer konservativen Behandlung des Problems und zur Anwendung teilweise oder überhaupt unbrauchbarer Methoden bei der analytischen Behandlung der Stabilität von Mauerwiderlagern.

In vielen Fällen weisen die geostatischen Bedingungen auf die Möglichkeit eines Bruches an bestimmten Flächen im Widerlager, die durch das Vorhandensein der geologischen Trennflächen bedingt wird, hin. Diese Trennflächen bestimmen die Größe dieser Bruchflächen, und von ihren charakteristischen Eigenschaften hängt größtenteils die Scherfestigkeit der Masse ab.

Die Brucherscheinungen der zerklüfteten Felsmasse, die man während ihres Entstehens bei Scherfestigkeitsuntersuchungen „in situ" einigermaßen beurteilen kann, erweisen sich bei der analytischen Behandlung der Felswiderlager als sehr komplizierte Erscheinungen. Gleichmäßig verteilte Spannungen entlang der potentiellen Scherflächen vorauszusetzen, wäre gewiß unreal.

Die Grundbrucherscheinungen sind in der Bodenmechanik bereits sehr gründlich erforscht worden. Grundbruch entsteht, wenn die Bodenbelastung die Scherfestigkeit des Bodens überschreitet. In diesem Fall bildet sich unter der Last in der Erdmasse die Gleitfläche mit kleinstem Widerstand aus. Über die Form dieser Gleitflächen mit kleinstem Widerstand bestehen verschiedene Annahmen.

Für die Berechnung der Bodentragfähigkeit gibt es mehrere Gleichungen, von denen sich die unten angeführte für die praktische Anwendung als vollkommen geeignet erwiesen hat:

$$q = c \cdot N_c + p_0 \cdot N_q + \gamma \frac{B}{2} N_\gamma, \tag{1}$$

worin:

q = kritische Bodenbelastung (Tragfähigkeitskapazität),

c = Kohäsion,

p_0 = Belastung auf der horizontalen Ebene,

γ = Raumgewicht,

B = Fundamentbreite,

N_c, N_q, N_γ = Halbraum — Tragfähigkeitsfaktoren.

Die Fundamenttragfähigkeit hängt von den mechanischen Bodeneigenschaften, von den Fundamentabmessungen (Größe, Tiefe usw.) und von der Art der Fundamentausbildung ab. Dieses Problem hat auch G. G. Meyerhof behandelt, der die Theorie auf Grund seiner Untersuchungen ergänzte und erweiterte. Der erwähnte Autor studierte auch den ziemlich häufigen Fall eines geneigten Fundamentes, d. h.

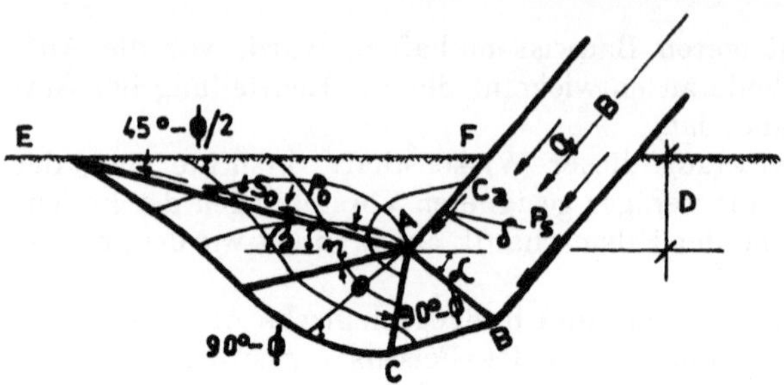

einer schrägen Basis mit darauf senkrecht wirkender Belastung (Abb. 1).

Die Tragfähigkeitskapazität eines schrägen Streifenfundamentes mit einer Basis, die auf die Belastung senkrecht steht, hat Meyerhof durch folgende Gleichung angegeben:

$$q = c \cdot N_{cq} + \gamma \frac{B}{2} N_{\gamma q}. \qquad (2)$$

Abb. 1. Die schräge Basis mit der senkrechten Belastung nach Meyerhof

Inclined base under vertical loading, acc. Meyerhof

Wegen der mathematischen Schwierigkeiten sind die Tragfähigkeitsfaktoren N_{cq} und $N_{\gamma q}$ in Diagrammen angegeben, wodurch das Berechnungsverfahren der Tragfähigkeitskapazität erheblich erleichtert und vereinfacht wird. Dies veranlaßte manche Autoren, diese von G. G. Meyerhof bearbeitete und erweiterte Methode auch zur Berechnung der Tragfähigkeit bei Widerlagern von Bogenmauern in einer zerklüfteten Felsmasse anzuwenden, und zwar in der Weise, wie es J. Talobre im Buch „La méchanique des roches" (S. 220—223) darstellt.

Die Methode von G. G. Meyerhof wurde für eine homogene Masse ausgearbeitet, d. h. für eine Masse, bei der die Scherfestigkeit in allen Richtungen gleich ist; demnach ist das Ergebnis schon von Anfang an mit dieser Einschränkung belastet.

Wir wollen auch den Fall behandeln, daß der Fels eine nur kleine Kohäsion hat, die vernachlässigt werden kann. Die Tragfähigkeitskapazität des schrägen Streifenfundamentes wird in diesem Fall durch folgende Gleichung angegeben (Abb. 2, 3, 4):

$$q = \gamma \frac{B}{2} N_{\gamma q}. \qquad (3)$$

Abb. 2. Blick flußaufwärts auf die Sperre — schematische Darstellung

Downstream view of the dam — schematic diagram

Um die Analyse der Fundamenttragfähigkeit in den Bogenmauerwiderlagern durchführen zu können, müssen wir das in Abb. 4 gegebene Schema benützen und alle Elemente, die die Tragfähigkeit beeinflussen, auf die Ebene $b—b$, die auf der Terrainlinie senkrecht steht, projizieren.

Demnach wird die Gl. (3) für die kritische Belastung folgende Form haben:

$$q' = \cos \beta \cdot \gamma \frac{B}{2} N_{\gamma q}, \qquad (4)$$

$$N_{\gamma q} = f\left(\varphi, \alpha, \frac{D'}{B}\right), \qquad (5)$$

wobei:

$$\varphi = \text{Reibungswinkel,}$$
$$\alpha = \text{Neigungswinkel der Fundamentbasis,} \qquad (6)$$
$$D' = D \sin \beta.$$

Wenn man die Tragfähigkeit der Widerlager nach den Gln. (4) und (6) in zwei extremen Fällen nachprüft, bemerkt man sofort, daß die Anwendung dieser Methode nicht berechtigt ist.

Fall I: Senkrechte Berglehne $\beta = 90^0$,

$$\frac{d'}{b} \neq 0, \quad q' = 0.$$

Fall II: Horizontale Berglehne $\beta = 0^0$,

$$\frac{d'}{b} = 0, \quad q' \neq 0.$$

Um die vorhergehende Meinung zu begründen, wird ein Rechenbeispiel angeführt, und zwar für:

$$\varphi = 45^0,$$
$$\alpha = 56^0,$$
$$\gamma = 2,6 \ \text{t/m}^3.$$
$$D = 9,8 \ \text{m,}$$
$$B = 22,0 \ \text{m.}$$

Neigung des Berghanges β	20^0	38^0	45^0	60^0
Kritische Belastung q' in kg/cm²	80,5	74 5	73,0	60,5

Wie man sehen kann, fällt die Tragfähigkeitskapazität des Fundamentes mit der Zunahme der Berghangneigung β ab, was im Gegensatz zur Tatsache steht, daß

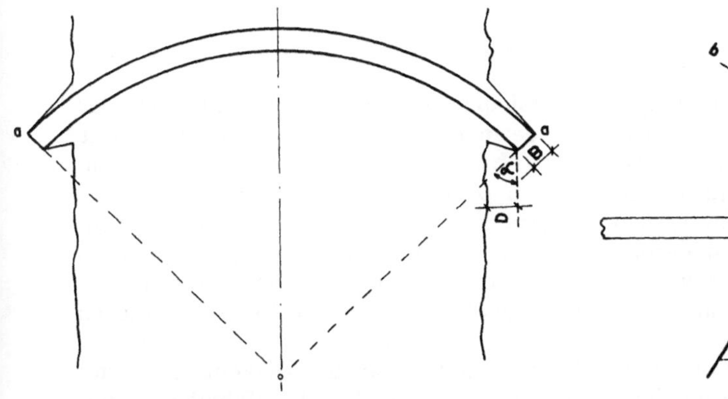

Abb. 3. Horizontaler Bogen (a—a), 1,0 m hoch
Horizontal arch (a—a), 1.0 ms high

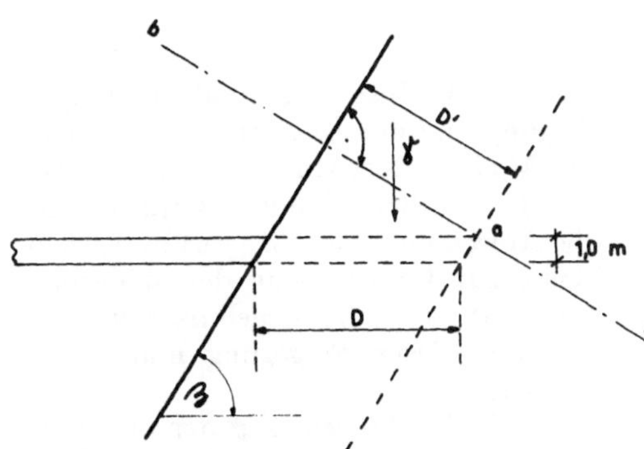

Abb. 4. Horizontaler Bogen (a—a) — Detail I
Horizontal arch (a—a) — detail I

die Kräfteübertragung in Bogenmauerfundamenten umso günstiger ist je größer die mitwirkende Felsmasse und je größer die auf die Schicht über den mitwirkenden Bereich wirkende Auflast ist.

Bei der Überprüfung der Anwendbarkeit der genannten Methode muß darauf aufmerksam gemacht werden, daß diese ein geneigtes Fundament für den Fall behandelt, daß die auf die Widerlager wirkende Kraft zentrisch und auf der Fundamentbasis senkrecht ist.

Es ist aber bekannt, daß die Fundamente von Bogenmauern wegen der Biegemomente und der horizontalen und vertikalen Kräfte, die auf das Bauwerk wirken, exzentrischen und schräg wirkenden Kräften unterworfen sind.

Für die Lösung eines Problems in der Praxis wurde auch dieser Einfluß analysiert. Unter der Voraussetzung, daß die Masse homogen ist und die Kohäsion nicht betrachtet wird, wurde eine Analyse der Stabilität in einem Schnitt, der auf die Fundamentebene (Schnitt $b-b$, Abb. 4) senkrecht steht, durchgeführt.

Die Bruchkräfte (P) wurden nach der Bedingung des Grenz-Gleichgewichtes durch eine Abgleichung der Momente der wirkenden Kräfte für verschiedene Spiralzentren berechnet; in allen untersuchten Fällen sind spiralförmige Gleitflächen vorausgesetzt.

Man kann diesen Fall auch mit den Annahmen des vorher angeführten Rechenbeispiels für $\beta = 38^0$ illustrieren:

1. Nach dem Meyerhofschen Diagramm:

$$P_1 = q \cdot B \cdot 1{,}0 = 16\,400{,}0 \text{ t/lfm.}$$

2. Nach der Spiral-Gleitfläche mit der Voraussetzung, daß die Belastung zentrisch wirkt und auf der Fundamentbasis senkrecht steht:

$$P_2 = 13\,000{,}0 \text{ t/lfm.}$$

3. Nach der Spiral-Gleitfläche mit der Voraussetzung, daß die Belastung zentrisch und schräg wirkt bzw. die Kraft mit der Vertikalen auf die Fundamentbasis einen Winkel $\delta = 11^0$ einschließt:

$$P_3 = 7180{,}0 \text{ t/lfm.}$$

4. Nach der Spiral-Gleitfläche mit einer exzentrischen und schrägen Belastung: $\delta = 11^0$, Exzentrizität $e = 1{,}60$ m (wobei eine effektive Fundamentbreite $B' = B - 2e$ angenommen wurde):

$$P_4 = 5170{,}0 \text{ t/lfm.}$$

Die Analyse zeigt, daß die Tragfähigkeitskapazität eines Fundamentes mit der Neigung und der Exzentrizität der Belastung geringer wird, was auch die Ergebnisse der Modelluntersuchungen, die G. G. Meyerhof durchgeführt hat, bekräftigen.

Die vorhergehenden Betrachtungen an den praktischen Beispielen zeigen, daß die Anwendung des Meyerhofschen Diagrammes auch im Falle der homogenen Masse zur Überprüfung der Widerlagerstabilität von Bogenmauern keine Berechtigung hat, weil die Ergebnisse weit von der Wirklichkeit abweichen können.

Bei dieser Folgerung muß auch die folgende Tatsache in Betracht gezogen werden:

Mit der Anwendung der erwähnten Methode kann man die Diskontinuität der Gefüge und die statische Felsenanisotropie nicht erfassen. Die Tragfähigkeitskapazität von solchen Felssystemen kann kleiner sein als bei einer durch keine Risse zerklüfteten Masse ($c = 0$) ohne bevorzugte Richtungen von kleinerer Festigkeit.

Mit der Anwendung der erwähnten Methode ist es nicht möglich, auch nur ungefähr die Auswirkung aller Kräfte, die sich durch die Errichtung einer Bogenmauer, durch den Aufstau und durch die anderen Baueingriffe im Bogenmauerprofil auf das Felsmassiv übertragen, anzugeben.

Literatur

[1] Hansen, J. B.: A General Formula for Bearing Capacity. Kopenhagen 1961.

[2] Meyerhof, G. G.: The Ultimate Bearing Capacity of Foundations. 1951.

[3] Meyerhof, G. G.: The Bearing Capacity of Foundations under Eccentric and Inclined Loads. Zürich 1953.

[4] Müller, L.: Das Kräftespiel im Untergrund von Talsperren. Geologie und Bauwesen, 1961.

[5] Talobre, J.: La méchanique des roches. Paris 1957.

Gebirgsbeschreibung aufgrund von Bohrergebnissen

Von

Karl Hoffmann*

Mit 5 Textabbildungen

Zusammenfassung — Summary — Résumé

Gebirgsbeschreibung aufgrund von Bohrergebnissen. Zur Gebirgsbeschreibung aufgrund von Bohrergebnissen gehören neben der Aufstellung von Schichtenverzeichnissen auch Feststellungen über die Festigkeit des Gesteins, seinen Zerteilungsgrad sowie die Durchlässigkeit und den Wasserstand im Gebirge. Beobachtungen hierüber können auf verschiedenartige Weise angestellt werden.

Für die Praxis hat es sich als besonders zweckmäßig erwiesen, die Beobachtungsergebnisse nicht nur zu beschreiben, sondern auch übersichtlich darzustellen. Das Bohrdiagramm enthält daher zweckmäßig:

Bohrprofil, stratigraphisches Profil und Diagramm der Kernfestigkeit (Brüchigkeit), der Klufthäufigkeit, der Öffnungsweite der Klüfte, des Wasserstandes, der Spülverluste, der Wasseraufnahmen bei Abpressungen sowie das Verrohrungsschema.

Das Bohrdiagramm erlaubt eine einfache Interpretation der Beobachtungsergebnisse sowie eine schnelle Erfassung der Gebirgsverhältnisse.

Die wesentlichsten Einzeldarstellungen des Bohrdiagrammes werden besprochen.

Description of Rock Mass on the Basis of Drill Holes. The description of rock based on boring results should consist not only of the boring logs, but should also include statements on the compactness of the rocks, their partition, permeability, and on the level of the groundwater. Observations thereon can be made by different methods.

It has been found useful, not only in practice, to map out the boring results, in addition to describing them. The author therefore devised a "drilling diagram" for all drillings supervised by himself. This diagram comprises the following important figures:

Geological drilling section, stratigraphical section, diagrams concerning the consistency of the cores, frequency and width of joints, losses of drilling fluid, amount of water absorbed during permeability tests, and the casing schedule.

Interpretation of the drilling results and rapid analysis of the caracteristics of the rocks are facilitated by the use of a drilling diagram.

The most important parts of this diagram are discussed.

La description des massifs rocheux d'après les résultats de sondage. La description des massifs rocheux d'après les résultats de sondage comporte le relevé descriptif des coupes et des constatations sur la résistance de la roche, sa fracturation, sa perméabilité et sur le niveau de l'eau dans le massif. Ces observations peuvent être établies par divers procédés.

En pratique, il ne suffit pas de décrire les résultats de ces observations et il faut les représenter clairement. C'est pourquoi l'auteur établit toujours un synoptique pour chaque sondage excécuté. Ce synoptique comporte les figures principales suivantes:

Coupe du sondage, coupe stratigraphique et graphique de carottage; graphique de fréquence des fissures, et d'épaisseur de ces fissures; graphique de niveau d'eau, de pertes d'eau de forage, d'absorption d'eau sous pression; enfin schéma de tubage.

Le synoptique de sondage permet l'interprétation des résultats d'observations et l'analyse rapide du comportement du massif rocheux, comme le montre la discussion des plus importantes parties de ce graphique.

---

* Dipl.-Geol. Dr. Karl H o f f m a n n, Koblenz—Metternich, Trierer Straße 99.

Eine Gebirgsbeschreibung kann auf die verschiedenste Art und Weise vorgenommen werden. An erster Stelle steht die Beschreibung des Gebirges im natürlichen Aufschluß. Sie wird ergänzt durch Beobachtungen in künstlichen Aufschlüssen, wie Schürfungen, Bohrungen, Stollen usw. Sowohl aus den natürlichen als auch den künstlichen Aufschlüssen können Gesteinsproben entnommen und laboratoriumsmäßig untersucht werden.

An dieser Stelle sollen nur die Möglichkeiten besprochen werden, die uns Bohrungen, vor allem Kernbohrungen, zur Gebirgsbeschreibung liefern. Hierbei wird jedoch nicht allein auf die normale, geologisch-petrographisch-stratigraphische Beschreibung eingegangen, sondern es werden speziell die Methoden behandelt, die zusätzlich zur Vervollständigung der Beschreibung des Gebirges angewendet werden können (Tab. 1).

So notwendig die textliche Beschreibung der Beobachtungsergebnisse, so unumgänglich ist aber auch eine übersichtliche, interpretierbare und auch für den

Tabelle 1. Methoden der Beobachtung und Darstellung der Ergebnisse von Bohrungen

Methods of observation and presentation of the results of drillings

Méthodes d'observation et de figuration des résultats des forages

Untersuchungsmethode	Beobachtung	Art der Darstellung
I. Ohne spezielle Methoden		
Bohrkernbeschreibung	Art des Gesteins	Bohrprofil
	Festigkeit, Bindung seiner Komponenten	Diagramme: Festigkeit des Kerns Kernverlust
Auszählung von Klüften	Klüftung, Schiefrigkeit	Diagramme: Klufthäufigkeit
	Alter	stratigraphisches Bohrprofil
II. Aufgrund des Bohrverfahrens		
Bohrfortschritt	Festigkeit des Gesteins	Bohrfortschrittsdiagramm
Verrohrung	Festigkeit, Brüchigkeit und Klüftung	Verrohrungsschema
Spülverlust	Durchlässigkeit	Spülverlustdiagramm
III. durch spezielle Beobachtungsmaßnahmen		
Messung des Wasserspiegels in der Bohrung	Wasserführung des Gebirges	Wasserstandsdiagramm
	Durchlässigkeit des Gebirges	
Abweichungsmessung	Einfluß des Gesteins auf die Bohrrichtung	Abweichungsdiagramm
Fernsehsondierung	Lage von Trennflächen im Gestein	Kluftrose, Lagekugeldarstellung
	Häufigkeit von Klüften etc.	Häufigkeitsdiagramm
	Öffnungsweite von Klüften	Diagramm: Öffnungsweite
IV. durch spezielle Maßnahmen		
Wasserverpressungen	Durchlässigkeit des Gebirges	Verpreßdiagramm

Nichtgeologen, insbesondere den Ingenieur, verständliche Darstellung der Beobachtungsergebnisse.

Die Bohrkernbeschreibung liefert die Schichtenverzeichnisse, aufgrund derer *die Bohrprofile* gezeichnet werden. Schichtenverzeichnis und Bohrprofil stellen die Grundlage für die Beurteilung des durchbohrten Gebirges dar.

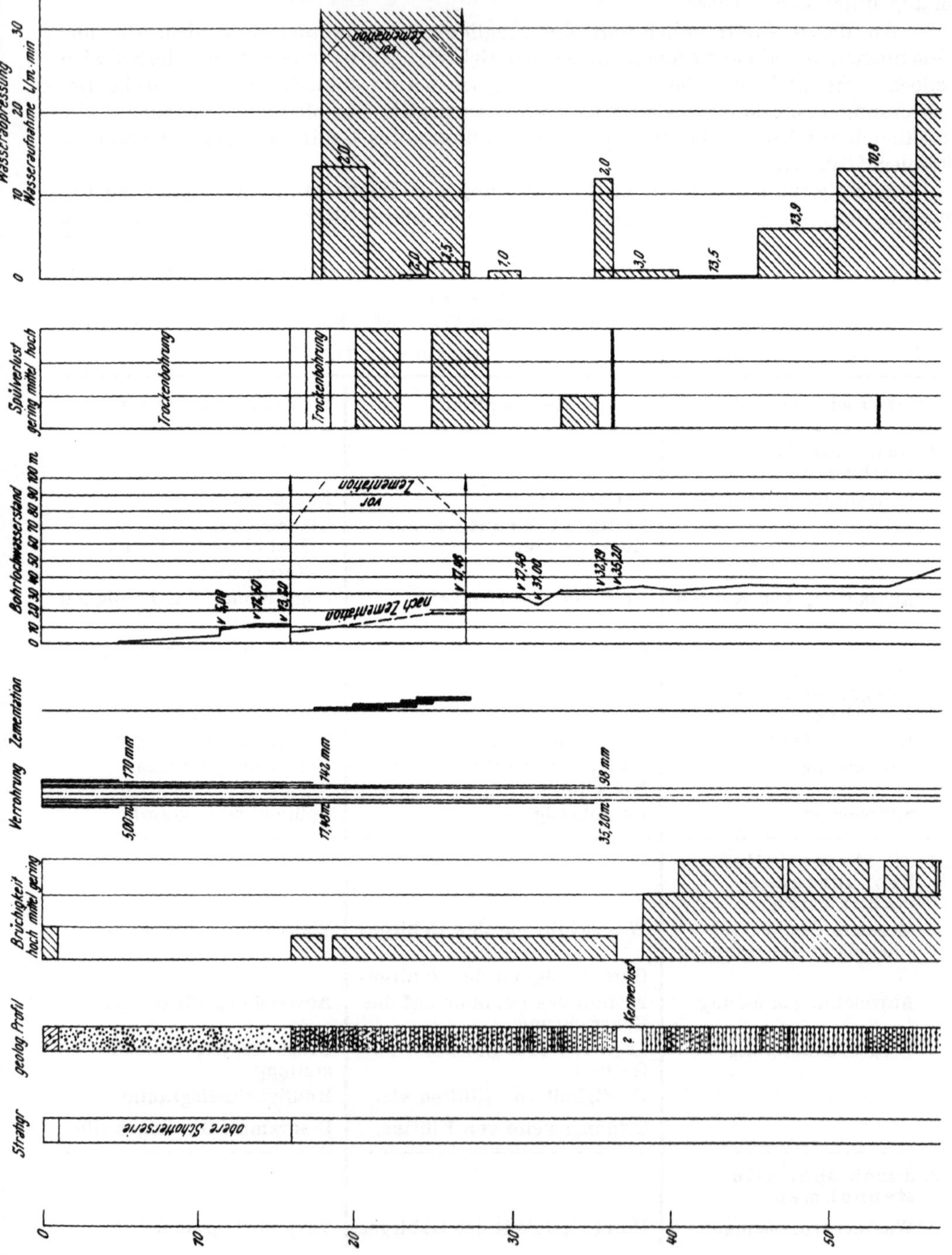

Damit auch Altersbeziehungen im Profil erkennbar werden, aus denen z. B. tektonische Vorgänge abgeleitet werden können, empfiehlt es sich häufig, neben das Schichtenprofil das *stratigraphische Profil* zu setzen. In anderen Fällen ist es zweckmäßig, ein vereinfachtes, zusammenfassendes Profil in die Darstellung der Bohrergebnisse mit aufzunehmen (Abb. 1). Alle weiteren Beobachtungen sind oft nur

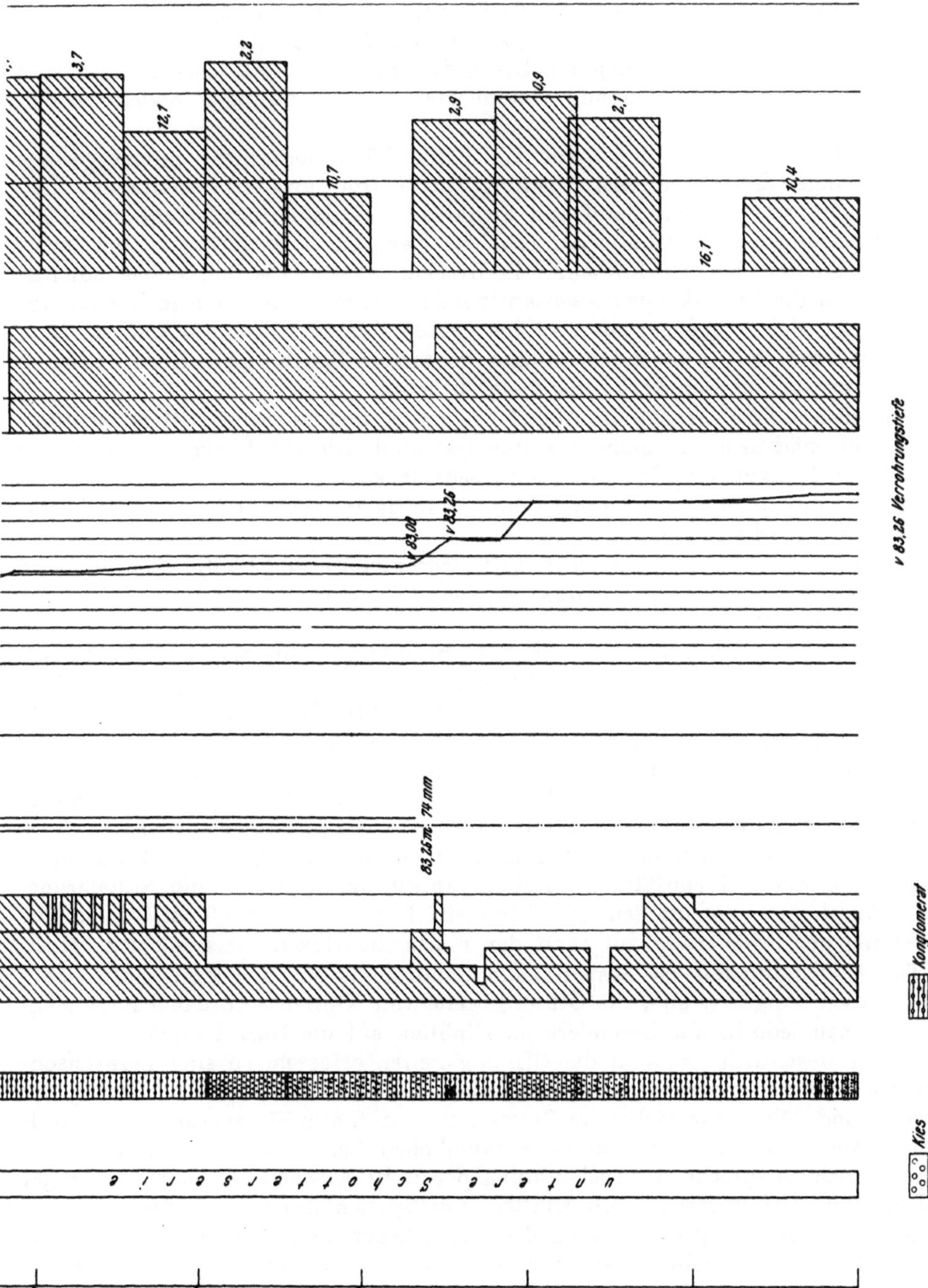

Abb. 1. Beispiel eines Bohrdiagrammes — Example of a drilling diagram — Exemple d'un diagramme de forage

aus der Bohrbeschreibung oder den Bohrberichten zu ersehen. Ohne graphische Darstellung ist es dann aber meist unmöglich, Einzelheiten der Beschreibung aufeinander zu beziehen und zu interpretieren. Deshalb muß das Bohrprofil durch weitere Darstellungen ergänzt werden.

Die Festigkeit des Gesteins läßt sich durch eine Klassifizierung nach der *Brüchigkeit des Bohrkernes* (Abb. 1) veranschaulichen. Eine solche Klassifizierung ist zwar subjektiv; wenn jedoch alle Kerne vom gleichen Geologen, Boden- oder Felsmechaniker beschrieben werden, so ist eine gleichmäßige Klassifizierung innerhalb eines Aufgabengebietes möglich. Die Erfahrung zeigt, daß auch bei mehreren Bearbeitern eine Übereinstimmung vereinbarter Klassifizierungen erzielt werden kann.

Im Beispiel der Abb. 1 ist von ca. 1 bis 15 m noch nicht einmal geringe Festigkeit eingetragen. In diesem Abschnitt wurde trotz größter Bemühungen kein Kern erbracht, sondern nur Sand und Feinkies im Schlammrohr zutage gefördert. Zwischen 15 und ca. 36 m hatte der erbohrte Kern einen so geringen Zusammenhalt, daß er sofort bei der Entnahme in Brocken auseinanderfiel, wobei lediglich teilweise noch die Einzelkomponenten aneinander hafteten. Nach einem kurzen Abschnitt ohne jeden Probengewinn war das Gestein bis in eine Tiefe von ca. 71 m fest. Es wurden, mit Ausnahmen einiger gebrächer Zonen, feste Kerne erbohrt. Bis ca. 75 m war der Bohrkern erneut sehr brüchig.

Aus dem Bohrdiagramm in Abb. 1 ist sofort und auf einen Blick sichtbar, daß nur die mächtigen Konglomeratbänke fest sind. Die Sandsteine und auch die Wechselfolge Konglomerat/Sandstein sind sehr locker.

Ein *Kernverlustdiagramm* ermöglicht ebenfalls in vielen Fällen Rückschlüsse auf die Gebirgsverhältnisse.

Die Aufzeichnung eines *Bohrfortschrittdiagrammes* ergibt häufig ebenfalls sehr gute Anhaltspunkte über die Gebirgsfestigkeit. Für sich allein betrachtet kann es jedoch zu Fehldeutungen führen.

Das *Verrohrungsschema* gibt ebenfalls brauchbare Hinweise auf die Standfestigkeit des Gebirges. In der Bohrung in Abb. 1 z. B. wurde nach Einbau eines Standrohres bis 5 m Teufe eine sehr brüchige Zone durchfahren, in der stellenweise, wie in Spalte 5 dargestellt, sogar zementiert werden mußte. Diese Zone wurde zweimal verrohrt. Das darunter folgende, feste Gebirge machte keine weitere Verrohrung notwendig, und erst nachdem ein beträchtliches Stück des lockeren Abschnitts unterhalb 70 m durchbohrt worden war, wurden erneut Rohre eingebaut.

Einen Überblick über den Zerteilungsgrad des Gebirges bietet die Darstellung der *Klufthäufigkeit.* Trennflächen schaffen Schichtung, Klüftung und Schieferung. Da die Anzahl von potentiellen Schieferungsflächen oft sehr groß ist, so ist eine Darstellung der Schieferung aufgrund der Flächenhäufigkeit meist nicht möglich.

Die Klufthäufigkeit im Gestein kann unmittelbar am Bohrkern zahlenmäßig erfaßt werden. Angaben über die Öffnungsweite von Klüften sind dabei nicht möglich. Will man jedoch, was besonders im Hinblick auf die Durchlässigkeit des Gebirges sehr wesentlich ist, auch die Öffnungsweiten erfassen, so sind Fernsehsondierungen sehr zu empfehlen. Sie ermöglichen eine unmittelbare Beobachtung der Bohrlochwand. Die Trennflächen im Gestein, aber auch alle Einlagerungen, Mineralgänge, Kavernen usw., können in ihrer räumlichen Lage unmittelbar eingemessen werden. Durch entsprechende Ausleuchtung der Bohrlochwandung kann bei einiger Erfahrung die Öffnungsweite von Klüften und Spalten gemessen werden.

In Abb. 2 ist unmittelbar neben dem eigentlichen Bohrprofil die Klufthäufigkeit pro Meter dargestellt. Man kann einigermaßen erkennen, daß die Quarzite wesentlich stärker geklüftet sind als die Schiefer. Wesentlich deutlicher zeigt dies das rechte Diagramm. Hier ist die mittlere Klufthäufigkeit pro Meter für jeweils eine Gesteinsart dargestellt.

Man könnte nach dem Vorstehenden annehmen, daß die Wasserdurchlässigkeit in dem tiefsten Quarzit fast so groß wie in dem Schieferhorizont (24,00 bis 33,25 m)

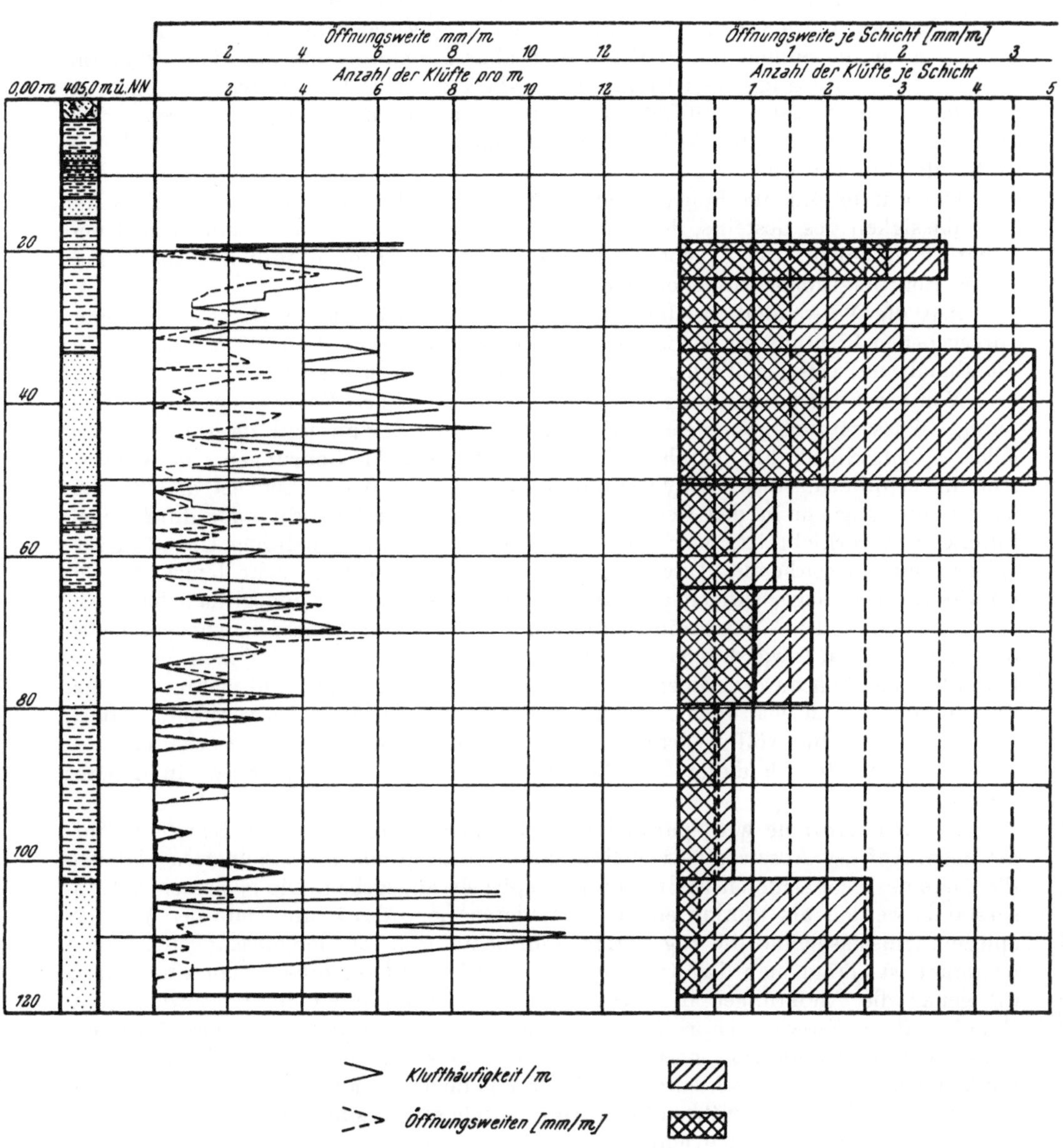

Abb. 2. Häufigkeit und Öffnungsweite der Klüfte in Abhängigkeit vom Gestein
Frequency and gaping of joints depending on the type of rock
Quantité de fissures et largeur de fentes

sei, bzw. ca. 50 % von derjenigen in den oberen Quarziten (33,25 bis 51,00 m) betrage. In Wirklichkeit drang aber bei den Wasserabpressungen in den unteren Quarzit wesentlich weniger Wasser als in die oberen Schichten ein.

Diese Beobachtung wird durch das Diagramm über die *Öffnungsweiten* der Klüfte und Spalten (Abb. 2) bestätigt. Sowohl in den oberen Schiefern und Quarziten als auch in der Zone mit einer engen Wechsellagerung von Schiefern und Quarziten lag die Summe der Spaltenweiten über 1,5 mm/m und z. T. über 2 mm/m. Unter 50 m Teufe ging die Öffnungsweite stark zurück. Im tiefsten Abschnitt, in dem der Quarzit stark geklüftet war, wurden nur noch 0,3 mm/m gemessen. Die Öffnungsweite der Klüfte bestimmt die Durchlässigkeit des Gebirges in stärkerem Maße als ihre Häufigkeit.

Die Darstellung im Bohrdiagramm erlaubt es, die Beziehungen zwischen der Durchlässigkeit des Gesteins, der Klufthäufigkeit und der Öffnungsweite schnell und sicher zu erkennen. Die Ergebnisse von Fernsehsondierungen sollten deshalb auch in die Bohrdiagramme aufgenommen werden.

Auf die Notwendigkeit, die *Richtungsabweichung* eines Bohrloches von seiner vorgegebenen Richtung zu beobachten, hat vor allem Heitfeld (1957) hingewiesen. Manchmal sind Gesetzmäßigkeiten in der Abweichung zu erkennen.

Von besonderer Bedeutung ist der *Wasserstand in den Bohrlöchern* (vgl. Stini, 1958). Wasserstandsmessungen sollten jeweils vor Beginn und nach Beendigung jeder Arbeitsschicht durchgeführt werden, da die meisten Bohrungen eine Wasserspülung erfordern. Auch bei Verwendung von klarem Wasser sickert die Spülung häufig nur langsam in das Gebirge ein, weil es z. B. sehr dicht oder durch einen Filterkuchen verklebt ist, oder weil infolge eines hohen, natürlichen Grundwasserspiegels ein ausreichender Überdruck fehlt. Häufig dauert es sehr lange, bis der Bohrlochwasserspiegel sich an den echten Grundwasserspiegel angeglichen hat.

Trägt man die Wasserstände, die bei Schichtbeginn gemessen wurden, neben dem Bohrprofil auf, wie dies z. B. die DIN 4023 vorschreibt, so wird man, besonders bei unterschiedlich durchlässigem Gebirge, bei tiefen Bohrungen und mehreren Rohrtouren, auch wenn man das zugehörige Datum neben dem Wasserstand notiert, häufig ein scheinbar völlig wirres, kaum mehr interpretierbares Bild erhalten.

Es empfiehlt sich daher zur Darstellung der Wasserspiegellagen ein anderes Verfahren.

Auf der Ordinate wird parallel zum Bohrprofil wie üblich die Bohrteufe aufgetragen. Auf der Abszisse ist von links nach rechts in einem verkleinerten Maßstab die Tiefe des beobachteten Bohrlochwasserspiegels angegeben. Die gemessenen Wasserstände werden als Punkte eingetragen, deren Lage durch die Tiefe des Wasserspiegels (Abszisse) und die zugehörige Bohrlochteufe z. Zt. der Messung (Ordinate) bestimmt ist. Trägt man Tag für Tag diese Punkte auf, so erhält man eine Kurve, die genau die Änderungen des Wasserspiegels während der Bohrzeit wiedergibt. Wandert diese Kurve nach rechts, so entspricht dies einem Sinken des Bohrlochwasserspiegels, wandert sie nach links, so zeigt dies ein Steigen des Wasserspiegels mit zunehmender Bohrlochteufe an. Bei vertikalem Verlauf der Kurve bleibt der Wasserspiegel konstant, während eine Horizontale einem sinkenden Spiegel bei konstanter Bohrlochteufe entspricht.

In der Bohrung in Abb. 3 z. B. wurde erstmalig bei 7,50 m Bohrteufe Wasser im Bohrloch in einer Tiefe von 5 m beobachtet. Die nächste Messung bei 10 m Bohrteufe lieferte den gleichen Wert. Beim Weiterbohren war kein Wasser mehr im Bohrloch. Die beiden Wasserstände dürften durch nicht abgelaufenes Spülwasser verursacht worden sein. Echtes Grundwasser konnte in diesem Falle nicht angenommen werden. Vielleicht ist die dünne Schieferschicht, die bei etwa 10 m angetroffen wurde, dafür verantwortlich, daß eine Zeit lang die Spülung nicht versickerte. Nach

Durchteufen von Kalken bis 30 m stand in den oben kalkigen, unten sandigen Schiefern bei knapp 30 m wieder Wasser im Bohrloch. Der Wasserspiegel sank kaum ab, obwohl die Bohrung weiter vertieft wurde. Erst unterhalb 50 m Bohrteufe fiel der Wasserspiegel plötzlich stark ab, nachdem durchlässiger Kalkstein (vgl. die Abpreßergebnisse im Bohrdiagramm, Abb. 3) erreicht worden war. Der Wasserspiegel innerhalb der, wie die Wasserabpressungen gezeigt haben, dichten Schieferfolge von ca. 30 bis 50 m ist sicherlich ebenfalls auf nicht versickertes Spülwasser zurückzuführen. Die Kalke unterhalb ca. 50 m waren wesentlich durchlässiger und der Wasserspiegel sank schnell ab. Beachtenswert ist die leichte Versteilung der Kurve ab ca. 63 m, die ein langsameres Absinken des Spiegels anzeigt. Ab ca. 63 m beginnen wieder weniger durchlässige Schichten. Von 70 m an blieb der Wasserspiegel konstant. Diese Teufe liegt wenige Dezimeter über dem Niveau eines ca. 50 m entfernten Flusses. Hier etwa liegt erst der echte Grundwasserspiegel im Gebirge. Bei der letzten Messung war der Bohrlochwasserspiegel aufgrund eines Hochwassers in dem benachbarten Fluß etwas angestiegen.

Ähnlich interessant ist der Verlauf der Wasserstandskurve in der Bohrung Abb. 1. Diese Bohrung wurde verrohrt, die jeweiligen Verrohrungsteufen wurden neben den Wasserständen eingetragen. Ohne deren Kenntnis ist eine Interpretation der Wasserstandsbeobachtung nicht möglich. Im oberen Abschnitt bis ca. 16 m sank der Wasserspiegel ziemlich gleichmäßig, bis das Bohrloch restlos trocken blieb. Nach verschiedenen Zementationen stellte sich dann wieder ein Wasserspiegel ein, der bis zu einer Bohrteufe von ca. 54 m ziemlich gleichmäßig bei ca. 30 bis 35 m erhalten blieb. Jetzt erst sank der Spiegel in den durchlässigen Konglomerathorizonten rasch ab. Ein ähnliches, rasches Absinken wiederholte sich noch zweimal. Gegen Ende der Bohrung blieb der Spiegel lange Zeit konstant und sank nur zum Schluß noch einmal leicht ab.

Auch in diesem Falle dürften die Wasserstände im Bohrloch keine echten Grundwasserspiegel repräsentieren. Es handelt sich sehr wahrscheinlich um nicht versickertes Spülwasser. Erst beim Antreffen sehr durchlässiger Schichten, wie z. B. der oberen Partie in der Konglomeratfolge bei ca. 55 m, sank das Wasser schlagartig tiefer. In dieser Teufe erfolgte übrigens auch totaler Spülwasserverlust. Eine Beziehung zwischen Spülverlust und Bohrlochwasserstand zeigt sich auch in dem Knick der Wasserstandskurve bei ca. 84 m. Der letzte Abfall des Wasserstandes kann möglicherweise auf eine offene Spalte zurückgeführt werden, die kurz nach dem plötzlichen Absinken des Wasserspiegels angebohrt wurde. Der Verlauf der Wasserstandskurve läßt keinen Schluß zu, daß der echte Grundwasserspiegel im Gebirge bereits erreicht ist. Dies war im vorliegenden Falle umso wichtiger, als die letzten Wasserstände Werte erbrachten, die unter dem Niveau eines ca. 400 m entfernten Flusses lagen.

Diese wenigen Beispiele zeigen die Bedeutung laufender Wasserstandsbeobachtungen und einer übersichtlichen Darstellung derselben. Stets ist eine Interpretation der Beobachtungen nur dann möglich, wenn die Beziehungen zwischen dem Gebirge, dem Wasserstand, den Spülverlusten und Abpreßergebnissen aufgespürt werden.

Dazu müssen in dem Bohrdiagramm auch die *Spülverluste* dargestellt werden. Auch diese helfen, die Gebirgsverhältnisse zu interpretieren. Für sich allein kann jedoch die Beurteilung der Spülverluste zu Mißdeutungen führen. So ist z. B. aus der Bohrung in Abb. 4 zu ersehen, daß das Spülwasser sehr wahrscheinlich nur bis zu einer Teufe von ca. 24 m in das Gebirge versickerte. Bis zu dieser Teufe sank der Bohrlochwasserspiegel stark ab, und die Wasserabpressungen erbrachten hohe Aufnahmewerte. Die darunter liegenden Schichten sind offensichtlich wesentlich dichter. Der Wasserspiegel war in diesem Abschnitt ziemlich konstant und die Wasserabpressungen ergaben geringe Aufnahmewerte. Hieraus darf gefolgert werden,

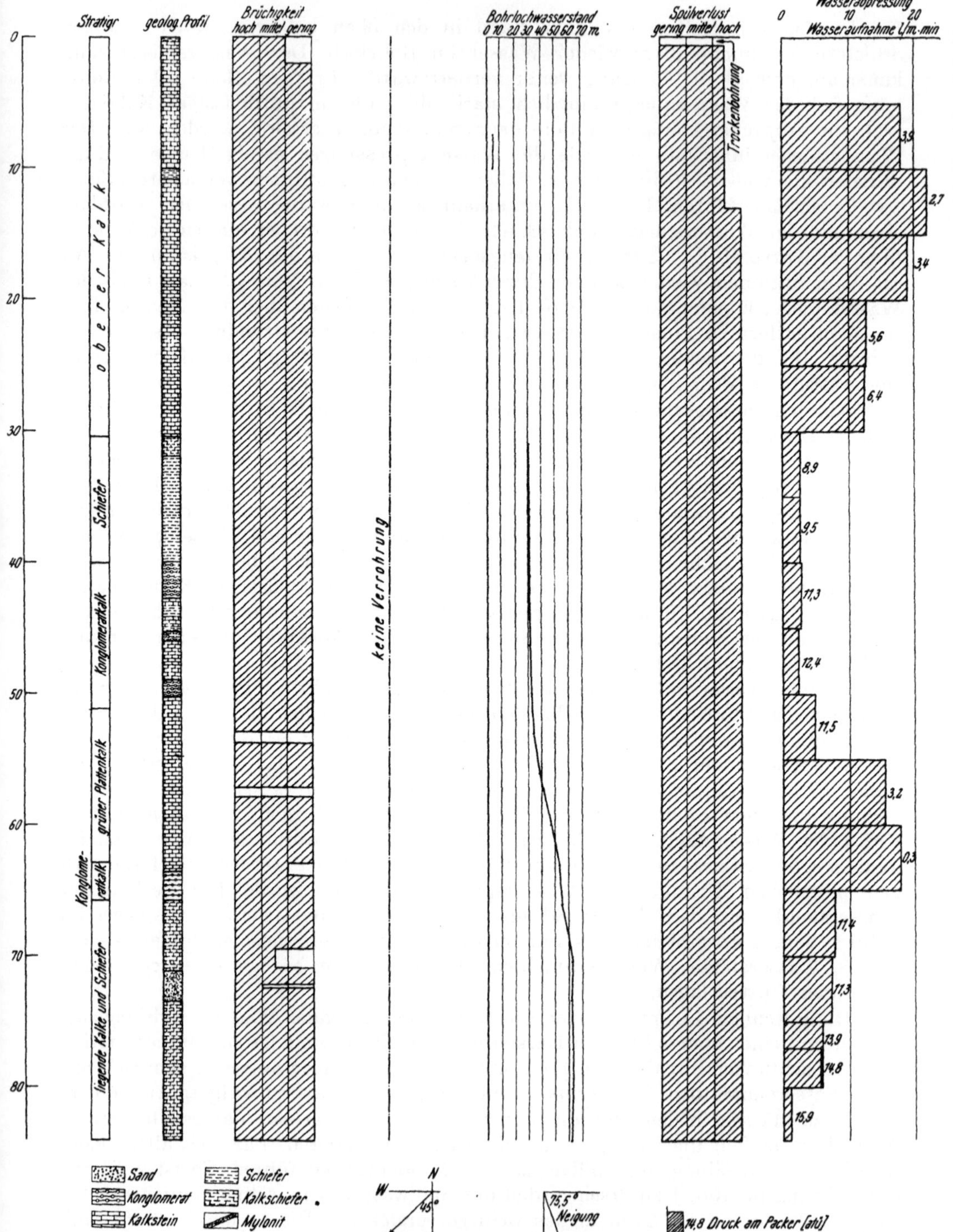

Abb. 3. Beispiel eines Bohrdiagrammes
Example of a drilling diagram
Exemple d'un diagramme de forage

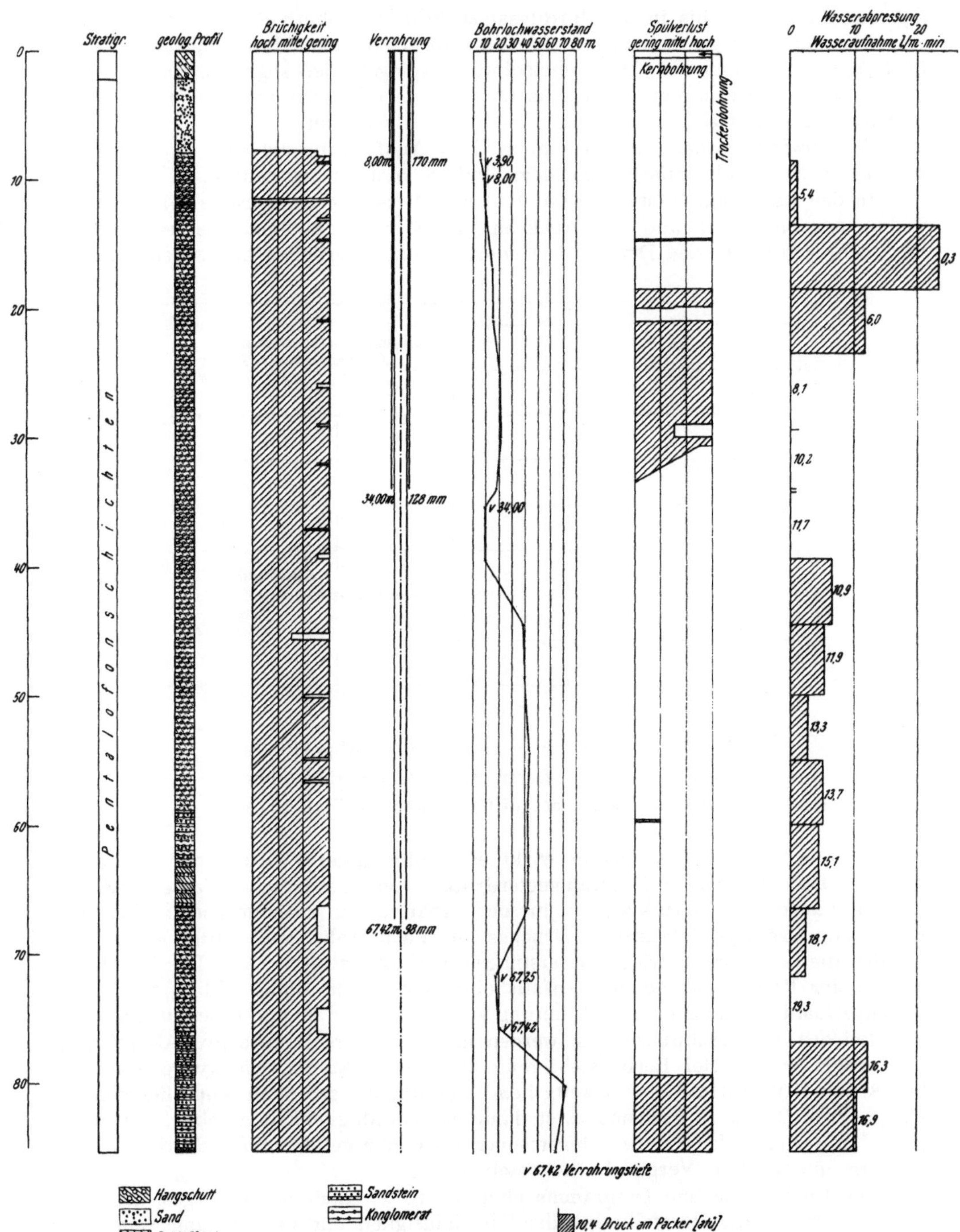

Abb. 4. Beispiel eines Bohrdiagrammes
Example of a drilling diagram
Example d'un diagramme de forage

daß das Spülwasser bis in die durchlässigen Schichten oberhalb 24 m anstieg und
erst dort ins Gebirge versickerte. Als eine Rohrtour bis 34 m eingebaut wurde,
hörte der Spülverlust auf. Der Bohrlochwasserspiegel stieg sogar an und fiel erst
wieder ab, als durchlässigere Schichten ab ca. 39 m erbohrt wurden. Erneute, totale
Spülverluste traten ab 78 m in durchlässigen Schichten auf.

   Wasserabpressungen sind sowohl in ihrer Ausführung als auch in ihrer Deutung oft recht problematisch. Trotzdem sind sie nicht zu entbehren.

   In den Bohrdiagrammmen (Abb. 1, 3 und 4) ist die eingepreßte Wassermenge
in l/m.min dargestellt und der Druck am Packer vermerkt. Leider ist es bis heute
noch sehr verbreitet, den Druck am Manometer als maßgebend anzusehen. Dies ist

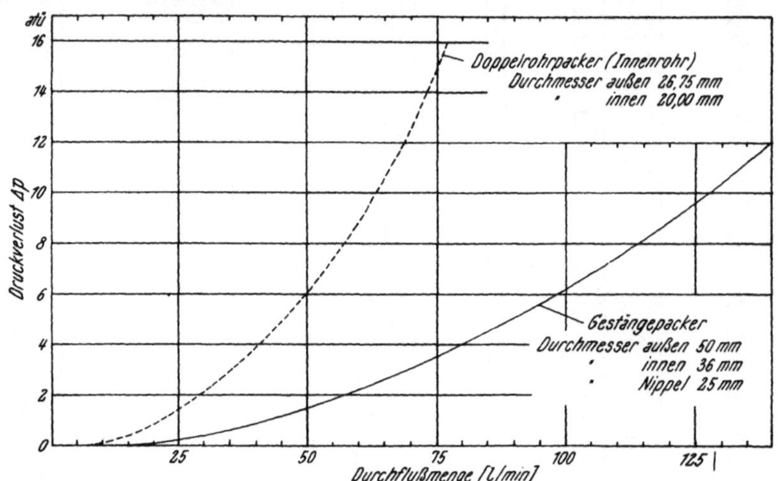

Abb. 5. Druckverluste in verschiedenen Abpreßrohren
Pressure loss in different types of injecting tubes
Pertes de charges des différentes tubes d'injections

jedoch völlig unrichtig. Im Packer wirkt ein völlig anderer Druck, der sich unter
anderem aus Pumpendruck (Manometerdruck), hydrostatischem Wasserdruck im
Packergestänge und Druckverlust in der Rohrleitung zusammensetzt. Bei der
Ermittlung des hydrostatischen Druckes im Packerrohr muß unter allen Umständen die Lage des Bohrlochwasserspiegels berücksichtigt werden. Der Druckverlust in der Preßleitung spielt ebenfalls eine sehr große Rolle. Abb. 5 zeigt zwei
Druckverlustdiagramme einer Rohrleitung von 36 mm und 20 mm Innendurchmesser.

   Bei den Wasserabpressungen werden meist mehrere Drücke pro Abpreßstufe
angewendet. Erst unter Berücksichtigung der o. g. Faktoren, wie hydrostatischem
Druck und Druckverlust, ist es weitgehend möglich, die gesamten Abpreßergebnisse
auf gleichen Druck zu beziehen. Erst damit ist die Möglichkeit gegeben, die Interpretation der spezifischen Aufnahmemengen nicht nur gefühlsmäßig durchzuführen,
sondern quantitative Vergleiche anzustellen.

   Die Bohrprofile und Diagramme über die Brüchigkeit des Gebirges, das Verrohrungsschema, die Darstellungen der Klufthäufigkeit und der Öffnungsweiten der
Klüfte, die Wasserstands-, Spülverlust- und die Verpreßdiagramme liefern zusammen ein Bild der Gebirgsverhältnisse, wie es eine reine Bohrkernbeschreibung oder
die eine oder andere Beobachtung für sich allein nie erbringen kann. Erst aus
der Zusammenschau aller Beobachtungen kann das Gebirge beschrieben werden.
Eine derartige Synthese ist aber nur möglich, wenn praktisch mit einem Blick die

Gebirgsverhältnisse in ihren Einzelheiten überschaut werden können. Dies sollen die Bohrdiagramme ermöglichen.

Die beschriebenen Methoden und die Darstellung der Ergebnisse sind in erster Linie für den auf der Baustelle tätigen Praktiker gedacht. Sie sollen aber auch theoretische Untersuchungen, insbesondere im Zusammenhang mit fels- und bodenmechanischen Aufgaben, erleichtern.

Literatur

Bendel, L.: Ingenieurgeologie. Bd. I, Springer, Wien 1948; Bd. II, Springer, Wien 1949.

Breddin, H.: Vorschläge zu einer international einheitlichen Darstellung auf lithologischen, tektonischen und hydrogeologischen Zeichnungen und Karten. 1. Teil. Geol. Mitt., *1*, H. 1, S. 1—120, Aachen 1960.

Gignoux, M. und R. Barbier: Géologie des barrages et des aménagements hydrauliques. Masson, Paris 1955.

Graupner, A.: Die Kartierung von Festgestein für geotechnische Zwecke. Geologie und Bauwesen, *26*, H. 4, S. 273, Wien 1961.

Heitfeld, K.-H.: Kleinlochbohrungen zur Klärung der geologischen Verhältnisse und für die Untergrunddichtung im Talsperrenbau. Erdölzeitschrift, H. 11, S. 3, Wien—Hamburg 1957.

König, H.-W. und K.-H. Heitfeld: Über Notwendigkeit und Ausmaß geologischer Untersuchungen im Talsperrenbau. Geologie und Bauwesen, *26*, H. 1, S. 63, Wien 1962.

Müller, L.: Der Felsbau. Enke, Stuttgart 1963.

Stini, J.: Soll die baugeologische Geländeaufnahme auch eine eingehende Untersuchung der Kluftwasserverhältnisse im Felsgebirge einschließen? Geologie und Bauwesen, *24*, H. 1, S. 31, 1958.

DIN-Normblatt 4022, Schichtenverzeichnis und Benennen der Boden- und Gesteinsarten.

DIN-Normblatt 4023, Baugrund- und Wasserbohrungen.